Genetics and
Product Formation in
Streptomyces

FEDERATION OF EUROPEAN MICROBIOLOGICAL SOCIETIES SYMPOSIUM SERIES

Recent FEMS Symposium volumes published by Plenum Press

1990 • MOLECULAR BIOLOGY OF MEMBRANE-BOUND COMPLEXES IN
PHOTOTROPHIC BACTERIA
Edited by Gerhart Drews and Edwin A. Dawes
(FEMS Symposium No. 53)

1990 • MICROBIOLOGY AND BIOCHEMISTRY OF STRICT ANAEROBES
INVOLVED IN INTERSPECIES HYDROGEN TRANSFER
Edited by Jean-Pierre Bélaich, Mireille Bruschi, and Jean-Louis Garcia
(FEMS Symposium No. 54)

1990 • DENITRIFICATION IN SOIL AND SEDIMENT
Edited by Niels Peter Revsbech and Jan Sørensen
(FEMS Symposium No. 56)

1991 • *CANDIDA* AND CANDIDAMYCOSIS
Edited by Emel Tümbay, Heinz P. R. Seeliger, and Özdem Anğ
(FEMS Symposium No. 50)

1991 • MICROBIAL SURFACE COMPONENTS AND TOXINS
IN RELATION TO PATHOGENESIS
Edited by Eliora Z. Ron and Shlomo Rottem
(FEMS Symposium No. 51)

1991 • GENETICS AND PRODUCT FORMATION IN *STREPTOMYCES*
Edited by Simon Baumberg, Hans Krügel, and Dieter Noack
(FEMS Symposium No. 55)

1991 • THE BIOLOGY OF *ACINETOBACTER:* Taxonomy, Clinical Importance,
Molecular Biology, Physiology, Industrial Relevance
Edited by K. J. Towner, E. Bergogne-Bérézin, and C. A. Fewson
(FEMS Symposium No. 57)

Genetics and Product Formation in *Streptomyces*

Edited by

Simon Baumberg

University of Leeds
Leeds, United Kingdom

Hans Krügel and *Dieter Noack*

Institut für Mikrobiologie und
Experimentelle Therapie
Jena, Germany

PLENUM PRESS • NEW YORK AND LONDON

Library of Congress Cataloging-in-Publication Data

Genetics and product formation in streptomyces / edited by Simon
  Baumberg, Hans Krügel, and Dieter Noack.
      p.   cm. -- (Symposium series ; no. 55)
    "Proceedings of a symposium held under the auspices of the
  Federation of European Microbiological Societies, held May 1-6,
  1990, in Erfurt, Germany"--T p. verso.
    Includes bibliographical references and index.
    ISBN-13: 978-1-4684-5924-1   e-ISBN-13: 978-1-4684-5922-7
    DOI: 10.1007/978-1-4684-5922-7

    1. Streptomyces--Congresses. 2. Bacterial genetics--Congresses.
  3. Microbial metabolism--Congresses. 4. Microbial biotechnology-
  -Congresses.   I. Baumberg, S. II. Krügel, Hans. III. Noack,
  Dieter. IV. Federation of European Microbiological Societies.
  V. Series: FEMS symposium ; no. 55.
  QR82.S8G46 1991
  589.9'2--dc20   91-12849
                                                                  CIP

Proceedings of a symposium held under the auspices of the
Federation of European Microbiological Societies,
held May 1-6, 1990, in Erfurt, Germany

ISBN-13: 978-1-4684-5924-1

© 1991 Plenum Press, New York
Softcover reprint of the hardcover 1st edition 1991
A Division of Plenum Publishing Corporation
233 Spring Street, New York, N.Y. 10013

L A U D A T I O

Professor Arnold Demain is known throughout the world as one of the preeminent scientists in modern microbiology and biotechnology. He has been responsible for a remarkable amount of work of the highest quality and importance on secondary metabolism in particular and microbial physiology in general during over 30 years at the world famous Massachusetts Institute of Technology in Boston, USA.

Professor Demain formed a whole generation of young microbiologists through his pioneering studies on regulation of production of penicillin and other antibiotics, the isolation and characterization of enzymes involved therein and, moreover, by numerous review papers and comprehensive volumes dealing with all aspects of antibiotic production as well as the formation of primary metabolites and enzymes. We appreciate his engagement in the organization of our scientific community, by establishing committees, meetings and schools as extraordinarily fruitful. His integrating efforts have facilitated the close development of industrial microbiology in America, Europe and Japan, and have provided a sound basis for full integration of Eastern Europe and the developing countries as well into this area of the scientific world. The series of Streptomyces meetings held at Weimar, and now in Erfurt, can be considered as a contribution to the development of closer international relationships. Today, 124 participants from 49 laboratories in 19 countries of Europe, Asia and America have come together at our meeting.

It is a great honour to the workers of the Central Institute of Microbiology and Experimental Therapy at Jena to welcome Arnold Demain here at this meeting. It is also a very personal pleasure for several colleagues among us, to meet him again here, as an outstanding and outgoing personality who has constantly stimulated our scientific work through his ideas as well as through his ever-ready help, when considering our scientific efforts. Hence we wish to express our appreciation of our sympathies with the colleague Arnold Demain by conferring upon him at this time the

Medal of Honour of the Central Institute
of Microbiology and Experimental Therapy.

We wish him, for his future work, continued success and never-ending pleasure in the new developments in our common field of interest, microbial genetics and metabolism.

ARNOLD L. DEMAIN

PREFACE

The Streptomycetes are industrially widely used microorganisms due to their ability to produce numerous different chemical compounds. These show very varied effects upon other living systems, and result from profound and subtle biochemical and morphological differentiation during the streptomycete life cycle. It is therefore not surprising that research on the genetics of antibiotic biosynthesis and differentiation in this group is currently progressing rapidly in many countries. Intimately connected with the production of antibiotics is resistance to them; analysis of this is giving further information about the origin and evolution of this class of genes and their hypothesized spread among other microorganisms. Another interesting feature of the Streptomyces group is their mycelial growth. Also, their ecologically important utilization of high molecular weight compounds requires enzymes to be transported outside the cell to hydrolyze non-diffusible substrates. Finally, we have as yet limited understanding of the various mechanisms of genome rearrangement observed in some of these species; deletions and/or amplifications of enormous amounts of DNA can occur without seriously affecting the viability of the organism under laboratory conditions.

The present volume, which includes contributions addressing the above subjects and others, originates from a meeting on "Genetics and Product Formation in Streptomyces" sponsored by the Federation of European Microbiological Societies in Erfurt on May 1-6 1990. Compared to previous ones of this kind held in 1979, 1983 and 1987 in Weimar, one can point to impressive progress in the study and applications of Streptomyces genetics. We may note some of the highlights. Sigma factors, transfer RNA, differentiation factors and other regulatory elements are all implicated in the cascade of gene expression during the life cycle. Ribosomal and transfer RNA are being increasingly studied for their own sake. Antibiotic biosynthesis was given detailed attention mainly in the context of tetracyclines as examples of the polyketide type of pathway and β-lactams as peptide antibiotics. Antibiotic resistance featured as protecting the producer from suicide; of the enzymes involved, antibiotic-modifying enzymes such as phosphotransferases, acetyltransferases and β-lactamases were instanced and their structure, regulation, possible origin and spread among microorganisms was discussed.

Protein synthesis and secretion focussed on the genes involved in utilisation of high molecular weight carbon sources and the possible use of streptomycetes for the production of exported foreign gene products. It is clearly tempting to consider the use of these non-pathogenic soil bacteria for the production of eukaryotic proteins; these have been shown to be expressible in streptomycetes with the help of appropriate plasmid vectors and expression cassettes. To assist this, investigation of the structural and segregational stability of these plasmids is of great value and some pioneering studies in this direction were presented.

Work on chromosomal instability and rearrangements featured mainly
S. lividans and showed several mechanisms operating in this species. End
points of the large deletions have been further characterized and further
genes affected by the deletions and amplifications were identified.

The contributions led to wide ranging and enthusiastic discussions and
the participants, of a broad range of nationalities, agreed that the
proceedings of the meeting testified that a particularly fruitful period in
Streptomyces studies had been reached.

The organizers of the meeting express their gratitude to the Federation
of European Microbiological Societies(FEMS), the Arzeneimittelwerk Dresden
GMBH (GERMED), and Blackwell Scientific Publications for their financial
support. They also wish to thank the representatives of the IGA in Erfurt
for their hospitality, and the many coworkers of the Central Institute of
Microbiology and Experimental Therapy for their commitment.

The Editors

CONTENTS

INTRODUCTION

K.F.Chater

John Innes Institute
Colney Lane
Norwich NR4 7UH, U.K.

Streptomycetes are morphologically among the most complex of bacteria: they grow as a branching mycelium, and characteristically they reproduce by the formation of chains of uninucleoidal spores from specialized aerial hyphae. The formation of aerial hyphae often coincides with antibiotic production and the occurrence of various pleiotropic mutants defective in both processes shows that the study of morphological differentiation may prove relevant to the fermentation industry.

In some strains, notably some isolates of *S.griseus*, sporulation can be made to occur abundantly and synchronously in liquid culture. Detailed biochemical analysis is therefore possible. Here, two papers exploit this attribute. Penyige et al. show that the membrane-bound protein-ADP ribosylation activity of *S.griseus* correlates with sporulation. The ADP ribosylating enzyme may be required for the low molecular weight hormone-like compound A-factor to exert its well-known stimulating effects on sporulation and antibiotic production. Also in *S.griseus*, Szeszak et al. describe work on the quite different sporulation and antibiotic production inducing compound, factor C. The addition of minute quantities of this 34KDa protein gives rise to a range of physiological effects, including induction of its own synthesis. Remarkably, factor C-like proteins seen to be produced by many microorganisms, as judged by immunological experiments.

In *S.coelicolor* A3(2), the formation of aerial hyphae and the biosynthesis of antibiotics are not known to involve A-factor, factor C or ADP ribosylation (though there is no proof of their non-involvement). Genetical studies, however, have shown that a tRNA for a rare leucine codon (UUA) plays a important part. This has given rise to the need to characterise tRNA molecules directly, as in the paper by Rokem, and to consider the potential importance of programmed translational ambiguity during development (the article by Weiser et al.). Several clusters of genes for other tRNAs are described in the paper by Redlimeier and Schmieger, and in the future it will be interesting to compare various aspects of their expression with expression of bldA (the gene for tRNA$_{UUA}^{Leu}$) and the ribosomal genes *(rrn)*: previous studies have shown that rrn expression is subject to stringent regulation, and as discussed in the paper by Strauch et al. that the stringent response may play a part in regulating the onset of differentiation.

Unlike *S.griseus*, *S.coelicolor* A3(2) does not sporulate in submerged culture. Nevertheless, the wide range of genetic and molecular techniques available for strain A3(2) has made it possible to investigate events occurring during sporulation of aerial hyphae, through the study of sporulation deficient mutants (the paper by Chater et al.). This has revealed that a sporulation-specific sigma factor is needed for the initiation of sporulation, and that the regulatory events during sporulation do not all fall into one simple cascade. It seems that both of the parallel sequences of regulatory events are necessary for the formation of mature pigmented spores, so the cloned spore pigment genes describe here by Chater et al. may provide a valuable investigative tool.

Over the next few years, we can confidently expect that the information coming from these wide-ranging approaches to the analysis of morphological and physiological differentiation will become more integrated. Probably it will provide insights that will not only be of interest to academic scientists but will also inspire new approaches to the industrial exploitation of streptomycetes.

GENE EXPRESSION DURING SPORULATION IN *STREPTOMYCES COELICOLOR* A3(2)

Keith F. Chater, Celia J. Bruton, Nigel K. Davis[*],
Kitty A. Plaskitt, Juan Soliveri and Huarong Tan

John Innes Institute, Colney Lane, Norwich NR4 7UH, England

INTRODUCTION

Colonies of *Streptomyces* spp growing on suitable agar medium are effectively multicellular organisms which, as they grow, show morphologically, physiologically and functionally distinct regions (reviewed by Chater, 1989a). The tips of hyphae growing out into the agar convert surrounding nutrients into biomass, whereas hyphae in the central region of the colony often show extensive lysis (Wildermuth, 1970), perhaps connected with the supply of their contents as nutrients to the growing aerial hyphae (Méndez et al., 1985). The aerial hyphae themselves have two kinds of cellular organisation: the multinucleoidal hyphae that result from the re-use of materials from the substrate mycelium for a second round of growth, and the strings of uninucleioidal compartments that are found at hyphal tips and which eventually form chains of spores (Wildermuth and Hopwood, 1970). In this paper we focus on current approaches to, and progress in, analysing the events in aerial hyphae that give rise to spore chains.

Morphological and Biochemical Events During Sporulation of Aerial Hyphae

When aerial hyphae form, colonies of *Streptomyces coelicolor* A3(2) change in appearance and surface properties: the aerial mycelium looks white and hairy, and is hydrophobic. At this time various new proteins of unknown function are deposited on the surface of the hyphae. They persist through sporulation, and can be washed from spores by mild detergent treatment, hence the term Sap, for spore-associated proteins (Guijarro et al., 1988). Conceivably they may contribute to the hydrophobicity of aerial hyphae, and help in protecting the hyphal contents from dessication. During the extension of aerial hyphae there are few - if any - cross-walls (Wildermuth, 1970; Wildermuth and Hopwood, 1970), and cytochemistry shows that glycogen granules are absent (even though they are usually present in the ageing substrate hyphae on which the aerial branches grow: Braña et al., 1986; K.A.P., unpublished). Eventually, the tips of many aerial hyphae undergo metamorphosis into long chains of spores. The first signs of this are the coiling of the tips, and the initiation of sporulation septation. Sporulation septa are morphologically distinct from the

---

[*]Present address: Glaxo Group Research, Greenford, Middlesex UB6 OHE, England

occasional cross-walls formed in vegetative hyphae, presumably because their eventual function is to allow cells to separate (Wildermuth and Hopwood, 1970; Hardisson and Manzanal, 1976). Within a hypha, they form more or less synchronously and at regular intervals, one chromosome (along with extrachromosomal elements) being partitioned to each compartment. At the same time, abundant glycogen deposits are formed (Braña et al., 1986; K.A.P., unpublished). Finally, the compartments change shape - from cylinders to ellipsoids - and undergo wall thickening to give rise to recognisable spores. As this happens, the glycogen deposits disappear and the aerial mycelium changes from white to grey, a colour associated with mature spores of *S. coelicolor*. Different spore colours are associated with many other species, suggesting that different pigments are produced.

Genetic Approaches to the Study of Sporulation

Generally, this series of events occurs only on colonies growing on solid surfaces, and not during submerged growth. This makes it difficult to use direct biochemical analysis as a primary tool in investigating *Streptomyces* sporulation. However, the use of genetics often permits the dissection of complex processes, and this approach has proved fruitful in analysing *Streptomyces* developmental biology. The need to adopt genetic analysis explains the choice of *S. coelicolor* A3(2) for the studies described here: genetic analysis is particularly well developed in this strain (Hopwood et al., 1985). Our studies over a number of years have depended on a collection of mutants for the *whi* genes, so called because the mutant colonies remain white on prolonged incubation (Hopwood et al., 1970). Genetic and phenotypic analysis has so far revealed at least eight *whi* loci scattered round the chromosome (Chater, 1972). Some of the *whi* mutants have blocks very early in sporulation: *whiA, B, G* and *H* mutants make few if any sporulation septa. Thus, grey pigmentation seems to be dependent on the proper completion of early sporulation events, implying that some kind of regulatory cascade is involved.

With the availability of cloning systems for *Streptomyces* (Hopwood et al., 1985), it has become possible to isolate some of the *whi* genes, to determine their nucleotide sequences and transcriptional regulation, and to examine the effects of increasing or decreasing their expression on the regulation of other sporulation genes and processes. Such studies, which we discuss in this paper, form the basis of our approach to *Streptomyces* developmental biology. Other laboratories have taken complementary approaches based on the use of random transcriptional fusions to *lux* genes to produce developmentally controlled light emission (Schauer et al., 1988), the employment of reverse genetics after the sequencing of spore-associated proteins (Guijarro et al., 1988), or on biochemical studies of sporulation in submerged cultures (Kendrick & Ensign, 1983).

RESULTS AND DISCUSSION

The Decision to Sporulate Depends on a Sporulation-Specific Sigma Factor

Studies on epistatic interactions between mutations in different *whi* genes suggested that *whiG* acted early in sporulation (Chater, 1975). The *whiG* gene was cloned by complementation of a *whiG* mutant with DNA inserted in a *Streptomyces* phage vector (Méndez and Chater, 1987). *whiG* was shown to be important in determining the developmental fate of hyphae, by observing the effects of increased copy number. Even a single extra copy caused perceptibly more abundant sporulation, and with *whiG* on a multicopy plasmid sporulation could be seen in the substrate hyphae and in submerged cultures (Chater et al., 1989). This effect was due to the *whiG* gene product, since a small in-frame deletion in *whiG* relieved hypersporulation.

The nucleotide sequence of *whiG* revealed that it specifies a sigma factor-like protein (sigma-*whiG*; Chater et al., 1989). This protein is specific for sporulation, since disruption of *whiG* or the presence of a *whiG* frameshift mutation both give rise to strains with no obvious defects in vegetative growth or in antibiotic synthesis, but which show the typical *whiG* sporulation-defective phenotype (Chater et al., 1989; C.J.B. unpublished).

Interestingly, the sigma-*whiG* protein is particularly similar (38% amino acid identity) to sigmaD of *Bacillus subtilis*, which plays no role in sporulation in that organism (Helmann et al., 1988). Probably the sporulation processes of *S. coelicolor* and *B. subtilis* have independent evolutionary origins. SigmaD directs *B. subtilis* RNA polymerase to promoters for motility-related genes, with a single known - but particularly interesting - exception: sigmaD mutants form filaments, giving a phenotype reminiscent of that of *whiG* mutants.

Although *whiG* is Epistatic to *whiB* in Morphological Tests, *whiB* is Transcribed in a *whiG* Mutant

Since sigma-*whiG* is required for sporulation, and *whiG* is epistatic in morphological tests to all other *whi* genes tested (*whiA, B, H* and *I*: Chater, 1975), it seemed possible that transcription of these genes might depend directly or indirectly on sigma-*whiG*. Recently, the *whiB* gene has been cloned, by using a low copy-number plasmid vector and seeking complementation of a *whiB* mutant (N.K.D., unpublished). Using a suitable fragment of *whiB*, S_1 nuclease protection studies of *whiB* mRNA during growth on agar medium have been carried out (J.S., unpublished). These studies have shown that even though *whiB* mRNA is detectable only when aerial hyphae are present, it is not dependent on *whiG*; detectable transcription takes place from the same start-site in a *whiG* mutant as in the wild-type. Consistent with this *whiG*-independence, the *whiB* promoter has -10 and -35 regions closely resembling those of "typical" prokaryotic promoters transcribed by RNA polymerase containing a principal sigma factor, but unlike promoters transcribed in *B. subtilis* by sigmaD. These findings suggest that more than one transcriptional cascade occurs in parallel during sporulation.

A Strategy for the Isolation of Promoters Dependent on Sigma-*whiG*

The observation that sigma-*whiG* and sigmaD of *B. subtilis* are closely similar in regions expected to contact the -10 and -35 regions of cognate promoters suggested an experiment in which a *B. subtilis* promoter dependent on sigmaD was introduced at high copy number into *S. coelicolor*. It had already been shown that the level of sporulation reflects the amount of sigma-*whiG*, so it was not surprising to find that the experiment led to reduced sporulation (a partial *whiG* mutant phenocopy), interpreted as resulting from a partial sequestration of the sigma-*whiG* holoenzyme (Chater et al., 1989). This result in turn suggested a further experiment, in which a library of small fragments of *S. coelicolor* DNA in a high copy-number *Streptomyces* plasmid was introduced into *S. coelicolor* in a search for clones with white colonies (H.T. unpublished). Several such clones were found. pIJ4083, the particular vector that was used, contains the *xylE* reporter gene (Ingram et al., 1989; Clayton and Bibb, 1990). Transcription of *xylE in vivo* is detectable by spraying with catechol, which is converted by the *xylE* gene product (catechol 2,3-dioxygenase), into yellow hydroxymuconic semialdehyde (Zukowski et al., 1983). This Ylo[+] phenotype was shown by each of the clones that inhibited sporulation. Thus, the fragments contained promoters. Further, the yellow colour appeared to be mainly in the aerial mycelium, as expected of *whiG*-dependent promoters. In order to provide more pertinent evidence of *whiG*-dependence, two of the sporulation-inhibiting plasmids were introduced into various *whi* mutants. There was no detectable expression in a *whiG* mutant, little in a *whiH* mutant, and full expression in representative *whiA*, *whiB* and *whiI* mutants. On the strength of these experiments, the further characterisation of the promoters is clearly warranted, using nucleotide sequencing, transcription studies, and characterisation of the gene(s) transcribed from these promoters. Eventually, we hope that the promoters will provide suitable templates for *in vitro* studies of RNA polymerase containing sigma-*whiG*.

Glycogen Accumulation and Spore Pigment Formation as Developmental Markers for Sporulation

Two potential sources of *whiG*-dependent promoters (*whi* genes and sporulation inhibiting fragments) have been described in the previous sections. A third potential source would be provided if biochemical processes peculiar to sporulating aerial hyphae could be discovered. Such an approach has already led elsewhere to the cloning and analysis of genes for spore-associated proteins, but so far there is no evidence that these genes are dependent on *whiG* (Guijarro et al., 1988 and personal communication). Earlier observations had shown that glycogen accumulates in sporulating hyphae (Braña et al 1986). We therefore investigated glycogen deposition in the aerial hyphae of *whi* mutants, using electron microscopy of suitably stained thin sections (K.A.P., unpublished). There was no evidence of localised deposition in the representative *whiA, B, G, H* and *I* mutants tested. Since none of these undergoes extensive sporulation septation, it seems likely that such septation may be a prerequisite for glycogen accumulation. If so, then ectopic sporulation (e.g. that observed in the substrate mycelium when multiple copies of *whiG* are present: Chater et al., 1989) should also be accompanied by glycogen accumulation. This was indeed observed (K.A.P., unpublished). We therefore believe that glycogen metabolism in sporulating hyphae is developmentally regulated directly or indirectly by sigma-*whiG*. The enzymes of glycogen metabolism may therefore provide a starting point for a "reverse genetics" approach to the cloning of genes specifically expressed during sporulation.

As pointed out earlier, perceptible grey spore pigment fails to develop in mutants blocked early in sporulation, including *whiB* and *whiG* mutants. Thus, like glycogen accumulation, spore pigment also seems to require the completion of early stages of sporulation (including events in each of two possibly separate regulatory pathways, as shown by the effects of *whiG* or *whiB* mutations). One class of *whi* mutants (*whiE*) produces morphologically normal spores that are devoid of grey pigment, suggesting that *whiE* may be a determinant of spore pigment (Chater, 1972; McVittie, 1974). The recent cloning and sequencing of *whiE* by complementation of a *whiE* mutant (N.K.D. and K.F.C., manuscript submitted) has strengthened this possibility, since the locus has proved to be a set of genes with extensive homology to biosynthetic genes for polyketide antibiotics such as granaticin (Sherman et al., 1989) tetracenomycin (Bibb et al., 1989) and actinorhodin (M. Fernández, personal communication). Even a low number of extra copies of this gene cluster causes an apparent increase in spore pigment intensity, but production of the pigment still depends on other *whi* genes. We believe that the cis-acting elements essential for correct expression of the *whiE* cluster must therefore reside in the cloned DNA, most probably in a 184 bp sequence upstream of the first open reading frame of what is possibly a single polycistronic transcription unit. The analysis of this 184 bp region, perhaps exploiting fusions to the *xylE* reporter gene mentioned earlier, should prove informative about the regulation of events inside the compartments destined to become spores.

A word of caution: so far, analysis of glycogen and pigment metabolism have been too superficial for us to be quite sure that they involve developmental regulation of transcription. Glycogen accumulation is the result of a net excess of glycogen synthesis over its degradation, and the balance of these two processes in other organisms is sensitively controlled at the level of enzyme activation or inhibition (e.g. Preiss, 1984). Synthesis of spore pigment may well be highly dependent on precursor availability, and the supply of precursors is itself likely to be influenced by the development of the organism. For example, glycogen could be the major precursor of the spore pigment, so that any circumstances impeding glycogen accumulation in sporulating cells would also limit pigmentation. There is circumstantial evidence compatible with this, since *whi* mutants blocked at early stages fail to accumulate both compounds; the two known *whiE* mutants both accumulate relatively large amounts of glycogen; and hyperpigmentation, caused by adding extra copies of *whiE*, results in severely reduced glycogen accumulation (K.A.P., unpublished). In short, we must be prepared to find regulation operating at levels other than transcription.

CONCLUSIONS AND PROSPECTS

The availability of good natural and artificial genetic tools in *S. coelicolor* has begun to yield valuable information about the regulation of gene expression during sporulation. The cloned genes already available pose important questions. How are *whiG* and *whiB* regulated (and what prevents the "vegetative-like" promoter of *whiB* from transcription during vegetative growth)? What is the function of the *whiB* gene product, deduced (from the DNA sequence) to be a small, highly charged protein without marked resemblances to any proteins in existing databases (N.K.D., unpublished)? What are the genes controlled by the promoters present in sporulation-inhibitory segments of DNA? What form of RNA polymerase transcribes *whiE*? In addition we would like to know whether glycogen metabolism influences osmotic pressure and turgor in sporulating aerial hyphae (Chater, 1989b) and, if so, whether changes in osmotic pressure might influence gene expression (Higgins et al., 1988). Finally, we can be sure that the cloning and sequencing of more *whi* genes will even further widen the scope and interest of these investigations.

ACKNOWLEDGEMENTS

We are grateful to D A Hopwood for helpful comments on the manuscript. We thank the John Innes Foundation for grants to N.K.D., K.A.P. and H.T., and the European Community for a Senior Fellowship to J.S.

REFERENCES

Bibb, M.J., Bíró, S., Motamedi, H., Collins, J.F., and Hutchinson, C.R., 1989, Analysis of the nucleotide sequence of the *Streptomyces glaucescens tcmI* genes provides key information about the enzymology of polyketide antibiotic biosynthesis, *EMBO J.*, 8:2727.

Braña, A.F., Méndez, C., Díaz, L.A., Manzanal, M.B., and Hardisson, C., 1986, Glycogen and trehalose accumulation during colony development in *Streptomyces antibioticus*. *J. Gen. Microbiol.*, 132:1319.

Chater, K.F., 1972, A morphological and genetic mapping study of white colony mutants of *Streptomyces coelicolor*, *J. Gen. Microbiol.*, 72:9.

Chater, K.F., 1975, Construction and phenotypes of double sporulation deficient mutants in *Streptomyces coelicolor* A3(2), *J. Gen. Microbiol.* 87:312.

Chater, K.F., 1989a, Sporulation in *Streptomyces*, *in*: "Regulation of

Procaryotic Development", I. Smith, R. Slepecky and P. Setlow, ed., American Society for Microbiology, Washington, pp.277.

Chater, 1989b, Multilevel regulation of *Streptomyces* differentiation, *Trends Genet.*, 5:372.

Chater, K.F., Bruton, C.J., Plaskitt, K.A., Buttner, M.J., Méndez, C., and Helmann, J., 1989, The developmental fate of *S. coelicolor* hyphae depends crucially on a gene product homologous with the motility sigma factor of *B. subtilis.*, Cell, 59:133.

Clayton, T.M., and Bibb, M.J., 1990, *Streptomyces* promoter-probe vectors that utilise the *xylE* gene of *Pseudomonas putida*, Nucl. Acids Res., in press.

Guijarro, J., Santamaria, R., Schauer, A., and Losick, R., 1988, Promoter determining the timing and spatial localization of transcription of a cloned *Streptomyces coelicolor* gene encoding a spore-associated polypeptide, *J. Bacteriol.*, 170:1895.

Hardisson, C., and Manzanal, M.B., 1976, Ultrastructural studies of sporulation in *Streptomyces. J. Bact.*, 127:1443.

Helmann, J.D., Márquez, L.M., and Chamberlin, M.J., 1988, Cloning, sequencing and disruption of the *Bacillus subtilis* σ^{28} gene, *J. Bact.*, 170:1568.

Higgins, C.F., Dorman, C.J., Stirling, D.A., Waddell, L., Booth, I.R., May, G., and Bremer, E., 1988, A physiological role for DNA supercoiling in the osmotic regulation of gene expression in *S. typhimurium* and *E. coli*, Cell, 52:569.

Hopwood, D.A., Wildermuth, H., and Palmer, H.M., 1970, Mutants of *Streptomyces coelicolor* defective in sporulation, *J. gen. Microbiol.*, 61:397.

Hopwood, D.A., Bibb, M.J., Chater, K.F., Kieser, T., Bruton, C.J., Kieser, H.M., Lydiate, D.J., Smith, C.P., Ward, J.M., and Schrempf, H., 1985, Genetic Manipulation of *Streptomyces*: A Laboratory Manual, The John Innes Foundation, Norwich.

Ingram, C., Brawner, M., Youngman, P., and Westpheling, J., 1989. *xylE* functions as an efficient reporter gene in *Streptomyces* spp.: use for the study of *galP1*, a catabolite-controlled promoter, *J. Bacteriol.*, 171:6617.

Kendrick, K.E., and Ensign, J.C., 1983, Sporulation of *Streptomyces griseus* in submerged culture, *J. Bacteriol.*, 155:357.

McVittie, A., 1974, Ultrastructural studies on sporulation in wild-type and white colony mutants of *Streptomyces coelicolor*, *J. gen. Microbiol.*, 81:291.

Méndez, C., and Chater, K.F., 1987, Cloning of *whiG*, a gene critical for sporulation of *Streptomyces coelicolor* A3(2), *J. Bacteriol.*, 169:5715.

Méndez, C., Braña, A.F., Manzanal, M.B., and Hardisson, C., 1985, Role of substrate mycelium in colony development in *Streptomyces*, *Canad. J. Microbiol.*, 31:446.

Preiss, J., 1984, Bacterial glycogen synthesis and its regulation, Ann. Rev. Microbiol., 38:419.

Schauer, A., Ranes, M., Santamaria, R., Guijarro, J., Lawlor, E., Méndez, C., Chater, K., and Losick, R., 1988, Visualizing gene expression in time and space in the morphologically complex, filamentous bacteria *Streptomyces coelicolor*, Science, 240:768.

Sherman, D.H., Malpartida, F., Bibb, M.J., Kieser, H.M., Bibb, M.J., and Hopwood, D.A., 1989, Structure and deduced function of the granaticin-producing polyketide synthase gene cluster of *Streptomyces violaceoruber* Tü 22, *EMBO J.*, 8:2717

Wildermuth, H., 1970, Development and organisation of the aerial mycelium in *Streptomyces coelicolor*, *J. gen. Microbiol.*, 60:43.

Wildermuth, H., and Hopwood, D.A., 1970, Septation during sporulation in *Streptomyces coelicolor*, *J. gen. Microbiol.*, 60:57.

Zukowski, M.M., Gaffney, D.F., Speck, D., Kauffmann, M., Findeli, A., Wisecup, A., and Lecocq, J.-P., 1983. Chromogenic identification of genetic regulatory signals in *Bacillus subtilis* based on expression of a cloned *Pseudomonas* gene, J. Bacteriol., 80:1101.

PRESENCE OF FACTOR C IN STREPTOMYCETES AND OTHER BACTERIA

F. Szeszák, S. Vitális, I. Békési and G. Szabó

Institute of Biology, University Medical School, Debrecen, Hungary

INTRODUCTION

Streptomyces have complex life cycle, routes of differentiation. The mycelium contains hyphae representing the vegetative and the reproductive phase of differentiation at the same time. A great deal of observation supports the idea that this differentiation is a highly ordered and genetically determined process. It is relevant that the regulation of morphological differentiation and secondary metabolism are intimately connected. An intriguing example of this relationship is the finding described by Chater's group[1,2] that the reproductive phase and secondary metabolism does not begin in the absence of a particular leu-tRNA reading the rare UUA codon. Considering that antibiotics are the products of the secondary metabolism, the importance of the studies concerning these regulatory processes cannot be overestimated[3]. A promising field of research is the investigation of those endogenously produced substances (autoregulators)

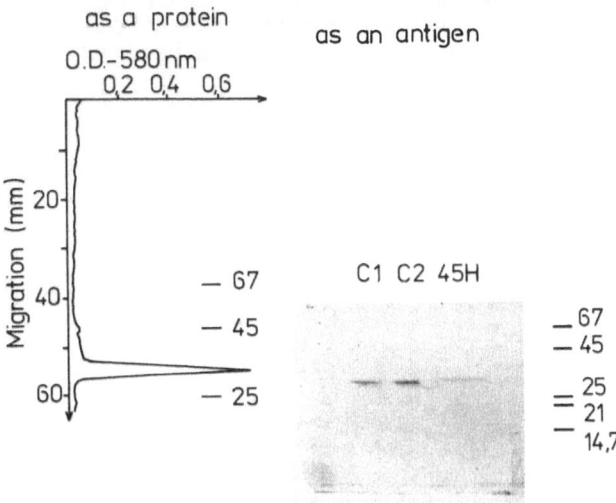

Fig.1. Molecular size of the purified, biologically active factor C and factor C antigen. Position of marker proteins are indicated by bars and their molecular mass given in kDa. A: Dodecyl sulphate-polyacrylamide gel electrophoresis. B: Immunoblotting made after electrphoresis of the same sort. C1 and C2: two preparations of purified factor C; 45H: Mycelial homogenate of the "producer" strain.

that were shown to influence the differentiation of *Streptomyces* (A-factor, B-factor, Pamamycin 607, and factor C, reviewed by Szabó et al.[4]).

Factor C was detected in and later isolated from the cultivation medium of the "producer" *S. griseus* 45H strain[5,6] as a regulatory protein. It induces the formation of preconidia characteristic for the reproductive phase of the life cycle[7] in submerged cultures of the "test" strain *S.griseus* 52-1 at concentrations as low as 0.5-1 ng/ml. The shift of the cultures to reproductive phase upon the effect of factor C was also detected by following the change of composition and synthesis of proteins in mycelial homogenates[8].

Studies on the production and mode of action of factor C got impetus from the introduction of new experimental methods such as measuring potassium release[9] in mycelial suspensions and elaborating ELISA and immunoblotting techniques with specific anti-factor C monoclonal antibodies[10].

FACTOR C AS A PROTEIN

Factor C was purified to electrophoretic homogeneity first by phosphocellulose and DNA-agarose chromatography[6] and was found to have molecular mass of 34.5 kDa. This value was corroborated by immunoblotting using specific anti-C monoclonal antibody (Fig.1.).

The only immunoreactive bands found in either two different preparations of purified factor C or in crude cell-homogenate of *S. griseus* 45H mycelium had the same molecular mass as factor C protein did. Further characteristics of factor C are: i) It loses biological activity by heating 10 min at 70 °C. ii) It binds both to single and double stranded heterologous DNA. iii) It showes hydrophobic property; It can be bound to and recovered from Xad-2 column. iv) It showed heterogeneity in isoelectric focusing. This heterogeneity was possible to indicate only before the introduction of quantitative ELISA and was a main obstacle in elaborating an efficient terminal purification method. By improving the ion-exchange chromatography steps and installing liquid phase isoelectric focusing in a novel type ampholyte carrier working along a pH range of 4-13, (Békési) a single main band of factor C was focused around pI 9.9 (Fig.2.). The paradox observed with the factor C preparation obtained by DNA agarose chromatography is something yet to clarify. It was homogenous as for molecular size but was heterogenous as for electric charge and at least three of its isoelectric bands also reacted with the spe-

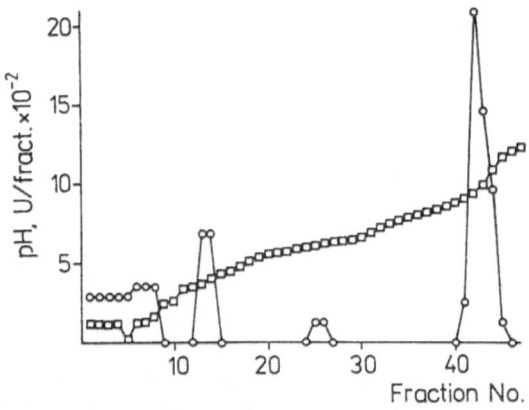

Fig.2. Liquid phase isoelectric focusing of purified factor C in ampholyte carrier of pH range between 4-13. Fractions were analysed by sandwitch ELISA technique using anti-C monoclonal antibody.
□ : pH: O : Factor C antigen.

cific monoclonal antibody. Differences in phosphorylation, ADP-ribosylation, different length of glucosamine side chain of the same molecular species are among the possibilities to study.

NEW DATA CONCERNING THE PRODUCTION OF FACTOR C

Fig.3. shows that factor C induces its own production. This assumption is supported by several observations. It was repeatedly controlled that factor C disappeared from cultures within 10 minutes after inoculation. The biological activity of factor C in the cultivation medium of 52-1 strain run higher than the input at the inoculation (Fig.3a.). The phenomenon appeared consistently on several subvariants of 52-1 strain (Fig. 3b.). It is also worth-mentioning that the auto-inducing activity was less effective at 5 U/ml than at 1 or 2.5 U/ml concentration.

In cultures grown on solid agar media, factor C antigen could readily be detected under or at some distances away from the colonies (Fig.4.).

These values should be considered, however, as qualitative indications only because of the high exoprotease activity present in the same system (data not shown). Even so, these data remarkably substantiated our previous assumption on the basic regulatory function of factor C in *Streptomyces* mycelium[4]. It was shown by immuncytochemistry that factor C is distributed unevenly, reproductive type of hyphae containing detectable amount of factor C appeared in the form of heavily stained foci of the mycelium. It seems as if the production of factor C would start at given points from

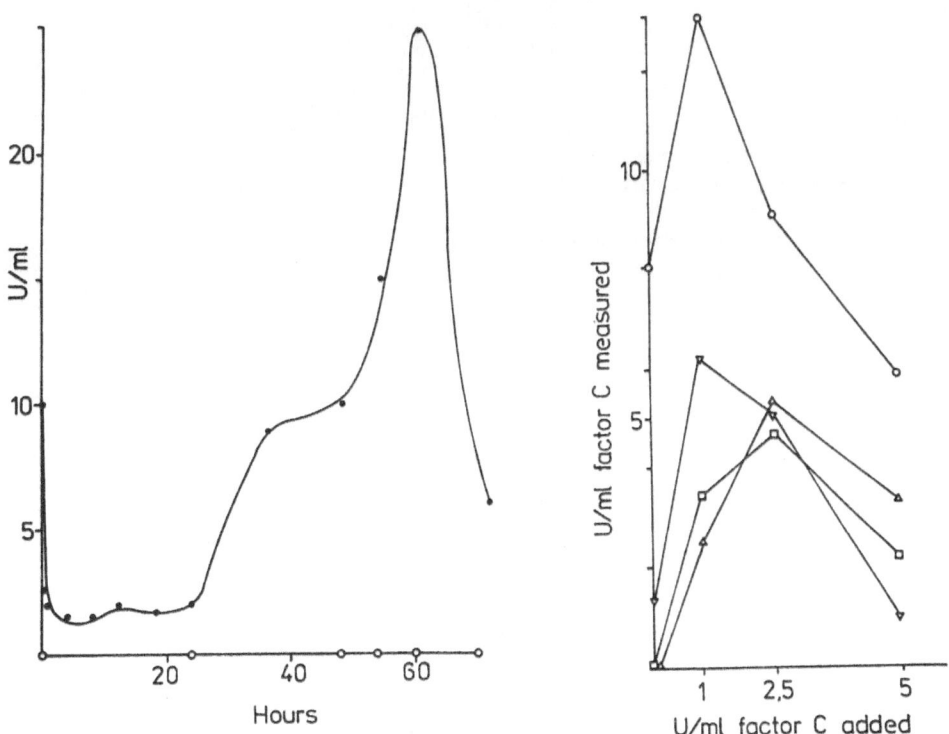

Fig.3. Inducing effect of factor C upon its own production. Factor C was added to the cultures at the time of inoculation. In the experiment shown in the right-hand side diagram, samples were tested after 72 h cultivation. Indicator strain subvariants: o :52-1/10; △ : 52-1/2; □ : 52-1/4; ▽ :52-1/1.

where it then spreads through an amplification process. If one takes also into account that factor C stimulated its own production as shown above, it can be concluded that the production of factor C begins in some centres of the mycelium, the active compound is excreted by the producer hyphae and induces other hyphae like a hormone. The induced hyphae would become factor C producers themselves.

EXAMINATION OF THE MODE OF ACTION OF FACTOR C

In washed mycelial suspensions factor C − in concentration as low as 0.7 U/ml (about 0.7 ng/ml) − shifted the equilibrium between *Streptomyces* mycelium and the medium toward higher external K^+ concentrations[9] (Fig.5.).

The K^+ release followed a characteristic "slow kinetics" (Fig.5a.) with about 3-10 min latency. It differed from the time course of the effect of compounds acting directly on the cytoplasmic membrane. Fig.5b shows K^+ release of "fast kinetics", ensuing within milliseconds as observed in an experiment in collaboration with Dr. U. Grafe, Jena in which a Pamamycin 607-621 mixture was added to the system. Several other compounds were also examined to see if they had a comparable effect to that of factor C on K^+

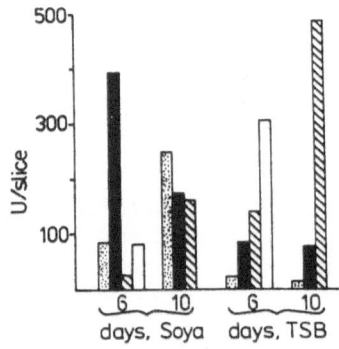

Fig.4. Factor C determined in agar cultures of S. griseus 45H. ▨ : Slice with the colony; ■ : Slice under the colony; ◩ : Slice beside the colony: □ : Slice 4 mm away from the colony.

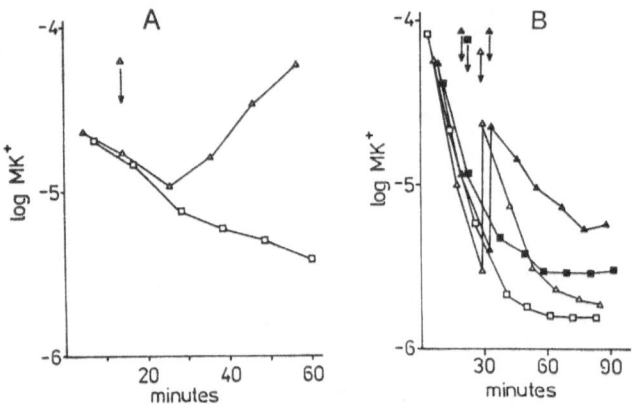

Fig.5. Release of potassium from S. griseus mycelium upon the effect of factor C and Pamamycin. A: Release of "slow kinetics" after the addition of factor C. □ : Control; △ : 28 U/ml factor C. B: Release of "fast kinetics" after the addition of 1.5 ug/ml of 1:1 mixture of Pamamycin 607-621. □: Control; ■ : 28 U/ml factor C; △ : Pamamycin; ▲ : Factor C then Pamamycin added to the same sample.

release. From among substances such as histones, poly-L-lysine, protamine, cytochrome C, heparin, poly-L-glutamic acid and chlorpromazine tested none of this character was found. Either they resulted in K^+ release of "fast kinetics" (histones and chlorpromazine) or were ineffective.

The effect of the other autoregulator A-factor on the K^+ release was also tested[4]. In contrast to factor C, it caused a decrease of extracellular K^+ in comparison with the control in *S. griseus* JA 5142/86. When the two autoregulators were added to the same mycelial suspension they did not synergize or inhibit each others effect.

Considering the assumptions[11] on the relationship between the effect of A-factor upon the change of activity of NADGH and the transition between the vegetative and reproductive phases of growth, we examined the effect of nicotinamide on the K^+ release[4]. To our surprise, this low molecular mass NADGH inhibitor induced a very similar K^+ release of "slow kinetics" as factor C did. It prompted us to examine the effect of factor C on NADGH activity.   (Table 1.).

Table 1

Effect of factor C on the NAD-glycohydrolase activity of Streptomyces

NAD-glycohydrolase activity was determined by measuring the consumption of NAD^+ after terminating the reaction with CN^- in mycelial homogenates and cultivation liquids of S. griseus strains as indicated.

Strain	Medium	Cult time h	NADGH activity Range[a] nmol/μg	Contr[b] Sup:myc	Ratio Sup[c] C:fC[e]	Myc[d] C:fC
a) Factor C added directly to the cultures.						
52−1	Synth[f]	16−22	0.2−2.5	1.61	4.07	2.49
52−1	Synth	19−25	0.4−7.8	0.27	4.41	6.54
52−1	Synth	16−23	0.13−0.55	1.04	7.01	2.14
52−1	Synth	19−21	−	0.15	2.08	3.27
2682	Soya[g]	19−21	0.5−4.6	0.60	4.06	1.30
52−1	Synth	17−19	0.42−2.0	2.37	2.85	2.05
52−1	Soya	20−26	0.06−1.13	0.83	6.66	5.95
2682	Soya	19−21	2.4−4.0	1.40	1.75	1.30
2682	Soya	19−21	3.0−18.0	2.02	1.57	1.03
b) Factor C added to washed mycelium suspensions.						
52−1	Synth	19−21	−	−	1.54	1.54
52−1	Synth	19−20	0.5−1.4	−	1.2	1.16
52−1	Synth	19−20	0.04−1.1	−	1.25	1.07
52−1	Synth	18−20	0.06−0.7	−	1.85	1.26

[a] The minimum and maximum of enzyme activity measured in the given experiment; [b] Control culture without factor C; [c] Cultivation medium; [d] Mycelial homogenate; [e] Ratio of control:factor C treated samples; [f] Synthetic medium; [g] Soy bean − corn steep medium.

30 U/ml factor C consistenly decreased NADGH activity in the cultivation medium and in the mycelial homogenate. The earliest differences were measured after 15-30 min and marked decrease developed after 2-3 h. (n.b. factor C had no effect when added directly to the enzyme assay). Quantitatively NADGH activity and its decrease was highly variable and did not seem to change in correlation with either the strain, medium or condition of incubation employed.

We do not know the details of relationship between the mode of action of A-factor and factor C although the fact that both affect NADGH activity indicate that the metabolic pathways regulated by them are at least in part common. Factor C induces a chain of events leading to differentiation. One of the steps in this chain might be the ribosylation of certain proteins. Both A-factor and factor C may act by affecting the level of NADGH in an opposite way, a key enzyme in the ADP-ribosylation.

Considering the pleiotropic effect of autoregulators by switching on the genes which coordinately direct the synthesis of clusters of enzymes necessary to development and antibiotic production makes us confident that research to understand the regulation of differentiation may lead to the discovery of genes of great practical importance.

OCCURENCE OF FACTOR C ANTIGEN IN DIFFERENT SPECIES

Our ELISA technique[10] was applied first to search for factor C antigen in different *Streptomyces* strains.

Table 2

Presence of factor C antigen in different Streptomyces strains

Strain	$Amy^{+/-}$	Time of harvest (h)[b]	Factor C antigen[a] supernatant	homogenate
S. griseus 45H	+	48	3240	3030
\[c] *S. griseus* 45-8702/16B	−	48	29.1	18.4
\ *S. griseus* 45-8702/16A	−	48	234	17.2
S. griseus 52-1	+	72	74	24
\ *S. griseus* 14 Fb 03	−	72	108	5.1
\ *S. griseus* 52-8509/44	−	72	74.5	n.d.[d]
S. griseus LS-1/8	−	24	224	n.d.
\ *S. griseus* LS-1/4052	−	72	175	n.d.
S. griseus 1439	−	72	185	n.d.
S. griseus JA 5142/86	−	72	32.5	n.d.
S. griseus NRRL B-2682	+	24	7.4	7.9
\ *S. griseus* Bald 10	−	18	n.d.	17.7
S. coelicolor 1438	+	24	494	486
\ *S. coelicolor* 145	+	60	33	21
S. lividans 66/1326	+	24	430	n.d.
S. fradiae LN 47-IG-1	+	120	2752	n.d.
S. flavofungini DV 147	+	120	1129	619
S. flavofungini AA	+	72	245	n.d.
S. chrysomallus ISP 5128	+	24	244	43
\ *S. chrysomallus* Col 02	+	24	37	43
S. antibioticus 11891	+	48	682	n.d.
S. odorifer FBUA 1161	+	24	191	23
S. levoris FBUA 1114	+	72	706	205

a   Units/mg protein
b   The highest factor C values determined during the cultivations are presented
c   Variants of some particular strain
d   Not detectable

As Table 2. shows, from among 23 *Streptomyces* strains belonging to 9 species investigated thus far there was none which, at least in some period of its cultivation, did not produce factor C antigen. This finding was concordant with earlier observations. i) Compounds showing factor C-like effect by inducing the formation of preconidia in *S. griseus* 52-1 were also detected in other *Streptomyces* strains[12]. ii) Factor C was shown to induce K^+ release from washed mycelium of nine *Streptomyces* strains[9].

The occurence of antigen was highly variable. Sometimes it was present in traces and the ELISA was successful from fermentation liquid concentrates only. A 1000 fold difference was found between the lowest and highest quantities determined.

By comparing its amount in mycelia and in fermentation liquids it is seen that most of factor C was usually secreted into the medium. Out of 23 strains, 9 contained factor C antigen only in their supernatants. 8 strains contained twice as much antigen at least in the supernatant as in the mycelium.

The observation that *Amy⁻* strains generally produced less antigen than those which sporulated well supports the assumption that factor C was an inducer of reproductive phase[5]. The well-sporulating *S. griseus* NRRL B-2682 was the only exception, producing very little factor C antigen.

In addition to *Streptomyces*, factor C antigen was also detected with immunoblotting[10] in a very broad range of different living things from archaebacteria to *B. subtilis*, *Neurospora crassa* and human tumour cell line (Table 3.).

Table 3

Molecular mass of factor C-like antigens in biological specimens of different origin

Strain	Cell homogenate	Culturing medium
	Molecular mass, kDa	
S. griseus 52-1	34	34
S. griseus 45H	34	34
S. flavofungini DV 147	34 and 70	34
S. coelicolor M-145	70	—*
S. antibioticus 77451	34	34
B. subtilis ATCC 6633	34 and 70	34
E. coli 3300	34	34
Halobacterium halobium NRL	34	34 and 70
Neurospora crassa RL-38 (wild)	34	34
Neurospora crassa FGSC-1118 (slime)	34	—
Mouse erythroleukemia DS-19	34	34
Human leukemia HL-60	—	34

* — not determined.

In all the specimens examined, 1 or 2 immunoreactive bands were found only having the same molecular mass as determined in *Streptomyces* (Mr 34500 and 70000). The presence of the same epitope in proteins of the same molecular mass makes it probable that all these species share a kind of protein closely related to factor C — an autoregulator of *S. griseus*.

Looking at the species examined, factor C would be among the most conserved proteins during evolution described until now.

References

1. E. J. Lawlor, H.A. Baylis and K. F. Chater, Pleiotropic morphological and antibiotic deficiencies result from mutations in a gene encoding a tRNA-like product in *Streptomyces coelicolor* A3(2), Genes Devt. 1:1305-1310 (1987)
2. K. F. Chater, E. J. Lawlor, C. Mendez, C. J. Bruton, N. K. Davis, K. Plaskitt, E. P. Guthrie, B. L. Daly, H.A. Baylis and K. Vu Trong, Gene expression during *Streptomyces* development, in: "Biology of Actinomycetes", Y. Okami, T. Beppu and H. Ogawara eds., Japan Scientific Society Press, Tokyo, pp. 64-70 (1988).
3. D. A. Hopwood, Antibiotics: opportunities for genetic manipulation, Phil. Trans. R. Soc. Lond. B 324:549-562 (1989).
4. G. Szabó, F. Szeszák, S. Vitális and F. Tóth, New data on the formation and mode of action of factor C, in: "Biology of Actinomycetes", Y. Okami, T. Beppu and H. Ogawara eds., Japan Scientific Society Press, Tokyo, pp. 324-329 (1988).
5. G. Szabó, T. Vályi-Nagy and S. Vitális, A new factor regulating life cycle of *Streptomyces griseus*. in: "Genetics of Microorganisms, Proc. Symp. on Heredity and Variability of Microorganisms", V. D. Timakova, ed., State Publishing House of Medical Literature, Moscow, pp. 282-292 (1962).
6. S. Bíró, I. Békési, S. Vitális and G. Szabó, A substance effecting differentiation in *Streptomyces griseus*. Eur. J. Biochem. 103:359-363 (1980).
7. S. Vitális, and G. Szabó, Cytomrphological effect of factor C in submerged cultures on the hyphae of *Streptomyces griseus* strain No. 52-1, Acta Biol. Acad. Sci. Hung. 20:85-92 (1969)
8. S. Vitális, G. Valu, I. Békési, F. Szeszák and G. Szabó, Changes of protein pattern of *Streptomyces griseus* strains that characterize differentiation of mycelia in submerged culture, J. Basic Microbiol. 28:393-407 (1988)
9. F. Szeszák, S. Vitális and G. Szabó, Factor C, a regulatory protein of *Streptomyces griseus* induces release of potassium from the mycelium, J. Basic Microbiol. 29:233-240 (1989)
10. F. Szeszák, S. Vitális, F. Tóth, G. Valu, J. Fachet and G. Szabó, Detection and determination of factor C — a regulatory protein — in *Streptomyces* strains by antiserum and monoclonal antibody, Arch. Microbiol. in press, (1990).
11. J. G. Ensign, M. J. McBride, L. J. Stoxen, A. Bertinuson, M. Pomplun and A. Ho, The life cycle of *Streptomyces*: germination and properties of spores and regulation of sporulation, in: "Biological, Biochemical and Biomedical Aspects of Actinomycetes, G. Szabó, S. Bíró and M. Goodfellow, Akadémiai Kiadó, Budapest, pp. 777-790 (1986).
12. G. Szabó, S. Bíró, L. Trón, G. Valu, S. Vitális, Mode of action of factor C upon the differentiation process of *Streptomyces griseus*. in: "Biochemical and Biomedical Aspects of Actinomycetes", L. Ortiz-Ortiz, L. F. Bojalil, V. Yakoleff, eds., Acad. Press, Orlando, pp. 197-214 (1984).

STEPTOMYCES GRISEUS PROTEINS POST-TRANSLATIONALLY MODIFIED BY

ADP-RIBOSYLATION DURING THE LIFE CYCLE

Andras Penyige[1], Gyorgy Barabas[1], and Jerald C. Ensign[2]

[1]Institute of Biology, Univ. Med. School, Debrecen, Hungary
[2]Department of Bacteriology, Univ. of Wisconsin
Madison, Wisconsin, 53706, USA

The physiology and regulatory processes of differentiation in Streptomycetes are still obsurce in many respects. Since the problem is very complex it can be approached from different sides. One of these is the investigation of the physiologicl role of post-translational regulation by ADP-ribosylation in Streptomyces griseus.

NAD, the well known respiratory cofactor, is used as a source of ADP-ribose (ADPR), in ADP-ribosyltransfcrase (ADPRT) reactions and the acceptor of ADPR is a protein molecule.

This post-translational protein modification process is an important mechanism to control many physiological processes both in eukaryotic and prokaryotic organisms.

ADP-ribosylation of proteins in eukaryotic cells plays a role in regulation of cell proliferation and differentiation, modification of chromatin structure and repair (6,9,16). several pathogenic bacteria produce exotoxins that catalyze ADP-ribosylation of specific eukaryotic proteins (10,8,17).

We have less information about ADP-ribosylation reactions in prokaryotic cells. From the few data available for prokaryotic ADP-ribosylations it is known that in E. coli the T4 and N4 phage infection results in ADP-ribosylation of the α subunit of the RNA-polymerase molecule (15). Nitrogenase activity in Rhodospirillum rubrum is regulated by ADP-ribosylation of the iron-protein subunit of the enzyme (14). Two cytoplasmic proteins of Pseudomonas maltophila were reported to be ADP-ribosylated (4). In this communication we present our results about the ADP-ribosyltransferase activity of Streptomyces griseus and discuss the role of this enzyme in the physiological regulation of sporulation.

The presence of the NAD-glycohydrolase (NADGH) activity in S. griseus is a well known fact (3,7). Several years ago our group began to investigate the possibility of the presence of the similar enzyme, the ADPRT in S. griseus strains (1).

ADPRT activity of S. griseus 52-1

The incubation of purified membrane preparation in the presence of ^{32}P-NAD$^+$ and the SDS-PAGE analysis of the reaction products followed by autoradiography revealed that one distinct protein band with a molecular weight of 32 kDa was labeled by ^{32}P-NAD. Longer exposure of the dried gel to X-ray film revealed less intensively labeled protein bands with molecular weights of 48 kDa, 84 kDa and 90 kDa. The autoradiogram of the SDS-PAGE analysis shown in Fig. 1.

In order to prove that the labeling of the membrane proteins was due to ADPRT activity, the nature of the transfer reaction and the modified protein products were investigated.

Identification of ADPRT activity

First, six inhibitors known to interfere with ADPRT activity were tested for their effect on incorporation of ^{3}H-NAD into proteins of S. griseus membrane preparations (13). Table 1 shows that each of the inhibitors partially blocked the incorporation of ^{3}H-ADP-ribose from ^{3}H-NAD$^+$ into the proteins. The most effective were 6-aminobenzamide and 3-aminobenzamide. It is important to be noted that these inhibitors at the concentrations used in our experiments had essentially no effect on the NADGH activity of the membrane preparations. These data clearly show that the incorporation of radioactivity from labeled NAD$^+$ to proteins is the result of ADPRT action.

The nature of the modifying group was determined by HPLC analysis. The bond between the ADP-ribose and the protein is alkaline labile; the stability depends on the type of the glycosidic bond. The HPLC analysis of the products of alkaline hydrolysis of the membrane sample prelabeled with ^{3}H-NAD$^+$ showed only one compound which was radioactive and the retention time of this UV absorbing peak was identical with that of the authentical ADP-ribose standard (data not shown) (13). Further alkaline hydrolysis of the material of this peak at higher temperature led to the formation of 5'-AMP. These results indicate that the modifying group on the protein molecules is ADP-ribose.

Evidence of NAD$^+$: arginine ADPRT activity in S. griseus

Table 1 shows the inhibitory effect of phenylglyoxal, an arginine blocking agent, on the ADPRT activity of the strain. Since phenylglyoxal is a specific blocker of guanidine groups and it is not an inhibitor of the enzyme itself, we concluded that the enzyme in the membrane is an NAD: arginine ADP-ribosyltransferase.

To confirm this possibility the purified membrane preparation was incubated in the presence of ^{32}P-NAD and agmatine, a small guanidine group containing molecule. Moss and co-workers showed that certain mono-ADPRT enzymes, for example the cholera toxin, were able to trasfer the ADP-ribose moiety to small guanidine compounds (11). We have analysed the reaction products by TLC. The result is shown in Fig. 2. Incubation of the membrane preparation in the absence of agmatine (lane 2) resulted in hydrolysis of all the NAD$^+$ to ADP-ribose by the NADGH activity of the membrane, but incubation of the preparation with agmatine and NAD$^+$ results in the formation of two labeled compounds which are stereoisomeric forms of ADP-ribose-agmatine, in addition to ADP-ribose (lanes 3,4). The activated cholera toxin hydrolysed NAD^+ to a small amount of ADP-ribose and formed the same labeled products. These data show clearly that the membrane preparation of S. griseus No. 52-1, like cholera toxin, contains NAD: arginine ADP-ribosyltransferase activity. Similar activity was detected in several other Streptomyces strains (2).

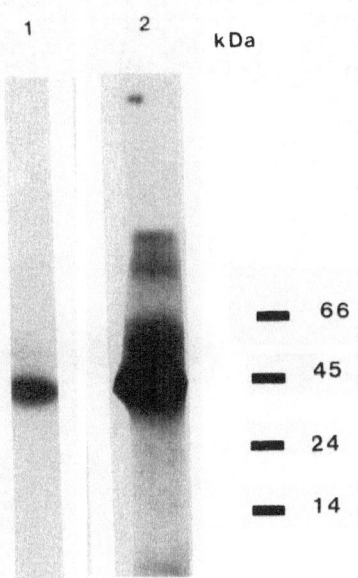

Fig. 1 Autoradiogram of SDS-PAGE separation of ADP-ribosylated membrane proteins of S. griseus No. 52-1

Table 1 The in vitro effect of inhibitors of ADPRT on incorporation of ^{3}H-NAD into proteins of S. griseus No. 52-1

Inhibitor	Concentration (mM)	Activity (% of control)
3-aminobenzamide	1	36
	10	49
Nicotinamide	1	20
	10	36
Isoniazide	1	27
	10	39
Phenylglyoxal	10	65
6-aminobenzamide	10	65

After the detection of ADPRT activity in <u>Streptomyces</u> strains we went on to study the role of this enzyme in the regulation of differentiation of <u>S. griseus</u>. We chose to work with the streptomycin producer wild type No. 2682 strain. J.C. Ensign and his co-workers developed a short life cycle culture system for this strain and their previous experiments revealed that the sporulation of this organism may involve a time determined sequence of events (5). Using valine as a repressor of sporulation and iso-leucine to reverse valine repression it was possible to determine that the cells reached the time of commitment to sporulation at 10-13 hours after germination in defined medium (5,2).

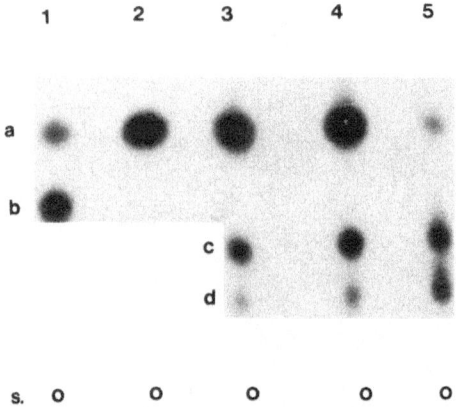

Fig. 2 Thin layer chromatographic analysis of products of ADP-ribosylation of agmatine by autoradiography.
Lane 1: position of authentic ADP-ribose (a) and NAD (b).
Lane 2: membrane sample incubated with ^{32}P-NAD without agmatine.
Lane 3 and 4: membrane sample incubated with ^{32}P-NAD and agmatine, c and d represents positions of the two isomeric forms of ADP-ribose-agmatine.
Lane 5: incubation of cholera toxin with ^{32}P-NAD and agmatine.

Using these culture conditions the specific activity of the ADPRT was determined in the function of the age of the culture. The results are shown in Fig. 3. Significant changes in the specific activity of the enzyme were observed during the life cycle. The specific activity displays a large maximum at 8 hours after germination, then after 12 hours the activity falls sharply and remains low during the rest of the life cycle except around 18 hours when it shows a smaller maximum. In the culture which was complemented with 20 mM valine at the inoculation time the second peak is missing.

In separate experiments we have also determined the pattern of ADP-ribosylated proteins present in the cells at different times during the life cycle. Aliqout samples were taken from the culture at certain times after germination and the proteins were labeled by ^{32}P-NAD. The proteins were then separated by SDS-PAGE and the pattern of labeled proteins were determined by autoradiography. Fig. 4. shows the autoragiogram of the ADP-ribosylated proteins.

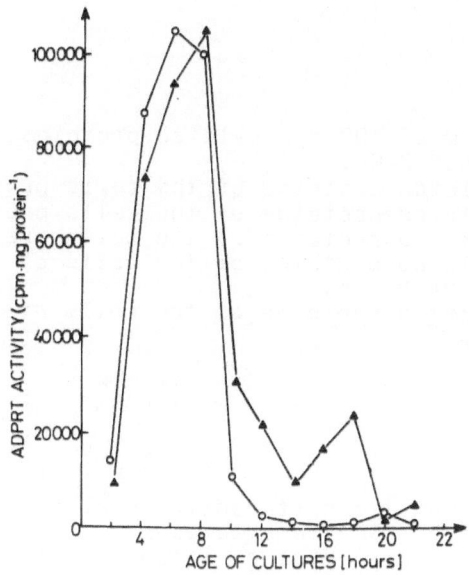

Fig. 3 ADPRT activity of strain No 2682 in DM-2 medium in the presence (○) and in the absence (▲) of valine.

The comparison of the pattern of ADP-ribosylated proteins present at 8 hours, 14 hours and 24 hours respectively after germination reveals a significant change in the protein pattern. At 8 hours only two ADP-ribosylated proteins are present in the cell with molecular weights of 47 kDa and 61 kDa respectively.

In the sample taken at 14 hours an additional higher molecular weight (71 kDa) protein appears and this pattern does not change during the rest of the life cycle. These data show that both the specific activity of the ADPRT and the pattern of the ADP-ribosylated proteins change significantly at the time of commitment to sporulation.

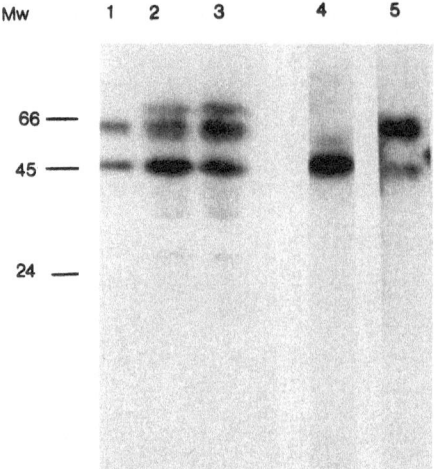

Fig. 4 Autoradiogram of ADP-ribosylated proteins of S. griseus No. 2682 separated by SDS-PAGE.
Lane 1: ADP-ribosylated proteins of the cells present at 8 hours.
Lane 2: ADP-ribosylated proteins of the cells present at 14 hours.
Lane 3: ADP-ribosylated proteins of the cells present at 24 hours.
Lane 4: ADP-ribosylated proteins of the cells of valine repressed culture present at 24 hours.
Lane 5: ADP-ribosylated proteins of the cells of the Bld mutant present at 24 hours.

To demonstrate that the differentiational state of the strain correlates with the ADP-ribosylation of specific acceptor proteins, the pattern of ADP-ribosylated proteins were determined in the valine repressed culture (lane 4. Fig. 4) of the No. 2682 strain at 24 hours after germination. It is clear that the late, highest molecular weight protein is missing (or not ADP-ribosylated) and because of this, the pattern is similar to the pattern characteristic to the control culture before the commitment time, except that the 47 kDa protein seems to be present in higher amount. In the possession of these results in similar experiments we have examined the ADP-ribosylated proteins present in a non-sporulating, so called bald (Bld) mutant of the wild type No. 2682 strain (lane 5. Fig. 4). The result is similar to that of the valine repressed culture. Again the same highest molecular weight band is missing from the pattern.

These data suggest that ADP-ribosylation of specific protein molecules in the S. griseus 2682 strain correlats with the events of sporulation.

To demonstrate that the enzyme plays an active regulatory role in the sporulation process, we have conducted in vivo experiments, using a specific inhibitor of the ADPRT, to inhibit the enzyme activity in the mycelia. The defined medium was supplemented with 10 mM 3-aminobenzamide, considered to be the most specific inhibitor, at the inoculation time or at 18

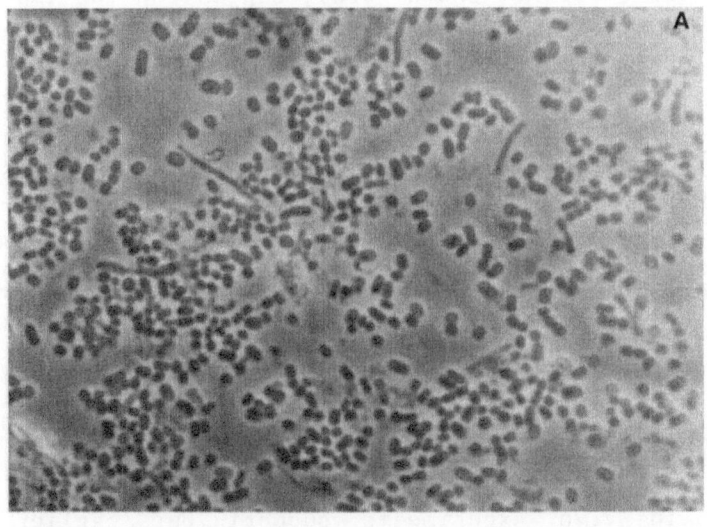

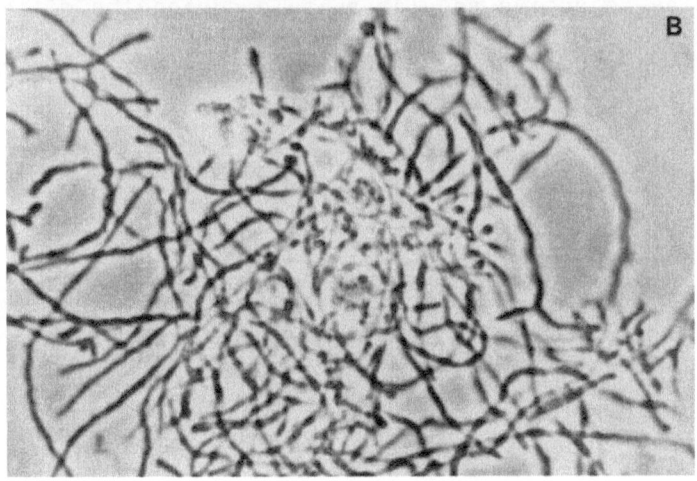

Fig. 5 The effect of 3-aminobenzamide on the sporulation of S. griseus No. 2682 grown in DM-2 medium.
A.: the control culture at 36 hours after germination.
B.: the culture was supplemented with 10 mM 3-aminobenzamide at the time of inoculation, the photomicrograph was taken at 36 hours after germination.

hours after germination, and the sporulation of the cultures was examined at 36 hours after germination. Fig. 5/A. shows the control sample containing only spores, and Fig. 5/B. shows that the 3-aminobenzamide added at 0 time inhibited the sporulation of the strain. 3-aminobenzamide added at 18 hours, however, had no effect at all (data not shown here), suggesting the importance of ADPRT involvment before commitment time in determining the fate of the mycelia.

To support this suggestion we also determined the in vivo effect of the same inhibitor on the non-sporulating Bld mutant. As it is well known, the spore forming ability of certain Bld mutants can be restored with A-factor, a morphogen produced by S. griseus. DM (defined medium) agar medium was inoculated with the Bld strain, and paper discs, containing the A-factor, were deposited on the surface at the time of inoculation. The inhibitor was added at 2 hours after inoculation in one case (Fig 6/A), and at 24 hours after inoculation in the other case (Fig 6/B), in 10 mM concentration, into the holes drilled into the medium at different distances from the paper discs to create different concentration gradients of A-factor and 3-aminobenzamide. Fig. 6 shows clearly that the inhibitor added early at life cycle inhibited the A-factor induced sporulation of the Bld mutant, when added one day later the 3-aminobenzamide had no effect at all on the spore formation.

These in vivo experiments support our earlier findings, suggesting that the ADPRT enzyme and ADP-ribosylation of certain proteins play an active regulatory role during the differentiation of S. griseus. Furthermore the in vivo experiments with the Bld mutant also suggest that the mechanism of action of A-factor possibly involves the ADPRT activity. The active enzyme is required for the A-factor induced sporulation process.

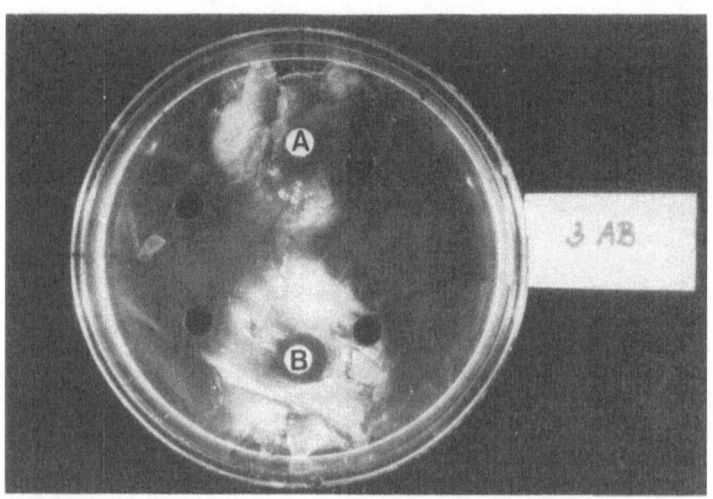

Fig. 6 The effect of 3-aminobenzamide on the A-factor induced sporulation of S. griseus Bld, a non-sporulating mutant. 3-aminobenzamide was added at 2 hours (A.) and 24 hours (B.) after the inoculation.

Literature

1. Barabás, Gy., A. Penyige, I. Szabó, and J. C. Ensign. 1985. ADP-ribosylation in Streptomyces griseus. In G. Szabó, S. Biró, M. Goodfellow. (ed.) Sixth Int. Symp. on Actinomycetes Biol. vol. 2. Hungarian Publishing House. Budapest. pp. 827-829.
2. Barabás, Gy., A. Penyige, I. Szabó, J. Barabás, and J. C. Ensign. 1988. Post-translational protein modification by ADP-ribosylation in Streptomyces. In Y. Okami, T. Beppu, H. Ogawara. (ed.) Biology of Actinomycetes. Japan Scientific Societies Press. Japan. pp. 439-444.
3. Bröker, M., J. Schindelmeiser, and H. Pape. 1979. A nicotinamide adenine dinucleotide(phosphate) glycohydrolase NAD(P)ase, EC 3.2.2.6 from Streptomyces griseus. FEMS Microbiol. Lett. 6:245-247.
4. Edmonds,C.,Griffin, E. G. and Johnstone, A. P. 1989.Demonstration and partial characterisation of ADP-ribosylation in Pseudomonas maltophila.Biochem. J. 261:111-118.
5. Ensign, J. C. 1988. Physiological regulation of sporulation of Streptomyces griseus. In Y. Okami, T. Beppu, H. Ogawara, (ed.) Biology of Actinomycetes. Japan Scientific Societis Press. Japan. pp.309-315.
6. Ferro, A. M., N. P. Higgins, and B.M. Olivera. 1983. Poly(ADP-ribosylation) of a DNA topoisomerase. J. Biol. Chem. 258:6000-6003.
7. Grafe, U., E. J. Borman, and G. Truckenbrodt. 1980. Control by phospho-adenosinediphospho-ribose of NADP-dependent isocitrate dehydrogenase and 6-phosphogluconatedehydrogenase in Strepromyces griseus. Z. Allg. Mikrobiol.20:607.
8. Lai, C-Y. 1986. Bacterial protein toxins with latent ADP-ribosyltransferase activities. Adv. Enzymol. 58:99-139.
9. Lucas, D. L., S. I. Tanuma, P. J. A. Davis, D. G. Wright, and G. S. Johnson. 1984. Maturation of human promyelocytic leukemia cells induced by nicotinamide: Evidence of regulatory role for ADP-ribosylation of chromosomal proteins. J. Cell. Physiol. 121:334-340.
10. Moss, J., and M. Vaughan. 1979. Activation of adenylate cyclase by choleragen. Ann. Rev. Biochem. 48:581-600.
11. Moss, J., and S. J. Stanley. 1981. Amino acid specific ADP-ribosylation. Identification of an arginine-dependent ADP-ribosyltransferase in rat liver. J. Biol. Chem. 256:7830-7833.
12. Payne, D. M., E. L. Jacobson, J. Moss, and M. K. Jacobson. 1985. Modification of proteins by (mono)ADP-ribosylation in vivo. Biochemistry. 24:7540-7549.
13. Penyige, A., Barabás, Gy., Szabó, I. and Ensign, J. C. 1990. ADP-ribosylation of membrane proteins of Streptomyces griseus strain 52-1. FEMS Microbiol. Lett.69:293-298.
14. Pope, M. R., A. Scott, and P. W. Ludden. 1985. Covalent modification of the iron protein of nitrogenase from Rhodospirillum rubrum by adenosine diphosphoribosylation of a specific arginine residue. Proc. Natl. Acad. Sci. U.S.A. 82:3173-3177.
15. Skorko, R. 1982. T4 and N4 phage-encoded ADP-ribosyl-transferase. In O. Hayashi. K. Ueda.(ed.) ADP-ribosylation Reactions. Academic Press, New York. pp.:647-694.
16. Tanigawa, Y., M. Tsuchiya, Y. Imai, and M. Shimoyama. 1983. ADP-ribosylation regulates the phosphorylation of histones by the catalytic subunit of cyclic AMP-dependent protein kinase. FEBS Lett. 160:122-129.
17. Ueda, K., and O. Hayaishi. 1985. ADP-ribosylation. Ann. Rev. Biochem. 54:73-100.

INDUCTION OF THE STRINGENT RESPONSE IN *STREPTOMYCES COELICOLOR* A3(2) AND

ITS POSSIBLE ROLE IN ANTIBIOTIC PRODUCTION

E. Strauch, E. Takano and M.J. Bibb

John Innes Institute, John Innes Centre for Plant Science
Research, Colney Lane, Norwich, England, NR4 7UH

SUMMARY

The stringent response has been implicated as one of the triggers for
antibiotic production and sporulation in *Streptomyces* (Ochi 1986; 1987b).
We have characterised the stringent response and studied its possible role
in the initiation of actinorhodin biosynthesis in the genetically well-
defined strain *Streptomyces coelicolor* A3(2). The stringent response was
elicited in exponentially growing cultures either by addition of the amino
acid analogue serine hydroxamate or by nutritional shiftdown. In both
cases, transient accumulations of ppGpp were observed after 10 to 15
minutes that correlated with a severe reduction in the level of
transcription of a previosly cloned rRNA gene set (*rrnD*; Baylis and Bibb,
1986a). The influence of the stringent response on actinorhodin
biosynthesis was less obvious. While transcripts of the actinorhodin
cluster were observed approximately one hour after a nutritional shiftdown,
no such transcripts were observed after induction of the stringent response
by serine hydroxamate. These results suggest that ppGpp accumulation alone
is not sufficient for the induction of antibiotic production.

INTRODUCTION

The stringent response is a well known regulatory mechanism that was
originally discovered in *Escherichia coli* (for reviews see Gallant, 1979,
and Cashel and Rudd, 1987). Starvation of exponentially growing *E.coli*
cultures for amino acids results in the accumulation of highly
phosphorylated guanine nucleotides, of which ppGpp (guanosine 5'-diphophate
3'-diphosphate) is the most abundant species. These unusual guanine
nucleotides are formed by a ribosome associated enzyme (stringent factor)
from ATP and GTP when uncharged tRNA molecules bind to the A site of the
ribosome. The accumulation of ppGpp leads to a drastic reduction in the
overall rate of RNA synthesis; the production of stable RNAs, which are the
most actively transcribed species in fast growing cells, is particularly
curtailed. Two classes of mutations that reduce ppGpp synthesis have been
studied in detail. In *relA* mutants the ribosome-associated stringent factor
is defective, while in at least some *relC* mutants the intact stringent
factor binds to the ribosome, but is non-functional because of alteration
or loss of ribosomal protein L11. These mutations result in continued RNA
synthesis during amino acid starvation. Besides the *relA* dependent pathway,

at least one other mechanism exists for ppGpp formation that is activated in response to energy starvation, but little is known of the enzymes involved (Cashel and Rudd, 1987). Whatever the means of ppGpp formation, it is clear that slower growth rates correlate with higher basal ppGpp levels in both relaxed and stringent strains (Sarubbi et al., 1988), and that ppGpp plays a central role in growth rate control.

The stringent response has been postulated as a trigger for the onset of antibiotic production in *Streptomyces* and presumptive *relC* mutants of several *Streptomyces* strains are impaired in antibiotic production (Ochi, 1986; 1987a; 1987b). Whether this reflects a causal relationship between ppGpp synthesis and the activation of antibiotic biosynthetic pathways remains to be determined.

We have investigated the stringent response in *Streptomyces coelicolor* A3(2), which is genetically the best studied member of the genus. We used two different methods to elicit the stringent response and studied its effect on growth rate, on total RNA synthesis, and at the molecular level, on the transcription of the rRNA *rrnD* gene set and of the *actIII* gene involved in actinorhodin biosynthesis. Furthermore, we monitored ppGpp levels after inducing the stringent response and during normal growth and attempted to correlate them with the onset of antibiotic production.

RESULTS

S.coelicolor A3(2) strain M145 was used throughout this study. The strain was grown in a liquid minimal medium with a low phosphate concentration (1 mM) which was supplemented with Casaminoacids to a final concentration of 0.2%
(w/v).

Serine hydroxamate induced stringent response

Serine hydroxamate is an analogue of serine that inhibits seryl tRNA synthetase (Tosa and Pitzer, 1971). The analogue was shown to provoke the stringent response in *Streptomyces hygroscopicus* (Riesenberg et al., 1984). The addition of serine hydroxamate to exponentially growing cultures of *S.coelicolor* resulted in transient accumulations of ppGpp. Addition of the analogue at final concentrations of 3 mg/ml and 6 mg/ml resulted in peak levels of ppGpp after 15 minutes of 47 pmol/mg and 75 pmol/mg dry weight, respectively; the overall RNA synthesis rates were reduced by 50% and 85%, respectively.

The stringent response is known to lead to a drastic reduction in the transcription of stable RNA genes. In *E.coli* the ribosomal RNA genes are expressed from two promoters (p1 and p2). p1 is the major promoter and is strongly inhibited during the stringent response, while p2, a weaker constitutive promoter, is relatively insensitive (Cashel and Rudd, 1987). The *rrnD* gene set of *S.coelicolor* is transcribed from four promoters (p1 - p4); while p1 and p2 possess sequences in their -10 and -35 regions that resemble the consensus sequences for the major class of procaryotic promoters, p3 and p4 show similarity only in their -10 regions (Baylis and Bibb, 1988b). RNA was isolated from cultures after addition of serine hydroxamate and high resolution nuclease S1-mapping experiments were performed using an end-labelled *AccI-BanI* fragment (uniquely labelled at the 5' end of the *AccI* site) that contained the *rrnD* promoter region. The relative levels of the promoter transcripts were determined by quantitative autoradiography. The serine hydroxamate induced stringent response caused a dramatic reduction in *rrnD* transcription. All of the promoters were subject to stringent control and apparently to equal degrees. Addition of the analogue at 3 mg/ml and 6 mg/ml reduced transcript levels to 35% and 10% of control levels, respectively, 20 minutes after addition.

Induction of stringent response by nutritional shiftdown

A sudden deprivation of amino acids can also be achieved by transferring exponentially growing cultures from an amino acid containing medium to minimal medium. This method was used by Ochi to demonstrate the occurence of the stringent response in several *Streptomyces* strains (Ochi, 1986; 1987a; 1987b). Nutritional shiftdown provoked the transient accumulation of ppGpp in *S.coelicolor*, giving a maximum of 192 pmol/mg dry weight after 10 minutes and a reduction in the overall RNA synthesis rate of approximately 75% after 30 minutes. Nuclease S1-mapping of the *rrnD* promoters again revealed that all promoters were subject to stringent control despite the differences in their primary sequences; transcript levels were reduced to 10% of pre-shiftdown levels 20 minutes after amino acid depletion.

ppGpp levels during normal growth

To investigate the possible role of ppGpp in antibiotic production during normal batch growth we studied the basal levels of ppGpp throughout the growth of a culture. While ppGpp was not detected during exponential growth, it was clearly observed during the transition from exponential to stationary phase (16 pmol/mg dry weight). A lower but significant level of ppGpp was found throughout stationary phase. The peak of ppGpp at the end of the exponential growth phase is intriguing since it coincides with the generally observed switch from primary to secondary metabolism. Production of the coloured antibiotics undecylprodigiosin and actinorhodin was assayed spectrophotometrically. While undecylprodigiosin was detected shortly after entering stationary phase, actinorhodin production could only be detected seven hours later.

Effect of the stringent response on induction of the *actIII* gene

We next studied the effect of inducing the stringent response on the transcription of an actinorhodin biosynthetic gene, *actIII*. An end-labelled *Sal*I-*Bam*HI fragment (uniquely labelled at the 5' end of the *Sal*I site) that contained the promoter and 5'coding region of the *actIII* gene (Hallam et al., 1988) was used in high resolution nuclease S1-mapping experiments. The *actIII* gene encodes an oxidoreductase that is believed to be responsible for catalysing an early step in actinorhodin biosynthesis, a β-keto reduction during assembly of the polyketide chain (Hallam et al., 1988). In normal batch cultures we could detect *actIII* transcripts four hours before the antibiotic was measurable in the culture medium. After nutritional shiftdown, the growth rates of the cultures were drastically reduced and they entered stationary phase with only a slight increase in optical density; *actIII* transcripts were detected one hour after shiftdown. After addition of serine hydroxamate (final concentration 3 mg/ml), the cultures continued to grow exponentially with a reduced growth rate and entered stationary phase about 12 hours later, in contrast to the 5 hours observed with untreated cultures. *actIII* transcripts were only detected when the cultures entered stationary phase.

DISCUSSION

The stringent response was provoked in *S.coelicolor* in two ways. Depletion of amino acids or addition of serine hydroxamate induced transient accumulations of ppGpp that were correlated with a dramatic reduction in the overall rate of RNA synthesis. This effect was studied in detail for the transcription of the *rrnD* gene set. In contrast to *E.coli*, where transcription from only the major promoter (p1) was reduced during the stringent response, transcription from all four of the *rrnD* promoters was inhibited, and apparently to the same degree.

Studies of normal batch cultures revealed a slight increase in the level of ppGpp at the end of exponential growth. This coincides with the generally observed switch from primary to secondary metabolism. However, studies on the transcription of an actinorhodin biosynthetic gene after provoking the stringent response by the two different methods gave different results. In nutritional shiftdown experiments *actIII* transcripts were observed one hour after inducing the stringent response. In serine hydroxamate treated cultures *actIII* transcription could only be detected when the cultures entered stationary phase. In the latter case, the transient accumulation of ppGpp far exceeded that observed during the transition from exponential to stationary phase, but yet did not switch on actinorhodin biosynthesis. This suggests that ppGpp alone is not sufficient for initiating antibiotic production.

Ochi has proposed that ppGpp plays a central role in the onset of secondary metabolism by isolating *relC* mutants of different *Streptomyces* strains that are impaired in antibiotic production (Ochi 1986; 1987a; 1987b). We have also isolated potential *relC* mutants that produce only 20% of the ppGpp observed in the wild type strain during nutritional shiftdown. However, these mutants, which are comparable in terms of ppGpp production to those of Ochi, are not deficient in actinorhodin and undecylprodigiosin biosynthesis. As our experiments using serine hydroxamate show, increased levels of ppGpp do not necessarily switch on antibiotic production genes. The observed increase in ppGpp levels at the end of exponential growth is an interesting feature. However, these ppGpp levels could reflect changes in growth rate, and may even be formed in a *relA* independent manner. In our opinion the role of ppGpp in the switch from primary to secondary metabolism remains uncertain. An elevation in ppGpp levels may well be a requirement, but on the other hand it seems unlikely that ppGpp alone is sufficient to trigger the onset of antibiotic production.

Acknowledgement: E. Strauch and E. Tanako acknowledge receipt of funds from the Deutscher Akademischer Austauschdienst and the John Innes Foundation, respectively.

LITERATURE

Baylis, H.A., and Bibb, M.J. (1988a) Organisation of the ribosomal RNA genes in *Streptomyces coelicolor* A3(2). Mol. Gen. Genet. 211: 191-196.

Baylis, H.A., and Bibb, M.J.(1988b) Transcriptional analysis of the 16S rRNA gene of the *rrnD* gene set of *Streptomyces coelicolor* A3(2). Mol. Microbiol. 2: 569-579.

Cashel, M., and Rudd, K.E. (1987) The stringent response. In *Escherichia coli* and *Salmonella typhimurium*: Cellular and Molecular Biology. Neidhart, F.C. (editor in chief). Washington DC: American Society for Microbiology, pp.1410-1438.

Gallant, J.A. (1979) Stringent control in *Escherichia coli*. Ann. Rev. Genet. 13: 393-415.

Hallam, S.E., Malpartida, F., and Hopwood, D.A.(1988) Nucleotide sequence, transcription and deduced function of a gene involved in polyketide antibiotic synthesis in *Streptomyces coelicolor*. Gene 74: 305-320.

Ochi, K. (1986) Occurence of the stringent response in *Streptomyces* sp. and its significance for the initiation of morphological and physiological differentiation. J. Gen. Microbiol. **132**: 2621-2631.

Ochi, K. (1987a) A *rel* mutation abolishes the enzyme induction needed for actinomycin synthesis by *Streptomyces antibioticus*. Agric. Biol. Chem. **51**: 829-835.

Ochi, K. (1987b) Metabolic initiation of differentiation and secondary metabolism by *Streptomyces griseus*: Significance of the stringent response (ppGpp) and GTP content in relation to A factor. J.Bacteriol. **169**: 3608-3616.

Riesenberg, D., Bergter, F., and Kari, C. (1984) Effect of serine hydroxamate and methyl α-D-glucopyranoside treatment on nucleoside polyphosphate pools, RNA and protein accumulation in *Streptomyces hygroscopicus*. J. Gen. Microbiol. **130**: 2549-2558.

Sarubbi, E., Rudd, K.E., and Cashel, M. (1988) Basal ppGpp level adjustment shown by new *spoT* mutants affect steady state growth rates and *rrnA* ribosomal promoter regulation in *Escherichia coli*. Mol. Gen. Genet. **213**: 214-222.

Tosa, T., and Pizer, L.I. (1971) Biochemical basis for the antimetabolite action of L-serine hydroxamate. J. Bacteriol. **106**: 972-982.

THE RELATIONSHIP BETWEEN PRIMARY AND SECONDARY METABOLISM IN STREPTOMYCETES

G. Padilla, Z. Hindle, R. Callis, A. Corner, M. Ludovice[*], P. Liras[*], and S. Baumberg

Dept. of Genetics, University of Leeds, Leeds, England, and [*]Dpto. de Microbiologia, Universidad de Leon, Leon, Spain

INTRODUCTION

Most of the work on control of gene expression in streptomycetes has been done up till now on pathways of secondary metabolism. Two primary metabolism systems, both catabolic, have been investigated in detail, namely those of glycerol (Smith and Chater, 1988) and galactose (Adams et al., 1988) catabolism. Biosynthetic pathways have however received less attention so far, although studies of the proline (D.A. Hodgson, personal communication) and histidine (Limauro et al., 1990) biosynthesis systems are in hand. This is perhaps surprising because of the obvious relationship of biosynthetic primary metabolism to secondary metabolism, in that the end products of the former constitute the starting materials for the latter. One potentially important aspect of this is that the availability of the primary metabolite precursors of secondary metabolism may be one factor in determining secondary metabolite yield. It might therefore be the case that where a primary metabolite is used in secondary metabolite synthesis, its own formation is subject to controls over and above those that apply when this is not so.

Few streptomycete systems seem to have been examined with a view to correlating secondary metabolite production with flux through pathways to and from primary metabolite precursors, for instance by looking at mutants that overproduce the latter. One system where interesting information of this type is available is that of biosynthesis of tylosin and related oligoketides from branched-chain amino acids (see Vancura et al., 1989, for references ; also Kralova et al. (1990).

We have chosen arginine as an example to study whether this might be the case. Regulation of the biosynthesis, and to a lesser extent the catabolism, of this amino acid has been studied intensively in *Escherichia coli* (Glansdorff, 1987) and *Bacillus subtilis* (North et al., 1989). Relevant pathways of arginine metabolism are shown in Fig. 1. Arginine is a precursor of streptomycin in *Streptomyces griseus*, donating two guanidino groups to substituted inositol for the formation of its streptidine moiety (Walker, 1975). (In this part of the work, we are also looking for possible effects of regulation of inositol metabolism on streptomycin yield). Arginine, via conversion to ornithine, is also probably a precursor of clavulanic acid in *S. clavuligerus* (Romero et al., 1986). A schematic depiction of these two secondary metabolite biosynthetic pathways is shown in Fig. 2. There is no evidence for arginine being a precursor of any secondary metabolite in *S. coelicolor* or *S. lividans*. Our eventual aims are therefore to elucidate the

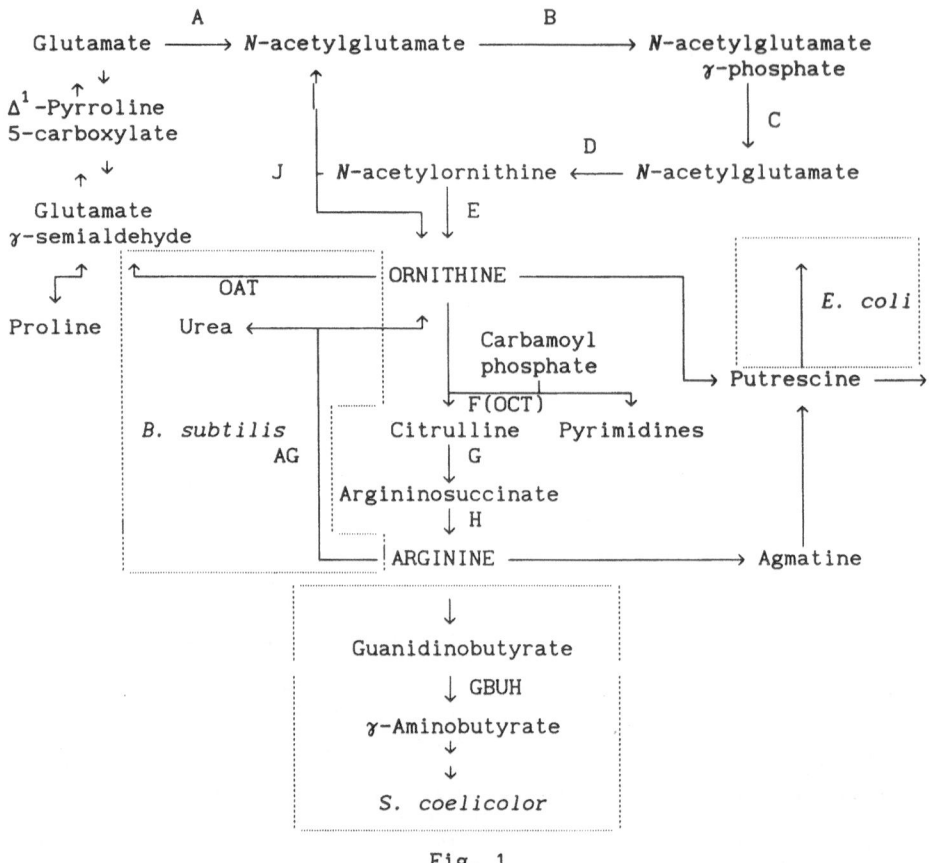

<u>Fig. 1</u>

Selected pathways of arginine metabolism in *E. coli*, *B. subtilis* and
Streptomyces. For abbreviations, see text.

controls of arginine metabolism in these streptomycetes, to see if there are
any differences in control between those organisms which do and those which do
not utilise arginine as a secondary metabolite precursor, and to find out the
effect (if any) of altering these controls on secondary metabolite yield.

EFFECTS OF MEDIUM SUPPLEMENTATION WITH PRIMARY METABOLITE

        It can be argued that the effect on secondary metabolite production
of an increase (e.g. following mutation) in the formation of a primary
metabolite precursor, can be mimicked simply by supplementing the production
medium with the primary metabolite. The literature contains examples where
such supplementation increases secondary metabolite yield, as well as examples
where it does not. Examples involving streptomycin are Majumdar and Kutzner
(1962) and Nomi (1963).

        We have examined the effect of inositol supplementation on
streptomycin production by *S. griseus* while characterising a mutant G126
isolated following nitrosoguanidine treatment of our standard *S. griseus*
strain, ATCC 12475. G126 behaves as a leaky inositol auxotroph (the first, as
far as we know, described for any prokaryote). On solid minimal medium, growth
was slowed only slightly by the omission of inositol, but sporulation was
completely prevented and probably no aerial mycelium was formed. Inositol did
not stimulate streptomycin formation by wild type grown in liquid R5 medium,

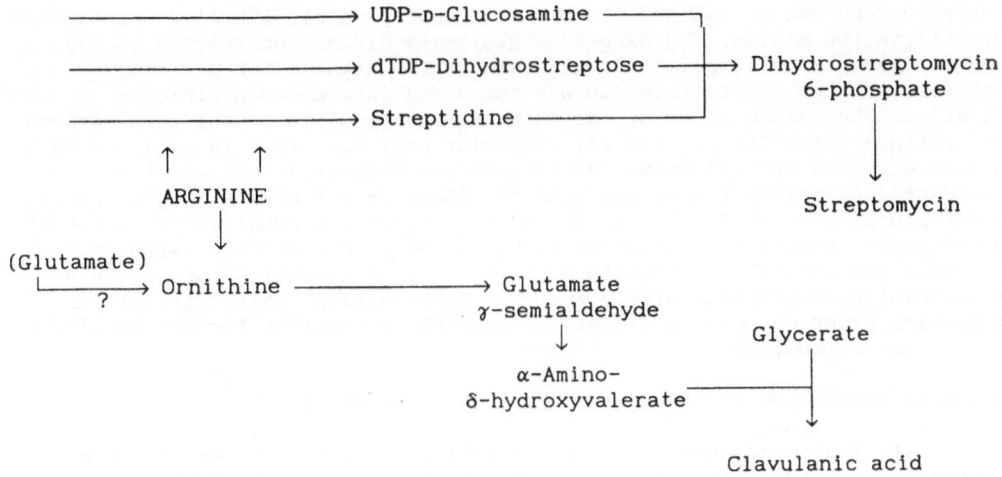

Schematic pathways of streptomycin and clavulanic acid biosynthesis showing
the involvement of arginine. For details, see for streptomycin Mansouri et al.
(1989) and for clavulanic acid Romero et al. (1986).

but did so in YEME medium. By contrast, streptomycin formation by G126 in
liquid R5 medium was barely detectable in the absence of inositol but was
comparable with the wild type in its presence ; in liquid YEME medium, G126
produced no streptomycin whether inositol was present or not.

But does a failure of supplementation with a primary metabolite
precursor to increase secondary metabolite yield unambiguously imply that
increase in endogenous formation of the primary metabolite would likewise have
no effect ? Clearly, there must be evidence that the primary metabolite is
effectively taken up. Even if that is so, there are two arguments suggesting
that the implication may not always hold. The first is that there may be
separate pools for endogenous and exogenous primary metabolite, with the
secondary metabolite being formed preferentially from the former. That such
separation of pools may occur in prokaryotes has been documented by Tabor and
Tabor (1969), following work by Sercarz and Gorini (1964). They measured
polyamine formation in *E. coli* arginine auxotrophs grown in a chemostat on
limiting amounts of citrulline-arginine mixtures, the added amino acids being
differentially labelled with either 3H or 14C From the 3H : 14C ratio in the
isolated polyamines, it could be inferred that more polyamines had been
produced from the exogenous arginine than from the arginine formed from the
endogenous citrulline. However, there was little or no preference for either
type of arginine in protein synthesized. These results imply that exogenous
and endogenous arginine are in functionally distinct pools, the former being
more accessible to arginine decarboxylase (the first enzyme in synthesis of
polyamines from arginine in *E. coli*) than the latter. Few other experiments of
this kind appear to have even carried out, but the possibility of separate
pools must be considered even though their physical basis is obscure.

The second argument, which may relate to or even be a rephrasing of
the first, is that exogenous primary metabolite may be preferentially
converted to another compound. An example is described by Inbar and Lapidot
(1988a) in their studies on actinomycin D formation by *S. parvulus*. In
fructose-glutamate medium, this is subject to strong catabolite repression by
the exogenous glutamate ; antibiotic formation begins only when the latter is
exhausted. Nevertheless, it can be shown that during formation of actinomycin
D, intracellular levels of glutamate remain high. The apparent contradiction

was resolved by use of 13C and 15N NMR with appropriately labelled fructose or glutamate in the medium. The exogenous glutamate did not contribute to the intracellular glutamate pool ; it was used preferentially for protein synthesis, but the excess taken up was converted into a new pyrimidine derivative. The carbon atoms of the intracellular glutamate pool were derived from fructose catabolism ; this intracellular pool gave rise to actinomycin D and also a second pyrimidine derivative. It was suggested that the two pyrimidine derivatives (Inbar and Lapidot, 1988b) may function as nitrogen storage products. Finally, it was found that choline, a putative precursor of the sarcosine component of actinomycin D, seemed to become associated with a more rigid and immobilized structure at the time of maximal accumulation of the antibiotic, consistent with a membrane localization. This extremely interesting paper suggests a variety of possible mechanisms for the regulation of secondary metabolism.

CONTROL OF ENZYMES OF ARGININE METABOLISM IN STREPTOMYCETES

As indicated above, we are interested in comparing the physiology and genetics of regulation of arginine metabolism as between streptomycetes in which this amino acid is (*S. griseus*, *S. clavuligerus*) and is not (*S. coelicolor*, *S. lividans*) used as a secondary metabolite precursor. Control can of course be exerted at the levels of either enzyme synthesis and/or enzyme activity. It has been reported (Udaka, 1966) that the streptomycetes possess the ArgJ enzyme, *N*-acetylglutamate:ornithine acetyltransferase ; and that as in other microorganisms with ArgJ, the ArgB enzyme *N*-acetylglutamate kinase is the one subject to feedback inhibition by arginine. We have preliminary data (not presented here) confirming ArgJ activity in streptomycetes. In our hands, however, ArgB activity is undetectable in Streptomyces extracts ; it may be that the enzyme specific activity is unmeasurably low (the assay is rather insensitive), or the enzyme is unstable when the cells are broken, or both.

We have attempted to determine specific activities of several biosynthetic activities, and representatives of arginine catabolic activities, in our four chosen streptomycetes. The most reliable estimations of arginine biosynthetic activities are for the ArgF enzyme, ornithine carbamoyltransferase (OCT) ; some other activities were too low to measure accurately or indeed at all. The specific activities of biosynthetic enzymes in many pathways, under conditions where full repression would not be expected (i.e. in cells grown in minimal media without added end product) appear to be appreciably lower in streptomycetes than in enterobacteria, pseudomonads or bacilli (D.A. Hodgson, personal communication). It can be argued that this reflects the diminished flux requirement through such pathways in the more slowly growing streptomycetes. For catabolic pathways, we have taken arginase (AG), whose activities in *S. clavuligerus* have already been reported in Romero et al. (1986), and guanidinobutyrate ureohydrolase (GBUH). The latter is the most readily assayed enzyme in the novel pathway of arginine utilization described for *S. griseus* by Thoai et al. (1966) which probably exists in most or all streptomycetes. Typical results are shown in Table 1. It is seen that under the conditions employed (see legend to Table 1), repression of OCT is by factors of 1.5 - 3.5, induction of AG by 1.2 - 2.5, and induction of GBUH by 1.8 (but see the proviso for M130) - 22. The small degree of apparent repression of OCT is in line with observations on other biosynthetic enzymes in streptomycetes, whose repression ratios almost never exceed 10 and are often much less (D.A. Hodgson, personal communication). (It has recently been found - P. Carranchas, personal communication - that if the exogenous arginine concentration is raised to 10 mM, repression of OCT in *S. clavuligerus* rises to 10-fold). The considerable induction of GBUH is consistent with its involvement in arginine utilization. The low induction and absolute levels of AG, already commented on by Romero et al. (1986) in regard to *S. clavuligerus*, are rather surprising if this enzyme is involved in arginine utilization. Romero et al. (1986) have suggested that in *S. clavuligerus* AG allows the

38

Table 1

Enzyme :		OCT	AG	GBUH
Strain				
M130	-arg	4.7	1.0	10 *
	+arg	3.1	1.2	18
TK54	-arg	5.5	0.8	6
	+arg	1.6	2.1	120
ATCC 12475	-arg	4.7	0.3	8
	+arg	2.0	0.7	130
T523	-arg	1.8	0.1	5
	+arg	0.8	0.2	160

The strains were : *S. coelicolor* A3(2) M130, *hisA1 proA1 uraA1* ; *S. lividans* TK54, *his leu* ; *S. griseus* ATCC 12475, prototroph ; and *S. clavuligerus* T523, *ura* (from Glaxo Group Research). 40 ml liquid minimal medium cultures without or with arginine (0.5 mM) were inoculated with 10^7 spores and incubated with shaking at $30°C$ for two days. Nitrogen sources for the data shown here were ammonium sulphate (M130, T523) or glutamate (TK54, ATCC 12475), but for all four strains use of ammonium sulphate, glutamate or glutamine gave essentially the same results. Mycelium was harvested by centrifugation, resuspended in 0.1 M potassium phosphate buffer pH 7.0 and broken by ultrasonic disintegration. The extract was clarified by centrifugation but small miolecules were not removed. Specific activities are in μmoles product formed/hr/mg protein. OCT, AG and GBUH were assayed as described in Mountain and Baumberg (1980), North et al. (1989), and Thoai et al. (1964) respectively.

*   Low GBUH activities such as this were characteristic of M130 in the presence of arginine under all conditions. Other *S. coelicolor* strains, however, gave specific activities in excess of 100 under the same conditions.

formation from arginine of ornithine needed as a precursor of clavulanic acid (see Fig. 2) ; this may well be true, but leaves unexplained the role of AG in the other streptomycetes.

We have isolated mutants of these four species resistant to the arginine analogue canavanine ; such mutants in *E. coli* and other bacteria often carry regulatory mutations affecting control of arginine enzyme synthesis. A number of canavanine-resistant mutants of *S. clavuligerus* were also kindly provided by Dr. B.A.M. Rudd (Glaxo Group Research), including some with phenotypes suggestive of changes in control of arginine metabolism. None of these appeared to possess altered specific activities of the above three enzymes or others of arginine metabolism that we attempted to assay. It will be interesting to see if any analogue-resistant isolates can be detected in these systems that correspond to those found by D.A. Hodgson (personal communication) among *S. coelicolor* mutants resistant to proline analogues, some of which are altered simultaneously in passage of proline across the cell membrane and in production of the secondary metabolite undecylprodigiosin, of which proline is a precursor.

CLONING OF STREPTOMYCETE GENES OF ARGININE BIOSYNTHESIS

        This has been pursued in parallel in our two laboratories, *S. coelicolor* and *S. clavuligerus* being studied in Leeds and Leon respectively.

CLONING OF *ARG* GENES FROM *S. COELICOLOR*

        We were kindly presented by Drs. S. Dowden and B.A.M. Rudd (Glaxo Group Research) with two plasmids pGX1 and pGX2, containing random fragments of *S. coelicolor* DNA cloned into pIJ922, which complemented *arg*⁻ mutations in *S. coelicolor* and *S. lividans*. Since the Streptomyces arginine auxotrophs were incompletely characterised, and assignment of enzyme deficiencies could be predicted to be hazardous because of the very low levels of most arginine biosynthetic enzymes even in prototrophs (see above), we attempted to subclone *arg* genes from pGX1 by complementation of well characterized *E. coli arg* mutants. We first used the pBR322-based expression vector pRK9, having in mind the difficulties generally encountered in obtaining expression of Streptomyces genes in *E. coli* (see discussion in Lim et al., 1989). Complementation was obtained of *argB, C, E* and *H* mutations ; the initial plasmids were however large (with duplications of material) and unstable. Eventually, from one of the large plasmids, pZC1, a smaller, apparently stable plasmid pZC177 was obtained in which much of the original *S. coelicolor* DNA, as well as part of the vector, had been deleted. pZC177 complemented *E. coli argB, C* and *E* mutations. Southern blotting showed that the arrangement of restriction sites

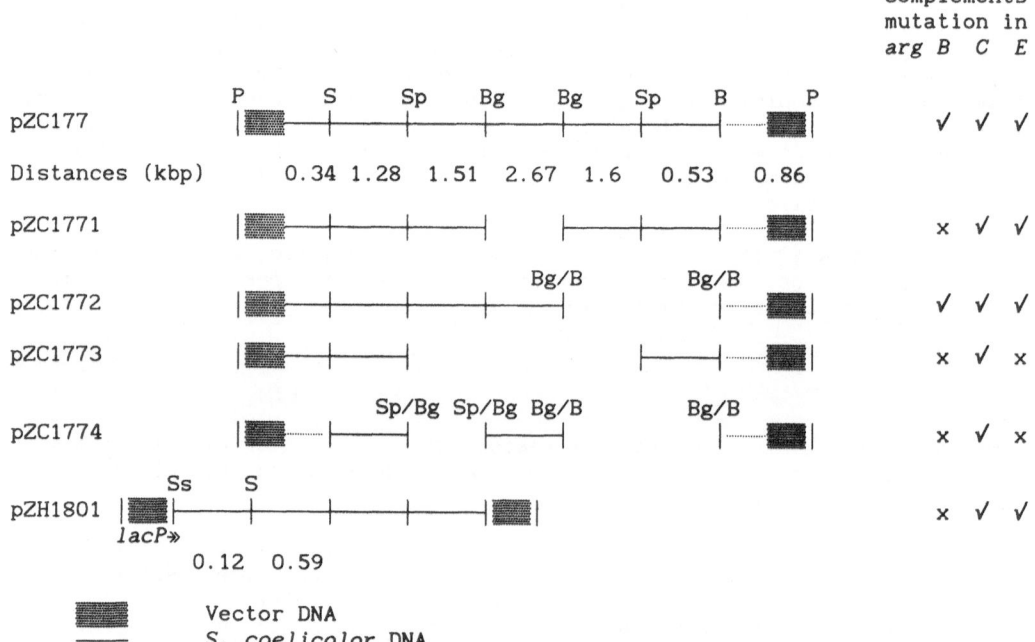

                                                    Complements
                                                    mutation in
                                                    arg B C E

pZC177

Distances (kbp)   0.34 1.28 1.51 2.67 1.6   0.53 0.86

pZC1771   x √ √

pZC1772   √ √ √

pZC1773   x √ x

pZC1774   x √ x

pZH1801   x √ √

        ▓▓▓   Vector DNA
        ─────   *S. coelicolor* DNA
        ┄┄┄┄   Region within which lies a junction between vector and *S. coelicolor* DNA, exact position unknown

Fig. 3

    Restriction map of pZC177 and derivatives, and pZH1801. Abbreviations of restriction sites : B, *Bam*HI ; Bg, *Bgl*II ; P, *Pst*I ; S, *Sal*I ; Sp, *Sph*I ; Ss, *Sst*I. There are additional *Sal*I sites to the one indicated. Complementation or failure thereof of *E. coli argB, C* and *E* mutations by the various plasmids is shown on the right.

was as in the *S. coelicolor* chromosome. In vitro deletion derivatives of
pZC177 were constructed ; Fig. 3 shows the restriction maps of the various
plasmids and the *E. coli arg* mutations that they complement. In addition, a
pGX1 SstI- BglII fragment known to overlap the left-hand end (Fig. 3) of the
insert in pZC177 was cloned in pUC18 to yield pZH1801 ; this is also shown in
Fig. 3. The same fragment was cloned in pUC19 to yield pZH1803, in which it is
in the reverse orientation with respect to the vector's *lac* promoter ;
nevertheless, pZH1801 and pZH1803 showed the same complementation pattern. The
results indicate a gene order *argC - E/J - B* ; the nomenclature *E/J* is used
because in theory complementation of an *E. coli argE* mutation could result
from the expression of *S. coelicolor argJ* (encoding
N-acetylglutamate:ornithine acetyltransferase ; see above) or of a possible *S.
coelicolor* equivalent of *argE* (encoding acetylornithinase).

We have determined the sequence of 1046 bp of DNA starting at the
leftmost pZH1801 SalI site shown in Fig. 3 and extending towards the *arg*
genes. Fig. 4 shows the first 360 bp of this. The ORF beginning at bp 301
extends to the end of the sequence ; Fig. 5 shows the deduced amino acid
sequence aligned with that of *E. coli* ArgC, to which it is clearly homologous.
Features of interest in the sequence shown in Fig. 4 are the *E. coli*
consensus-type promoter -35 and -10 boxes at bp 246-251/269-274, and the
two-fold symmetrical sequence at bp 254-269 involving a pair of 6 bp motifs.
It is tempting to speculate that the latter may represent a regulatory site. In
terms of comparison between this sequence and regulatory regions of *E. coli*
arg genes, a remarkable feature (whose significance if any remains to be
established) is the sequence TTGCATAAA at bp 246-254, overlapping the -35
box : this is identical to a sequence within the "ARG box" (putative operator,
in the sense of arginine repressor-binding site) of the *E. coli argR* gene
(Glansdorff, 1987).

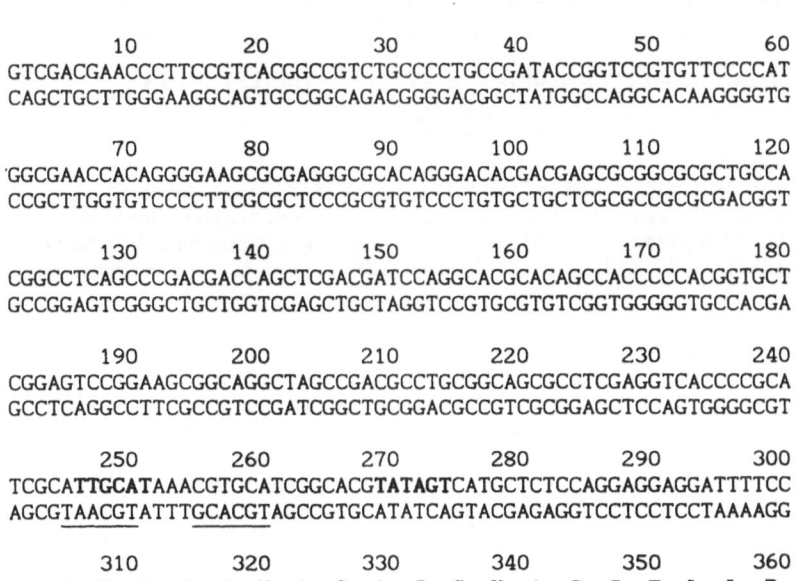

<u>Fig. 4</u>

Sequence of *S. coelicolor* DNA reading rightwards from the leftmost *Sal*I site
shown in Fig. 3. Sequencing was by the dideoxy method using standard methods.
The putative -35/-10 consensus promoter boxes at bp 246-251/269-274 are in
bold type, and the inverted repeat at bp 254-269 is underlined.

```
                 10        20        30        40        50
S.c. ArgC 1-54   MAVRAAVAGASGYAGGELLRLLLTHPEVEIGALTGNSAGQRLGALQPHLLP-LA
                 . . . *******.**.  .  **...* *** *   .. * * ..* * *
E.c. ArgC 1-53   -MLNTLIVGASGYAGAELVTYVNRHPHMNITALT-VSAQSNDAGKLISDLHPQLK

                 60        70        80        90        100       110
S.c. ArgC 55-103 DRVLEATTP-----EVLGGHDVVFLALPHGQSAAVAEQLGPDVLVVD-MGADFRL
                 . *    *      *  * ****** .*  *  .* *.    **  ... **.
E.c. ArgC 54-108 GIVDLPLQPMSDISEFSPGVDVVFLATAHEVSHDLAPQFLEAGCVVFDLSGAFRV

                 120       130       140       150       160
S.c. ArgC 104-153 KDAGDWERFYGSPHAGTWPYGLPELPGARAALEGSK-----RIAVPGCYPTAVSL
                 .**   *..** *    * *  ...  * *.*    ********** *
E.c. ArgC 109-161 NDATFYEKYYGFTHQ--YPELLEQAAYGLAEWCGNKLKEANLIAVPGCYPTAAQL

                 170       180       190       200       210       220
S.c. ArgC 154-206 ALFPAYAASLAE--PEAVIVAASGTSGAGKAAKPHLLGSEVMGSMSPYGVGGGHR
                 ** *   * * .   .** *.** ****. *      ** *. **** **
E.c. ArgC 162-213 ALKPLIDADLLDLNQWPVINATSGVSGAGRKAAISNSFCEV--SLQPYGV-FTHR

                 230       240       250       260
S.c. ArgC 207-248 HTPEMIQNLGA-VAGEPVTVSFTPTLAPMPRGILAYCTAKAKP
                 * **.  .*** *   *   .*     ***** *  . *.
E.c. ArgC 214-249 HQPEIATHLGADVIFTPHLGNF-------PRGILETITCRLKS...
```

Fig. 5

Computer-aided alignment of partial amino acid sequences of *S. coelicolor* and *E. coli* ArgC (*N*-acetylglutamate γ-phosphate reductase) deduced from DNA sequences. The (complete) *E. coli* ArgC sequence was reported by Parsot et al. (1988). Asterisks and full stops indicate identities and conservative changes respectively.

The direction of transcription of *S. coelicolor argC* is therefore rightwards with the orientation of Fig. 3 and towards *argE/J-B*. Since the junction between vector and *S. coelicolor* DNA lies in the truncated *tet* gene of pRK9 (data not shown), the supposition that transcription of *argC* is by readthrough from a vector promoter would imply that in pBR322 there is significant transcription of *tet* in the anti-sense direction. In view of the complexity of transcription in pBR322 (Balbas et al., 1986), this cannot be ruled out. However, as noted above both pZH1801 and pZH1803, with the *S. coelicolor arg* genes in opposite orientation to the *lac* promoter, complement *argC* and *E* mutants of *E. coli*. It is therefore probable that at least *argC* and *E*, and perhaps a putative *argC-E/J-B* operon, can be transcribed in *E. coli* from an *S. coelicolor* promoter. It will be interesting to determine whether this is the same as the promoter used in *S. coelicolor*. It may be noted that Meade (1986) suggested that the cloned *S. cattleya argG* gene is expressed in *E. coli* from its own promoter.

Organization of similar clusters of *arginine* biosynthesis genes is known for *E. coli* (Glansdorff, 1987) and *B. subtilis* (Smith et al., 1986), the gene orders being *argECBH* and *argCAEBD–cpa–F* respectively. To the extent that an *argJ* gene is in principle the functional equivalent of a combination of *argA* and *E*, the *S. coelicolor* gene order resembles that of *B. subtilis*. In contrast, the order of histidine biosynthesis genes in *S. coelicolor* is the same as that in *E. coli* (Limauro et al., 1990).

Cloning of *arg* genes from *S. clavuligerus*

The hitherto uncharacterized *arg* mutation in *S. lividans* auxotrophic strain 1674 is complemented by a 1.8 kbp fragment of *E. coli* DNA containing *argC* but no other complete *arg* gene (Crabeel et al., 1979). A partial Sau3A digest of total *S. clavuligerus* DNA was cloned in the shuttle vector pIJ699 ; transformation into 1674 yielded a prototrophic transformant containing a hybrid plasmid, pULML30, with a 3.2 kbp insert. pULML30 in addition complements XC33, an *E. coli argC* mutant, and also *S. clavuligerus* 328, an arginine auxotroph which fails to produce clavulanic acid ; 328(pULML30) transformants both are prototrophic and produce clavulanic acid. The insert in pULML30 was found to hybridize with a fragment containing the *S. coelicolor argC* gene (see above), but (unsurprisingly) not the *E. coli argC* gene. Further results (data not shown) are consistent with a gene order *argC–E/J–B* in *S. clavuligerus* as in *S. coelicolor*. Preliminary sequencing has revealed a region with strong homology to the *S. coelicolor argC* sequence described above.

A surprising property of 1674(pULML30) transformants was noted. Whereas ArgC (*N*-acetylglutamate γ-dehydrogenase) activity is undetectable in extracts of wild type strains of several Streptomyces species and also in *E. coli* XC33(pULML30), measurable activity (0.65 umoles/min/mg protein) was found in 1674(pULML30) extracts. It will be interesting to determine the basis of this phenomenon.

ACKNOWLEDGEMENTS

The provision of strains, plasmids, and unpublished information by S. Dowden, N. Glansdorff, D.A. Hodgson, B.A.M. Rudd, and members of D.A. Hopwood's Department at the John Innes Institute, is gratefully acknowledged. We thank R. Nicholson for the ArgC sequence alignment. Support was received by G.P. from the Colombian Government, the University of Leeds and the WHO, and by A.C. from the Wellcome Trust in the form of a Vacation Scholarship. M.L. acknowledges support from CIPAN, Lisbon, and from the Junta Nacional de Investigacao Cientifica e Tecnologia of the Portuguese Government. Collaboration between the Leeds and Leon groups was facilitated by an EC grant under the Acciones Integradas programme.

REFERENCES

Adams, C.W., Fornwald, J.A., Schmidt, F.J., Rosenberg, M. and Brawner, M.E., 1988, Gene organization and structure of the *Streptomyces lividans gal* operon, J. Bacteriol., 170:203.

Balbas, P., Soberon, X., Merino, E., Zurita, M., Lomeli, H., Valle, F., Flores, N. and Bolivar, F., 1986, Plasmid vector pBR322 and its special-purpose derivatives - a review, Gene, 50:3.

Crabeel, M., Charlier, D., Cunin, R. and Glansdorff, N., 1979, Cloning and endonuclease restriction analysis of *argE* and of the control region of the *argECBH* bipolar operon in *Escherichia coli*, Gene, 5:207.

Glansdorff, N., 1987, Biosynthesis of arginine and polyamines, in "*Escherichia coli* and *Salmonella typhimurium* : cellular and molecular biology", Ingraham, J.L., Low, K.B., Magasanik, B., Schaechter, M and Umbarger, H.E., eds., ASM, Washington, DC.

Inbar, L. and Lapidot, A., 1988, Metabolic regulation operating in *Streptomyces parvulus* during actinomycin D synthesis, studied with 13C and 15N labeled precursors, by 13C, 15N NMR and GC-MS, J. Bacteriol., 170:4055.

Inbar, L. and Lapidot, A., 1988, The structure and biosynthesis of new tetrahydropyrimidine derivatives in actinomycin D producer *Streptomyces parvulus* : use of 13C, 15N glutamate and 13C, 15N NMR spectroscopy, J. Biol. Chem., 263:16014.

Kralova, S., Kolinska, L., Vancura, A., Marsalek, J., Kristan, V. and Basarova, G., 1990, Metabolism of branched-chain amino acids and tylosin production in *Streptomyces fradiae*, Abstr. P13 Int. Symp. "Genetics and Product Formation in Streptomyces" (Erfurt, GDR, May 1-6 1990).

Lim, C.-K., Smith, M.C.M., Petty, J., Baumberg, S. and Wootton, J.C., 1989, *Streptomyces griseus* streptomycin phosphotransferase : expression of its gene in *Escherichia coli* and sequence homology with other antibiotic phosphotransferases and with eukaryotic protein kinases, J. Gen. Microbiol., 135:3289.

Limauro, D., Avitabile, S., Cappellano, M., Puglia, A.M. and Bruni, C.B., 1990, Cloning and characterisation of the histidine biosynthetic gene cluste of *Streptomyces coelicolor* A3(2), Gene, 90:31.

Majumdar, S.K. and Kutzner, H.J., 1962, Studies on the biosynthesis of streptomycin, Appl. Microbiol., 10:157.

Mansouri, K., Pissowotzki, K., Distler, J., Mayer, G., Heinzel, P., Braun, C., Ebert, A. and Piepersberg, W., 1989, Genetics of streptomycin production, in "Genetics and molecular biology of industrial microorganisms, C.L. Hershberger, S. Queener and G. Hegeman, eds., ASM, Washington, DC.

Meade, H., 1985, Cloning of *argG* from Streptomyces : loss of gene in Arg⁻ mutants of *S. cattleya*, Bio/technology, 3:917.

Nomi, R., 1963, Streptomycin formation by intact mycelium of *Streptomyces griseus*, J. Bacteriol., 86:1220.

North, A.K., Smith, M.C.M. and Baumberg, S., 1989, Nucleotide sequence of a *Bacillus subtilis* arginine regulatory gene and homology of its product to the *Escherichia coli* arginine repressor, Gene, 80:29.

Parsot, C., Boyen, A., Cohen, G.N. and Glansdorff, N., 1988, Nucleotide sequence of *Escherichia coli argB* and *argC* genes : comparison of *N*-acetylglutamate kinase and *N*-acetylglutamate-γ-semialdehyde dehydrogenase with homologous and analogous enzymes, Gene, 68:275.

Romero, J., Liras, P. and Martin, J.-F., 1986, Utilization of ornithine and arginine as specific presursors of clavulanic acid, Appl. Environ. Microbiol., 52:892.

Romero, J., Liras, P. and Martin, J.-F., 1988, Isolation and biochemical characterization of *Streptomyces clavuligerus* mutants in the biosynthesis of clavulanic acid and cephamycin C, Appl. Microbiol. Biotechnol., 27:510.

Sercarz, E.E. and Gorini, L., 1964, Different contribution of exogenous and endogenous arginine to repressor formation, J. Mol. Biol., 8:254.

Smith, C.P. and Chater, K.F., 1988, Structure and regulation of controlling sequences for the *Streptomyces coelicolor* glycerol operon, J. Mol. Biol., 204:569.

Smith, M.C.M., Mountain, A. and Baumberg, S., 1986, Sequence analysis of the Bacillus subtilis *argC* promoter region, Gene, 49:53.

Tabor, H. and Tabor, C.W., 1969, Partial separation of two pools of arginine in *Escherichia coli* ; preferential use of exogenous rather than endogenous arginine for the biosynthesis of 1,4-diaminobutane, J. Biol. Chem., 244:6383.

Thoai, N.V., Thome-Beau, F. and Olomucki, A., 1966, Induction et specificité des enzymes de la nouvelle voie catabolique de l'arginine, Biochim. Biophys. Acta, 115:73.

Udaka, S., 1966, Pathway-specific pattern of control of arginine biosynthesis in bacteria, J. Bacteriol., 91:617.

Vancura, A., Vancurova, I., Kopecky, J., Marsalek, J., Cikanek, D., Basarova, G. and Kristan, V., 1989, Regulation of branched-chain amino acid biosynthesis in *Streptomyces fradiae*, a producer of tylosin, Arch. Microbiol., 151:537.

Walker, J.B., 1975, Pathways of biosynthesis of the guanidinated inositol moieties of streptomycin and bluensomycin, Methods Enzymol., 43:429.

ISOLATION AND SEQUENCE OF A tRNA^{Gly} (CCC) FROM *Streptomyces coelicolor* A3(2)

J.Stefan Rokem[1], Astrid Schön[2] and Dieter Söll

Department of Molecular Biophysics and Biochemistry, Yale University, New Haven CT 06511 U.S.A. Present address:[1]Department of Applied Microbiology, The Hebrew University of Jerusalem Jerusalem, Israel, and [2]Laboratorium für Biochemie Universität Bayreuth, 8580 Bayreuth, F.R.G.

SUMMARY

A tRNA species from *Streptomyces coelicolor* A3(2) was isolated and sequenced. The tRNA has a glycine specific anticodon CCC. The GC-content of the sequenced tRNA 55 %, is much lower than the average of 74 % GC for *Streptomyces* genomes, but similar to glycine tRNAs from other bacteria.

INTRODUCTION

Streptomyces are important Gram positive mycelial soilbacteria. They undergo morphological differentiation, produce many economically important antibiotics and have a GC-content of 74 % (Gladek and Zakrevska, 1984). Very little is known about gene structure and regulation of gene expression in *Streptomyces*. Sofar only initiator tRNAs have been sequenced from *Streptomyces* (Kuchino et al., 1982;

Gamulin and Söll, 1987). The genetically most thoroughly studied species, *Streptomyces coelicolor,* is known to produce four different antibiotics. To the best of our knowledge no tRNA sequence has been described for this species even though it was indicated that the bldA gene is coding for a tRNA like sequence (Lawlor et al., 1987). A Glycine-specific tRNA from *Streptomyces coelicolor* A3(2) was isolated and sequenced and the results are presented here.

MATERIALS AND METHODS

   Streptomyces coelicolor A3(2) J650 was obtained from J.Piret (Northeastern University, Boston, MA. U.S.A.). Solid media for growth of the bacterium was R2YE (Thompson et al., 1982). Liquid cultures were grown in aureothricin medium (AM: Okanishi et al., 1970) with the following composition (per l): Difco Beef Extract, 5 g; Difco Bacto Peptone, 5 g; NaCl, 5 g; Glycerol, 10 ml (added after autoclaving). For tRNA isolation the cells were incubated for 48 hours at 30oC at 150 rpm in a gyratory shaker. Cells were harvested and washed (Tris-HCl buffer, 0.01 M, pH 7.0) were disrupted by N2-cavitation (Osmani et al., 1981) passed directly into phenol, extracted and nucleic acids were precipitated. The tRNAs were deacylated by incubation in 0.01 M Tris-HCl buffer pH 9.0 at room temperature for one hour. The crude tRNAs were obtained by DEAE cellulose column chromatography using a NaCl step gradient (Roe, 1975). The resulting tRNAs were separated on a RPC-5 column (Pearson et al., 1971). The tRNAs were further purified on a 20 % polyacrylamide/8M urea the tRNA is 55 %, considerably lower than the average value of 74 % for *Streptomyces* genomes (Gladek and Zakrewska, 1984).

Only a few tRNA sequences have been determined, mainly on the DNA level, for this group of bacteria. This is the first report where the RNA sequence including some of the modified bases has been determined for *Streptomyces coelicolor.*

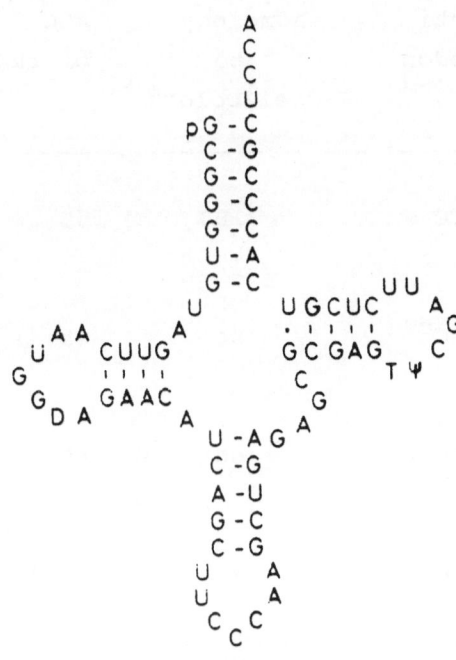

Fig.1 . Sequence of tRNA^{Gly} (CCC) of *Streptomyces coelicolor*.

Comparison of this sequence with sequences of tRNAs with an anticodon for glycine involved in cell wall synthesis in another grampositive bacterium, *Staphylococcus epidermidis* (Roberts, 1974) showed comparatively low homology, suggesting that the glycine tRNA sequenced from *Streptomyces coelicolor* is involved in protein synthesis (Table 1).

Table 1. Homology between different glycine accepting bacterial tRNAs

Organism	anti codon	%homology to S.coelicolor[a]	%GC of tRNA[a]	Average% GC in organism[b]
S. coelicolor	CCC	100	55	74
Staphylococcus epidermidis	UCC[c]	54	45	30-37
Staphylococcus epidermidis	UCC[c]	60	45	30-37
Bacillus subtilis	UCC	86	50	42-43
Escherichia coli	CCC	81	57	50-51
Escherichia coli	UCC	74	55	50-51
Pseudomonas aeruginosa	UCC	86	57	67

[a]Based on data presented by Sprinzl et al. (1989).
[b]Data from Bergey's manual (1974).
[c]Two different glycine tRNAs are involved in cell wall biosynthesis (Roberts, 1974).

No tRNAs from Gram positive organisms with the anticodon CCC
have been reported sofar. However, the homology with the
tRNA with an UCC anticodon of *Bacillus subtilis* is 86 %
(Table 1). Gram negative bacteria (*Escherichia coli* and
Pseudomonas aeruginosa) with CCC and UCC anticodons also
show high homology. It is of interest to note that
independent of the GC-content of the organism the GC-content
of the glycine tRNAs are similar (Table 1) except for the
Staphylococcus epidermidis tRNA not involved in protein
synthesis.

REFERENCES

Bergey's Manual of Determinative Bacteriology. R.E.Buchanan
    & N.E.Gibbons (eds.) 8th edition, 1974, Williams and
    Wilkins, Baltimore.

Gamulin,V., and Söll, D. 1987, The initiator tRNA genes from
    Streptomyces rimosus, Nucleic Acids Res. 15: 6747.

Gladek,A., and Zakrewska, J. 1984, Genome size of
    Streptomyces. FEMS Microbiol.Lett. 24: 73.

Krupp, G., and Gross, H.J. 1983, Sequence analysis of *in
    vitro* ^{32}P-labeled RNA. in: "The modified nucleotides in
    transfer RNA,II." P.F. Agris and R.A.Kopper (eds.)
    Alan R. Liss, New York.

Kuchino, Y., Yamamoto, I., and Nishimura, S. 1982,
    Nucleotide sequence of *Streptomyces griseus* initiator
    tRNA. Nucleic Acid Res. 10: 6671.

Lawlor, E.J., Baylis, H.A., and Chater, K.F. 1987,
    Pleiotropic, morphological and antibiotic deficiencies
    result from mutations in a gene encoding a tRNA like
    product in *Streptomyces coelicolor* A3(2). Genes &
    Developm. 1: 1305.

Okanishi, M., Ohta,T., and Umezawa, H. 1970, Possible
    control of formation of aerial mycelium and antibiotic
    production in *Streptomyces* by episomic factors.
    J.Antibiot. 23: 45.

Osmani, S.A., Marston, F.A.O., Selmes, I.P., Chapman A.G.,
    and Scrutton, M.C. 1981, Pyruvate carboxylase from
    Aspergillus nidulans. Eur.J.Biochem. 118: 271

Pearson, R.L., Weiss, J.F., and Kelmers, A.D. 1971, Improved separation of transfer RNA's on polychlorotriflouroethylene supported reversed phase chromatography columns. <u>Biochim.Biophys.Acta</u> 228: 770.

Roberts, R.J. 1974, Staphylococcal transfer ribonucleic acids. <u>J.Biol.Chem</u>. 249: 4787.

Roe, B.A. 1975, Studies on human tRNA I. <u>Nucleic Acid Res</u>. 2: 21.

Sprinzl,M., Hartmann,T., Weber,J. Blank, J., and Zeidler, R. 1989, Compilation of tRNA sequences and sequences of tRNA genes. <u>Nucleic Acids Res</u>. 17: r1.

Thompson, C.J., Ward, J.M., and Hopwood, D.A. 1982, Cloning of antibiotic resistance and nutritional genes in *Streptomyces*. <u>J.Bacteriol</u>. 152: 668.

Waldman,R., Gross, H.J., and Krupp,G. 1987, Protocol for rapid chemical RNA sequencing. <u>Nucleic acids Res</u>. 15: 7209.

MEASUREMENT OF TRANSLATIONAL ACCURACY IN *STREPTOMYCES*

J. Weiser[1], J.-L. Pernodet[2], M. Cassan[2], M. Ehrenberg[3], J. Náprstek[1], M. Guerineau[2], M. Picard[2]

[1]Institute of Microbiology, Czechoslovak Academy of Sciences, Prague, Czechoslovakia,
[2]Université de Paris-Sud, Centre d'Orsay, Orsay, France, and
[3]University of Uppsala, Biomedical Center, Uppsala, Sweden

INTRODUCTION

Bacteria as well as eukaryotic cells developed several mechanisms which prevent accumulation of errors in proteins they synthetize. Nevertheless, in both the types of cells the natural level of translational accuracy achieved is far from its attainable maximum since it can be increased by mutations affecting mostly ribosomal proteins. This indicates that a certain optimal level of translational fidelity was selected during evolution rather than its maximum level (Ehrenberg *et al.*, 1986). The reason for this could be that translational errors may play an important regulation role.

Many examples of organisms that take advantage of translational errors have been shown (Yoshinaka *et al.*, 1985, Wilson *et al.*, 1986, Craigen and Caskey, 1986). They use either a read-through event where the system does not respect the termination codon and the reading frame is extended, or a frame shift event which causes opening of a new reading frame superposed to the one in use. In both cases it is a consequence of an increased translational ambiguity. Among these organisms, the lower eukaryote *Podospora anserina* seems to use inaccuracy in translational control of gene expression during cellular differentiation. Results of analysis of several ribosomal mutations in *Podospora* indicate that there is a relationship between translational ambiguity and sporulation (Dequard-Chablat and Coppin-Raynal, 1984). That means that in *Podospora* errors are necessary for the spore development while other data show that high level of errors is not suitable for sexual reproduction (Picard-Bennoun, 1982).

Among bacteria, streptomycetes are very good candidates for using translational ambiguity as means of regulation of differentiation. They undergo fairly complicated life cycle and produce numerous antibiotics in complex biosynthetic pathways whose regulation is apparently interconnected with regulation of their cell cycle.

In order to be able to measure translational accuracy in *Streptomyces* we looked for suitable *in vitro* and *in vivo* assays. Here we describe an *in vitro* measurement of translational accuracy of ribosomes isolated from wild-type *Streptomyces granaticolor* and a streptomycin-resistant mutant derived from this strain. For *in vivo* measurements we developed a genetic system employing the firefly luciferase gene with a TGA stop codon incorporated in its proximal part as a reporter gene. The results of the *in vitro* measurements show that in the streptomycin-resistant mutant of *Streptomyces granaticolor* ribosomes confer to the resistance and that these mutant ribosomes are more accurate in *in vitro* poly(U) translation than are those from the wild-type strain. *In vivo* measurements revealed that the increase in the translational ambiguity of ribosomes caused by sublethal concentration of streptomycin in the culture can be determined by increased nonsense suppression of the TGA codon in *Streptomyces lividans*.

MEASUREMENTS *IN VITRO*

Streptomycin and some other aminoglycosides were first shown by Gorini and Kataja (1964) to cause misreading of mRNA. Subsequently Gorini (1970) described mutations in ribosomal protein genes which can influence the accuracy of protein biosynthesis. Therefore it is not surprising that most of the ribosomal mutants of *E. coli* exhibiting increased accuracy in protein synthesis are also streptomycin-resistant. A similar bias was described for sporulation defects and resistance to several aminoglycoside antibiotics in *Bacillus subtilis* (Campbell and Chamblis, 1977).

These were the reasons why we tried to isolate similar mutants from *Streptomyces granaticolor*. After UV light treatment of *S. granaticolor* spores we selected many mutants resistant to concentrations of streptomycin as high as 1000 μg/ml. Unfortunately none of the mutants had lost the ability to sporulate. Nevertheless, we selected one clone which was resistant to 1000 μg/ml of streptomycin during vegetative growth and sporulating on plates containing not more than 50 μg/ml of streptomycin. This strain, referred to as R-21, was then used for isolation of ribosomes and subsequent analysis of their accuracy characteristics.

The measurement of translational accuracy *in vitro* is rather a complicated problem since the maintenance of translational fidelity on the ribosome itself is a kinetic problem. Therefore the system in use has to exhibit kinetic parameters, *viz.* the rate and accuracy, very similar to the situation observed *in vivo*. That means for *E.coli* the rate of 15 − 20 amino acids incorporated per second on one ribosome and the error frequency in the range of 10^{-3} to 10^{-4}. Such a system for accuracy measurements was developed using purified *E. coli* components. It translates poly(U) with the rate of 12 amino acids per second per ribosome and error frequency for leucine close to 4 x 10^{-4} (Jelenc and Kurland, 1979, Wagner *et al.*, 1982). This system allows to measure kinetic parameters of individual components of translational system, especially those bearing mutations in their structure.

Ribosomes from *Streptomyces granaticolor* wild-type and mutant R-21 were isolated on a Sephacryl S-300 column as described by Jelenc (1980) and used in the above described *in vitro* system to measure their accuracy. By using the *E. coli* system we could not obtain the real value describing their accuracy in corresponding strains, but we wanted to see the possible difference in accuracy between the mutant and wild-type ribosomes in this system. This comparison is shown in Fig. 1 where the frequency of leucine incorporation into the polyphenylalanine growing chain is plotted against the ratio of tRNA^{Leu} to tRNA^{Phe} present in the system. The results show, first, that the relative accuracy of the *Streptomyces* ribosomes is approximately in the same order of magnitude as for *E. coli* ribosomes homologous with the system, and second, that the ribosomes from the streptomycin-resistant mutant R-21 are about two-fold more accurate than are those from the wild-type strain.

Table 1 shows the comparison of the same measurement performed in the presence or absence of streptomycin. The results show that while the wild-type ribosomes become infinitely inaccurate in the presence of streptomycin, ribosomes from the R-21 mutant are affected only very little. This indicates that the ribosomes confer to the resistance of the strain towards the streptomycin, most likely due to very low binding of the drug.

In order to look for a possible alteration in mobility of the ribosomal proteins as a consequence of the above functionally characterized mutation, we extracted all ribosomal proteins from the ribosome particles by the method of Barritault *et al.* (1976) and analyzed them in a two-dimensional electrophoretic system according to Geyl *et al.* (1981). The result of this analysis (Fig. 2) reveals that there are no principal differences in mobilities of the proteins from the wild-type (Fig. 2A) and R-21 (Fig. 2B) ribosomes. However, we identified three proteins which appeared consistently in a very low concentration in protein profiles of the R-21 mutant ribosomes. They are referred to as GCR 7, 14 and 30 in Fig. 2A. This observation indicates that the mutation is present either in the ribosomal protein whose mobility was not changed by the mutation or in rRNA where apart from conferring to resistance to streptomycin it can be a cause of lower stickiness of the missing proteins in the ribosomal particle. Thus they can be lost during the preparation of the ribosomes.

The above measurements show that a specially designed *in vitro* poly(U) translation system derived from *E. coli* can be successfully used for measurements of translational accuracy parameters of the components isolated from different translation mutants of *Streptomyces*.

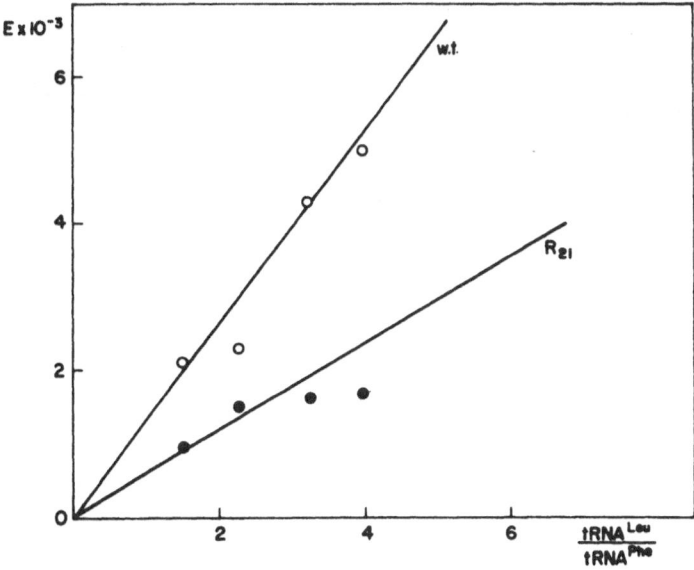

Fig. 1. Comparison of accuracy of ribosomes from the wild-type strain of *Streptomyces granaticolor* and the streptomycin-resistant mutant R-21 expressed as error frequency of Leu/Phe incorporation (E) in poly(U)-directed *in vitro* translation system plotted against the ratio of tRNA^{Leu}/tRNA^{Phe}.

Table 1. Comparison of accuracy of ribosomes from *Streptomyces granaticolor* wild-type strain and the streptomycin-resistant mutant R-21 in *in vitro* poly(U) translation system expressed as error frequency of Leu/Phe incorporation in the presence and absence of streptomycin.

Ribosomes	Error frequency[a]	
	Sm[b]	
	−	+
wild-type	1.3×10^{-3}	0.24
R-21	6.0×10^{-4}	3.6×10^{-3}

[a] the ratio Leu incorporated/Phe incorporated
[b] + denotes 100 μg/ml of streptomycin

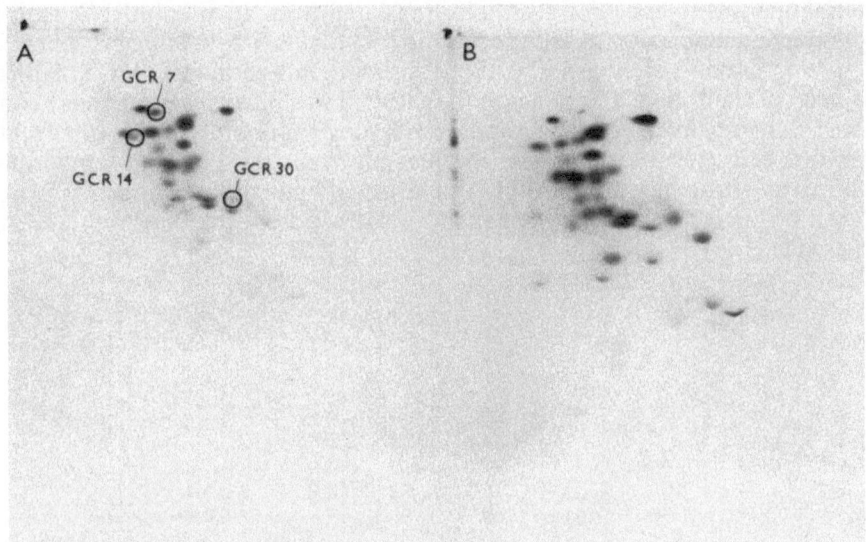

Fig. 2. Two-dimensional polyacrylamide gel electrophoresis of ribosomal proteins. Ribosomes were isolated from wild-type *Streptomyces granaticolor* (A) and from the streptomycin-resistant mutant R-21 (B). The proteins marked GCR 7, 14 and 30 on the wild-type profile appear in a very low concentration in the R-21 profile.

MEASUREMENTS *IN VIVO*

There are several ways of detecting missense or nonsense errors during translation. The most frequently used one follows up the frequency of suppression of a termination codon inserted in a defined position in the gene. In our measurement we employed the same principle which can be used in different organisms to make a minor gene products whose amount and time of appearance are automatically coupled to the expression of the major gene product.

The reporter gene we have chosen for this purpose was *luc* coding for luciferase in the North American firefly *Photinus pyralis*. This gene has been recently cloned in *E. coli* and mammalian cells (de Wet *et al.*, 1985, de Wet *et al.*, 1987) and since then used in many cases as a highly sensitive reporter gene. Firefly luciferase, in the presence of ATP, converts luciferin to oxyluciferin with a concomitant production of yellow-green light which can be quantitated using a luminometer or a scintillation counter (Nguyen *et al.*, 1988). The assay for luciferase is very rapid and inexpensive and moreover, it was estimated to be 30 to 1000 times more sensitive than the chloramphenicol-acetyltransferase assay (de Wet *et al.*, 1987). Very high sensitivity of the assay is especially important for the measurement of accuracy where both increase and decrease from the basal level are considered. This means that in this particular case sensitivity must be high enough for measuring the natural level of suppression of the TGA termination codon. The reason for use of an eukaryotic gene instead of a bacterial one was to assay for a protein translated from an alien mRNA as the transcriptional regulation signals are concerned.

We obtained the plasmid pRSV-L bearing this gene from Dr. S. Subramani, University of California. The gene on the plasmid was intronless and under control of the RSV long terminal repeat promotor allowing expression in *E. coli* as well as in eukaryotic cells. In order to make the plasmid transformable into *Streptomyces*, a small intron and a poly(A) stretch was deleted from the plasmid DNA, and was replaced by the 6.2 kb *Streptomyces* vector pIJ487 bearing the thiostrepton resistance marker (Fig. 3).

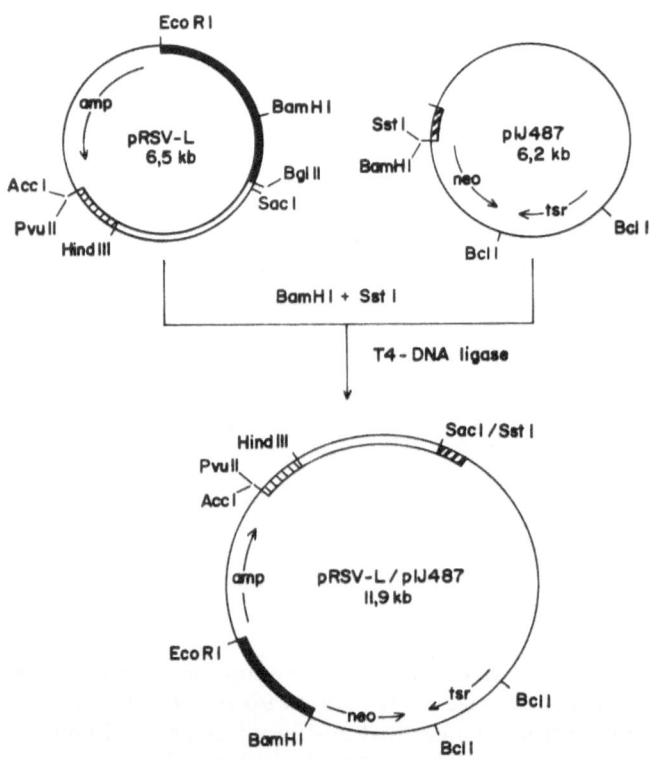

Fig. 3. Construction of plasmid pRSV-L/pIJ487 containing the *luc* gene from *Photinus pyralis*.

　　　　　　 luc gene,　　　　　　 SV 40 DNA,
　　　　　　 RSV promotor,　　　　　 terminator.

Then a TGA stop codon was introduced into the proximal part of the *luc* gene on this plasmid by means of site directed mutagenesis using a commercial site directed mutagenesis kit. Comparison of primary sequences of the proximal part of the wild-type *luc* gene and the construction of the *luc* gene containing the TGA stop codon is in Fig. 4.

```
(A)                    met glu asp ala lys asn ile lys lys gly
                       ATG GAA GAC GCC AAA AAC ATA AAG AAA GGC....


(B)        met glu asp ala ser stop ser asn ile lys lys gly
           ATG GAA GAC GCT AGC TGA  TCA AAC ATA AAG AAA GGC....
```

Fig. 4. Primary sequence of the proximal part of the *luc* gene in plasmid pRSV-L/
/pIJ847. (A), Sequence of the wild-type gene; (B), sequence of the construction with the TGA termination codon.

Both types of plasmid DNA were replicated and expressed in *E. coli* strain
DH1 (Hanahan, 1983). Plasmid DNA isolated from *E. coli* by the alkaline lysis
method (Kieser, 1984) was used for transformation of protoplasts from *Streptomyces
lividans* strains TK 21, TK 54 and TK 64 as described by Bibb *et al.* (1978). Strains
used for transformation were selected according to the markers they bore. TK 21 and
TK 54 are streptomycin-sensitive while TK 64 is streptomycin-resistant. All of the
strains were transformed with both plasmids carrying the wild-type *luc* gene and the
TGA stop codon construction. Luciferase production was tested in the obtained thiostrepton-resistant transformants. Selected clones were grown in liquid complex medium in the presence of 30 μg/ml thiostrepton analogue and vegetative mycelium was
harvested at the late log phase. The cells were resuspended in buffer, disrupted by
sonication in an ice bath, and luciferase was quantitated in the resulting cell-free
homogenates in the presence of luciferin and ATP. The light emitted in the reaction
was measured in a scintillation counter.

Table 2 shows the comparison of the production of luciferase from the wild-
type *luc* gene and from the *luc* gene bearing the TGA stop codon in 20 and 46 h old
vegetative cells of the TK 64 strain. The results show that the expression of luciferase
from the intact *luc* gene in *S. lividans* is quite high in spite of the presence of eleven
TTA codons that are very rare in the structural genes in *Streptomyces* and that probably play an important role in the regulation of differentiation (Chater, 1989). The
amount of the enzyme represented by produced light of 10^9–10^{10} cpm/mg of total
protein is equal to about 10–100 nanograms of luciferase and it is quite comparable
to the amounts of the enzyme expressed from the same plasmid in yeast cells
(M. Cassan, unpublished results).

For this study, however, the amount of functional luciferase which can be
translated from the *luc* gene containing the TGA stop codon in its proximal part is
more important. Although the amount of the enzyme translated from the stop codon
construction was almost four orders of magnitude lower than that from the wild-type
gene, it was high enough to be measurable. As shown in Table 1, the amount of light
produced by luciferase translated from the gene containing the TGA stop codon was
10-fold higher than the background in both 20-h and 46-h cultures. This apparently
reflects the natural level of nonsense suppression of the TGA termination codon in
Streptomyces cells and thus it should be possible to measure both the increase and
decrease of such a suppression also as a consequence of changes in translational
accuracy in both directions.

Table 2. Comparison of light production by luciferase expressed from both the wild-type and the TGA termination codon-bearing firefly *luc* gene in cell-free homogenates from *Streptomyces lividans* strain TK-64.

clone[a]	age of the culture (h)	emitted light (cpm)	cpm/mg of total protein
37	20	2.84×10^9	1.62×10^{10}
2*	20	2.02×10^5	2.50×10^6
0	20	2.10×10^4	–
37	46	0.58×10^9	1.13×10^9
2*	46	2.34×10^5	1.34×10^6
0	46	2.40×10^4	–

[a]clone 37 – wild-type *luc* gene
clone 2* – *luc* gene with the TGA stop codon
clone 0 – control strain without plasmid

In order to confirm this hypothesis, we grew two strains of *S. lividans*, the streptomycin-resistant strain TK 64 and the streptomycin-sensitive strain TK 21 in the absence and presence of 25 µg/ml and 2 µg/ml streptomycin, respectively, and measured the production of luciferase from both plasmid constructions containing the wild-type *luc* gene or *luc* containing the TGA stop codon. Streptomycin in sublethal concentrations ought to increase misreading of the ribosomes in this experiment and thus influence the expression of luciferase from the gene containing the TGA stop codon.

The results of this experiment are listed in Table 3. They show that the presence of streptomycin in the cultures indeed increased the production of luciferase from the gene interrupted by the TGA termination codon up to 5-fold in the TK 64 strain and about two-fold in the TK 21 strain while in both strains the expression of luciferase from the wild-type gene was about 2-fold lower due to the general inhibition effect of streptomycin on translation. Thus the overall increase of expression of luciferase from the stop codon-containing gene is almost 10-fold during increased misreading of ribosomes caused by streptomycin compared to the level of expression of the enzyme from an intact gene.

This shows that the luciferase reporter gene system using the nonsense suppression can be used for *in vivo* translational accuracy measurements in *Streptomyces*. It opens the possibility to measure translational accuracy in *Streptomyces* during their life cycle especially in the stages preceding the formation of aerial mycelium and spores as well as during the late stationary growth phases when production of secondary metabolites starts. This will include comparative measurements of asporogenous as well as non-producing mutants or special translational mutants under physiological conditions favouring the changes in translational accuracy (e. g. starvation for amino acids, various stress conditions and possibly also the effects of small regulatory molecules).

Table 3. Expression of firefly luciferase from the wild-type and TGA termination codon-interrupted *luc* gene in the presence and absence of sublethal concentrations of streptomycin in Sm-sensitive (TK-21) and Sm-resistant (TK-64) strains of *Streptomyces lividans*.

clone[a]	Sm[b]	light emitted cpm/mg	luciferase ng/mg of total protein
TK-64 (37)	−	4.50×10^9	5.3
TK-64 (37)	+	2.60×10^9	3.06
TK-64 (2*)	−	0.26×10^6	0.0003
TK-64 (2*)	+	1.20×10^6	0.0014
TK-21 (37)	−	7.10×10^{10}	83.5
TK-21 (37)	+	3.25×10^{10}	38.2
TK-21 (2*)	−	2.10×10^6	0.0025
TK-21 (2*)	+	5.00×10^6	0.0059

[a]clone 37 − wild-type *luc* gene
clone 2* − *luc* gene with the TGA stop codon
clone 0 − control strain without plasmid
[b]streptomycin concentration used was 2 and 25 μg/ml for the strains TK-21 and TK-64, respectively

CONCLUSIONS

As we have mentioned in the introductory part of this contribution, streptomycetes are good candidates for use of the mechanisms based on induced errors in translation in their regulatory pathways like it was previously shown for *Podospora*. In this way, streptomycetes could switch on expression of certain regulatory genes whose products are required in small amounts and in an exact time sequence along with major genes they are supposed to regulate. Opening up, this problem in studies of regulation of morphological and physiological differentiation in *Streptomyces* requires to build up methodological background for that, part of which we tried to demonstrate here.

Both methods described here should enable us to gain more insight into the problem of translational accuracy in *Streptomyces*. With a set of suitable mutants we should be able to find out whether translational accuracy in *Streptomyces* is regulated during their life cycle. Furthermore, we should be able to show if regulatory mechanisms directing either sporulation or biosynthesis of secondary products require changes in translational accuracy and, in particular, if these processes employ readthrough or frame shift events during translation of special regulatory genes. The availability of an *in vitro* defined translation system then could help us to measure accuracy characteristics of individual components of the system involved in such regulatory steps.

REFERENCES

Barritault D., Expert-Bezançon A., Guérin M.-F. and Hayes D., 1976, The use of acetone precipitation in the isolation of ribosomal proteins, Eur. J. Biochem. 63: 131.

Bibb M.J., Ward J.M. and Hopwood D.A., 1978, Transformation of plasmid DNA into *Streptomyces* at high frequence, Nature 274: 398.

Campbell K.M. and Chamblis G.H., 1977, Streptomycin-resistant, asporogenous mutant of *Bacillus subtilis*, Mol. Gen. Genet. 158: 193.

Chater K.F., 1989, Multilevel regulation of *Streptomyces* differentiation, Trends Genet. 5: 372.

Craigen W.J. and Caskey C.T., 1986, Expression of peptide chain release factor 2 requires high-efficiency frameshifting, Nature 322: 273.

Dequard-Chablat M. and Coppin-Raynal E., 1984, Increase of translational fidelity blocks sporulation in the fungus *Podospora anserina*, Mol. Gen. Genet. 195: 294.

Ehrenberg M., Kurland C.G. and Blomberg C., 1986, Kinetic costs of accuracy in translation, in: "Accuracy in Molecular Processes", Kirkwood T.B.L., Rosenberg R.F. and Galas D.J., eds., Chapman and Hall, London, New York.

Geyl D., Bock A. and Isono K., 1981, An improved method for two-dimensional gel electrophoresis: Analysis of mutationally altered ribosomal proteins of *Escherichia coli*, Mol. Gen. Genet. 181: 309.

Gorini L., 1970, The contrasting role of *strA* and *ram* gene products in ribosomal functioning, Cold Spring Harbor Symp. Quant. Biol. 34: 101.

Gorini L. and Kataja E., 1964, Phenotypic repair by streptomycin of defective genotypes in *E. coli*, Proc. Natl. Acad. Sci. USA 51: 487.

Hanahan D., 1983, Studies on transformation of *Escherichia coli* with plasmids. J. Mol. Biol. 166: 557.

Jelenc P.C., 1980, Rapid purification of highly active ribosomes from *Escherichia coli*, Anal. Biochem. 105: 369.

Jelenc P.C. and Kurland C.G., 1979, Nucleoside triphosphate regeneration decreases the frequency of translation errors, Proc. Natl. Acad. Sci. USA 76: 3174.

Kieser T., 1984, Factors affecting the isolation of ccc DNA from *Streptomyces lividans* and *Escherichia coli*. Plasmid 12: 19.

Nguyen V. T., Morange M. and Bensaude O., 1988, Firefly luciferase luminescence assays using scintillation counters for quantitation in transfected mammalian cells, Anal. Biochem. 171: 324.

Picard-Bennoun M., 1982, Does translational ambiguity increase during cell differentiation? FEBS Lett. 149: 167.

Wagner E.G.H., Jelenc P.C., Ehrenberg M. and Kurland C.G., 1982, Rate of elongation of polyphenylalanine *in vitro*, Eur. J. Biochem. 122: 193.

de Wet J.R., Wood V.K., Helinski D.R. and de Luca M., 1985, Cloning of firefly luciferase cDNA and the expression of active luciferase in *Escherichia coli*, Proc. Natl. Acad. Sci. USA 82: 7870.

de Wet J.R., Wood K.V., de Luca M., Helinski D.R. and Subramani S., 1987, Firefly luciferase gene: Structure and expression in mammalian cells, Mol. Cell. Biol. 7: 725.

Wilson W., Malim M.H., Mellor J., Kingsman A.J. and Kingsman S.M., 1986, Expression strategies of the yeast retrotranspozon Ty: a short sequence directs ribosomal frame shifting, Nucl. Acids Res. 14: 7001.

Yoshinaka T, Katoh I., Copeland T.D. and Orozlan S., 1985, Murine leukemia virus protease is encoded by the *gag-pol* gene and is synthesized through suppression of an amber termination codon, Proc. Natl. Acad. Sci. USA 82: 1618.

t-RNA GENES IN *STREPTOMYCES LIVIDANS*: STRUCTURE, ORGANIZATION AND CONSTRUCTION OF SUPPRESSOR t-RNA

Reinhard Sedlmeier and Horst Schmieger

Institut für Genetik und Mikrobiologie
Universität München
D-8000 München 19

INTRODUCTION - SUPPRESSOR- AND NONSENSE MUTANTS AS GENETIC TOOLS

The successful study of bacteriophages require a collection of various conditionally lethal mutants, preferably nonsense mutants. Nonsense mutants have a mutationally caused stop codon inside a reading frame which causes the premature termination of a peptide chain in translation. Nonsense mutants, on the other hand, can only be obtained and handled if bacterial suppressor strains are available. Such strains are characterized by a mutated t-RNA which has an alteration in the anticodon enabling it to insert an aminoacid at the nonsense codon position. Therefore, the first step in starting genetics with a phage is the search for suitable suppressor strains.

THE SEARCH FOR SUPPRESSOR MUTANTS IN *S. LIVIDANS*

We are interested in the actinophage ΦC31 which is the most intensively studied *Streptomyces* phage with regard to its biological behavior and from which a family of useful cloning vectors derives. Unfortunately, very little is known about its genome organization and almost nothing is known about its gene functions and their regulation. A more detailed understanding of this phage could contribute to its even better utilization in genetic engineering. We tried to isolate suppressor mutants of *S. lividans*, the bacterial host for this phage. However, the selection of a nonsense suppressor by means of a plasmid carrying a resistance marker with an *amber*-mutation was unsuccessful (Schmieger et al., 1988). Paradiso et al. (1987) using a similar approach with a different resistance gene did not succeed either. Several reasons could account for this failure, some of which are connected with the extremely biased codon usage in streptomycetes. The most reasonable explanations seemed to be the following: (1) The decision to use an *amber*-mutation for selection was wrong. The UAG codon is very rarely used as a stop signal for genes of the species

E. coli. Therefore, suppression of this codon by overreading and inserting an amino acid would not disturb many proteins. However, by analyzing a series of published streptomycetes sequences we found that many genes end with this *amber*-codon. Therefore, its suppression could frequently lead to the production of fusion proteins in the cells which might be deleterious. (2) Another possibility might be the extremely biased codon usage in *Streptomyces*. Codons with G or C at the third position are strongly favoured compared to synonymous codons with A or T at the third position. A compilation of eight protein coding sequences of streptomycetes (Sedlmeier, 1987) showed that for example 93% of all valines are encoded by GTG or GTC, whereas only 7% are encoded by GTA or GTT. This bias should be reflected by the different concentrations of the corresponding tRNAs in the cell (Ikemura, 1981). In Fig. 1 all tRNAs are listed which can be converted to suppressor tRNAs by a single step mutation.

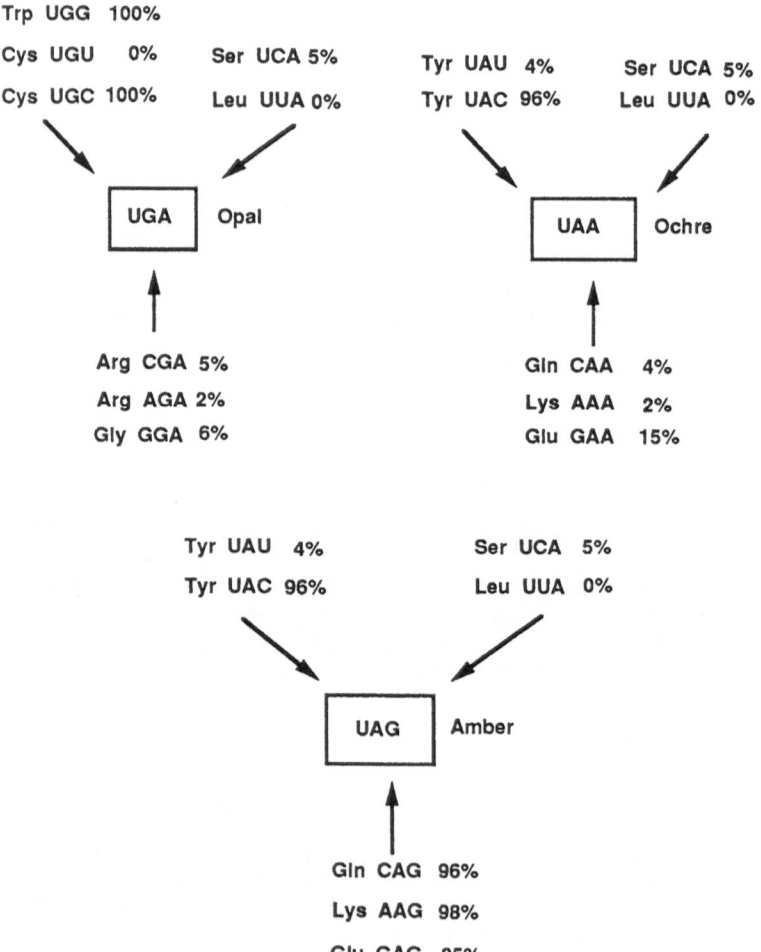

Fig. 1.tRNAs able to mutate by a single base exchange to suppressor tRNA. "%" indicates the percentage of that triplet within a family of synonymous codons used in streptomycetes.

If the gene for a minor tRNA is hit by chemical mutagenesis, the concentration of mature tRNA might be not sufficient for effective suppression. In cases where the gene for a major tRNA is hit, the mutation would be deleterious to the cell, if the tRNA is encoded by a unique gene.

ISOLATION OF tRNA GENES FROM *S. LIVIDANS*

To circumvent these difficulties we decided to add an additional tRNA gene to the genome of *S. lividans* which has an anticodon able to recognize a nonsense codon. In this way detrimental effects to the cell should be minimized. Since the new suppressor gene should be expressed and correctly processed in streptomycetes, it was necessary to start with a tRNA gene which derives from this genus. However, no information was available about location, nucleotide sequence and expression of tRNA genes in this group of organisms. This lead us first of all to study the situation of tRNA genes in *S. lividans* thoroughly.

The first step was to isolate tRNA from *S. lividans* 66. For this purpose total RNA was isolated and separated by PAA gelelectrophoresis. A tRNA mixture from *Saccharomyces cerevisiae* served as a size marker. Just as in the RNA marker, several bands of different intensity can be observed in the *Streptomyces* DNA. The major bands were isolated. They were labelled with ^{32}P and used as a probe for screening a gene library of *S. lividans* cloned in a phage Lambda-vector. By plaque hybridization with the main tRNA fraction positive plaques were identified. Several clones were recovered, and their inserts were characterized by restriction enzyme analysis. Two phage inserts, clones 4.21 and 15.21 in Fig. 2, were found which overlap by about 7 kb and which represent a total of more than 15 kb. Their analysis will be given as an example.

By Southern blotting and hybridization with the main fraction of labelled tRNA three positive subfragments could be identified which were subcloned and sequenced. A computer search showed that on each of these three fragments one tRNA gene was located: A Lys-tRNA recognizing the codon AAG (on a *Sst*I fragment), an absolutely identical copy of this gene about 12,5 kb away (on a *Bam*HI-fragment), and a Gly-tRNA for the codon GGA (on a *Xho*I-*Bam*HI fragment) between the Lys-tRNA genes. All these genes are transcribed in the same direction (Fig. 2).

All genes are followed by inverted repeats which are putative terminators. For example in the case of the Gly-tRNA a hairpin can be formed with a stem of 15 base pairs and only one mismatch.

As shown in Fig. 3 all these genes are close to Open Reading Frames; two of them are transcribed opposite to the adjacent tRNA gene (clones 44.1 and 47.4). Only the Lys-tRNA could be cotranscribed with the preceding ORF (50.2). These ORFs have the typical *Streptomyces* codon usage (Bibb et al., 1984), but the search for protein sequences in the EMBL gene bank with homology to the translated ORFs gave no results.

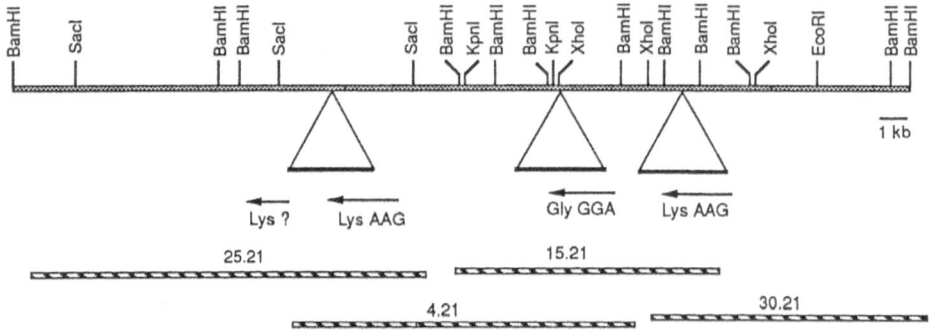

Fig. 2. Restriction map of a genomic region of *S. lividans* 66 isolated by plaque hybridizing with labelled tRNA. Arrows indicate the location and orientation of tRNA genes identified by nucleotide sequence analysis.

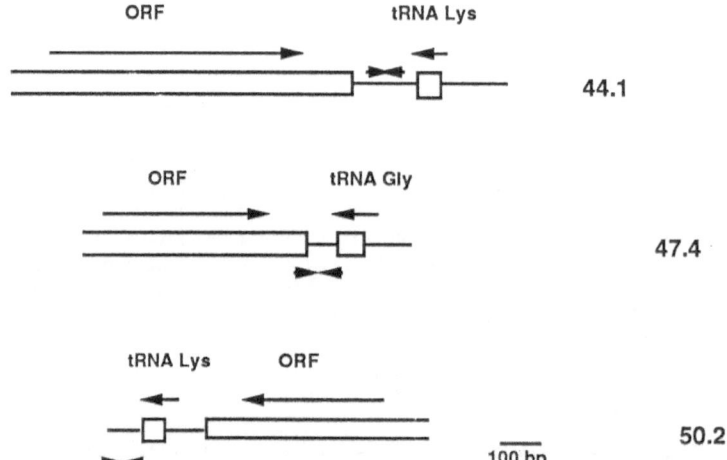

Fig 3. Schematic illustration of the sequenced region. Arrows indicate the direction of transcription, ▶◀ shows the position of the possible terminators.

COPY NUMBER OF tRNA GENES AND CLUSTER FORMATION

To determine the copy number the labelled tRNA genes were hybridized against genomic DNA of *S. coelicolor* A3(2) and two strains of *S. lividans*. The Southern blot showed that the Gly-tRNA is encoded by a unique gene. The Southern blot with the Lys-tRNA probe is shown in Fig. 4.

In the *Bam*HI- and the *Xho*I-digest the expected fragments light up. However, in the *Sst*I-lanes a third signal appeared, indicating that there must be a third gene for Lys-tRNA closely upstream of the isolated DNA, since it is obviously on the same *Bam*HI-fragment.

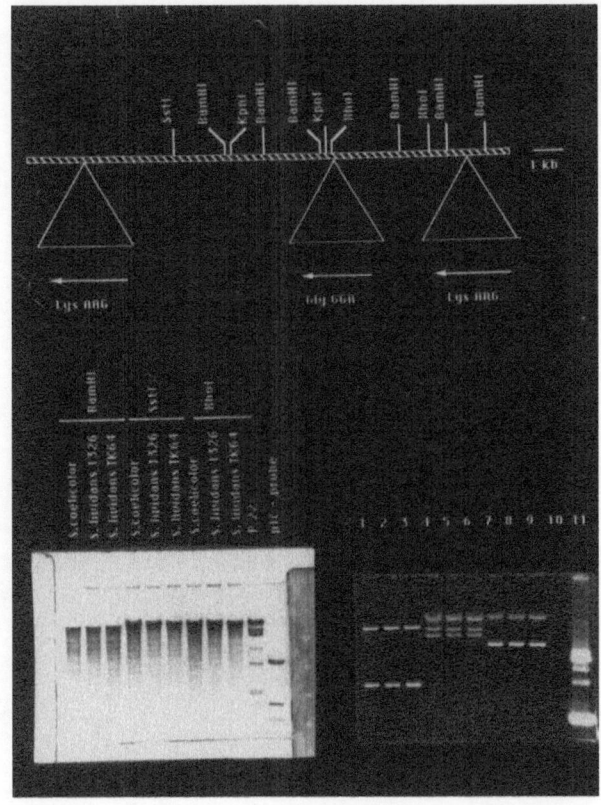

Fig. 4. Southern hybridization of
labelled Lys-tRNA gene against
genomic DNAs of *S. coelicolor*
and *S. lividans*.

To test whether there are additional tRNA genes in the
adjacent region we walked along the chromosome up- and downstream of the already isolated sequence and found the clones
25.21 and 30.21 (Fig. 2). Southern hybridization with labelled
tRNA revealed, however, that there are no other tRNA genes in
a region now spanning about 33 kb except the third copy of the
Lys-tRNA gene (on the *Bam*HI-*Sst*I-fragment).

OTHER tRNA GENES ON NON OVERLAPPING CLONES

The mentioned tRNA genes appeared to be scattered along
the chromosome with relatively large pieces of DNA between the
single genes.

However, three other positive plaques from the library
screening showed clusters of tRNA genes (Fig. 5).

On clone 63.1 a cluster of five tRNA genes was identified
containing one copy of a Gly-tRNA GGC and a Cys-tRNA GUC, and
three copies of Val-tRNA recognizing the anticodon GUC. One of
the valine tRNA genes differs from the other two in two base

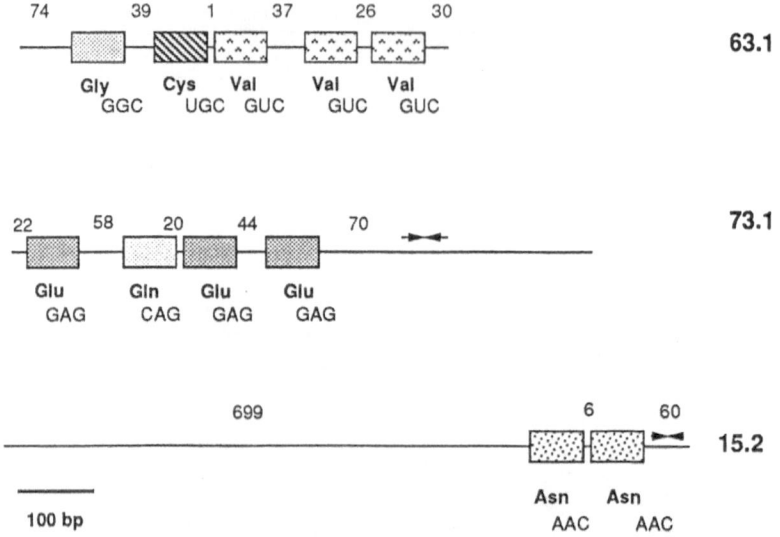

Fig.5. Organization of three tRNA gene clusters.
Numbers above the genes indicate nucleotides.
➤◄ marks possible transcriptional
terminators.

pair exchanges in the aminoacyl stem (Fig. 6 A). All genes
have the same orientation and are closely clustered.

Clone 73.1 also shows tightly clustered genes for Glu-
tRNA for the codon GAG which are absolutely identical.
Inserted between two Glu-tRNA genes is a Gln-tRNA gene for the
codon CAG. These genes also have the same orientation. Downstream an inverted repeat could act as a transcription terminator.

A third clone, 15.2, carries two Asn-tRNA genes for the
codon AAC which differ in one base pair in the aminoacyl stem
(Fig. 6 B). Both are orientated in the same way and followed
by an inverted repeat which again may be a terminator signal.

COMPARISON WITH OTHER BACTERIAL SYSTEMS AND EXPRESSION SIGNALS

1. Having identified tRNA clusters it would be reasonable
to compare these with clusters in other bacterial species. In
Bacillus subtilis, which is also a gram-positive
microorganism, tRNA genes are located in only two regions of
the chromosome (Vold, 1985). Clusters may comprise as many as
21 tRNA genes, but never more than one copy for the same tRNA
gene has been found. Another interesting observation is that
these genes are generally flanked by rRNA genes.

Asn-tRNA   ## Val-tRNA

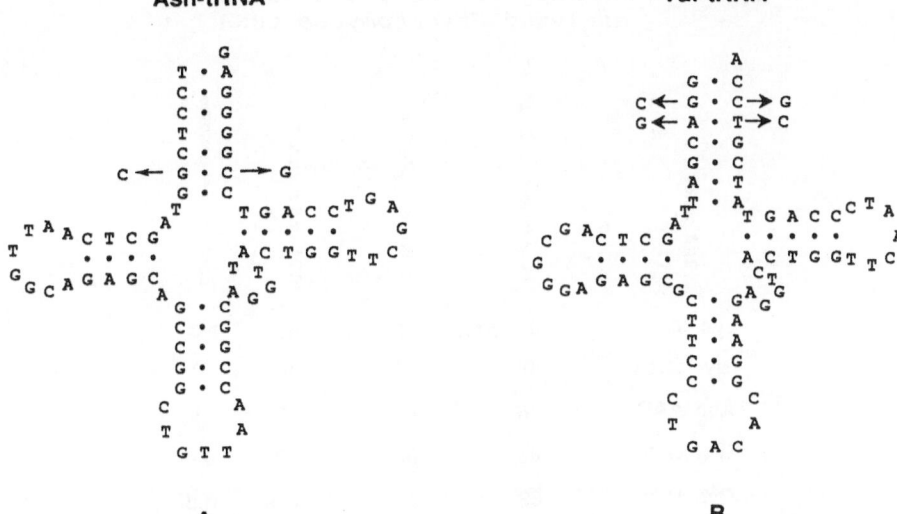

Fig. 6. Cloverleaf structure of Asn- and Val-tRNAs. Arrows mark the deviations between different gene copies.

In *E. coli*, a gram-negative species, tRNA genes are combined in smaller cluster which have no more than seven tRNA genes (Fournier and Ozeki, 1985). These clusters are scattered all over the chromosome. In some cases single tRNA genes or clusters are asssociated with rRNA genes.

Our results do not yet indicate whether *Streptomyces* reflects the situation of *B. subtilis* or of *E. coli* or of neither of them. A computer search for rRNA genes in the regions we have sequenced so far was negative. Whether this is also true for the adjacent regions has yet to be tested.

2. A comparison of the nucleotide sequences of *Streptomyces* tRNA genes with the corresponding genes of *B. subtilis* and *E. coli* does not give a uniform picture (Table 1).

There are genes with a high degree of homology to the corresponding genes of both species (Sprinzl, 1989), such as Gly-tRNA GGC. Some genes could only be compared to a related tRNA with another anticodon; nevertheless, they also show high homology. However, there are two genes, Gly GGA and Gln CAG, which exhibit only a weak homology. This gains even more importance if we consider that about 30% of the nucleotides in all prokaryotic tRNAs are conserved or semiconserved.

It is interesting that only one of the identified tRNA genes encodes the 3´-CCA-end of the mature tRNA. 12 genes out of 51 genes in *B. subtilis* and all tRNA genes of *E. coli* encode the 3´-CCA-end.

Table 1. Summary of the features of the analyzed *Streptomyces* tRNA genes

tRNA genes	% homology to *B. subtilis*	% homology to *E. coli*	CCA-end encoded	copy number/genome
Lys AAG	82 AAA	75 AAA	no	≥ 3
Gly GGA	49	51	no	1
Asn AAC	76	n.d.	no	2
Gly GGC	83	85	no	n.d.
Cys UGC	69	n.d.	yes	n.d.
Val GUC	72 GUU	73 GUU	no	3
Glu GAG	72 GAA	71 GAA	no	≥ 3
Gln CAG	50 CAA	62	no	n.d.

3. If we have a look at the copy number of the tRNA genes (Table 1) we can see that one of the possible explanation for the failure to isolate suppressor mutants in *S. lividans* is not valid. The Lys-tRNA for the codon AAG which can mutate in a single step to an *amber*-suppressor is encoded by at least three gene copies. The same is true for the Glu-tRNA which reads the codon GAG.

Since the aim of this investigation is the construction of a suppressor-tRNA, it is important that the gene modified in the anticodon can be expressed permanently in *S. lividans* which means that it is controlled by a suitable promoter. However, there were no sequences found which represented an entire *E. coli* like *Streptomyces* promoter region. We also tested upstream sequences of some tRNA genes for promoter activity in an *E. coli* promoter probe vector. But we didn't get any expression of the reporter gene.

We are currently cloning tRNA genes in promoter probe plasmids for streptomycetes to make sure that the tRNA gene is expressed during the whole life cycle. At the moment we are also introducing all three stop codons into a hygromycin resistance gene by site directed mutagenesis. These gene variants should act as reporter genes for our mutated suppressor tRNAs.

ACKNOWLEDGEMENTS

We thank Dr. J. Altenbuchner, Stuttgart, for the gene library and various useful cloning vectors. This work was supported by the Deutsche Forschungsgemeinschaft.

REFERENCES

Bibb, M.J, Findlay, P.R., and Johnson, M.B., 1984, The relationship between base composition in codon usage in bacterial genes and its use for the simple and reliable identification of protein-coding sequences. Gene, 30: 157-166.

Fournier, M.J. and Ozeki, H., 1985, Structure and organization of transfer ribonucleic acid genes of *Escherichia coli* K-12. Microbiol. Rev., 49: 379-397.

Ikemura, T., 1981, Correlation between the abundance of *Escherichia coli* transfer RNAs and the occurrence of the respective codons in its protein genes: A proposal for a synonymous codon choice that is optimal for the *E. coli* translational system. J. Mol. Biol., 151: 389-409.

Paradiso, M.J., Roberts, G., Streicher, S.L., and Goldberg, R.B., 1987, Characterization of suppressible mutations in the viomycin plasmid pVE138. J. Bacteriol., 169: 1325-1327.

Schmieger, H., Knerr, R., and Köberlein, M., 1988, Codon usage as a reason for unsuccessful search for amber-suppressor mutants in *Streptomyces lividans*? Genet. Res., 52: 163-167.

Sedlmeier, R., 1987, Charakterisierung einer Quecksilber-Resistenz aus *Streptomyces lividans*. Dipl. Thesis, Regensburg.

Sprinzl, M., Hartmann, T., Weber, J., Blank, J., and Zeidler, R., 1989, Compilation of tRNA sequences and sequences of tRNA genes. Nucleic Acids Res., 17: r1 - r172.

Vold, B.S., 1985, Strucutre and organization of genes for transfer ribonucleic acid in *Bacillus subtilis*. Microbiol. Rev., 49: 71-80.

CHARACTERIZATION OF EIGHT GROUPS (SPECIES) OF *SACCHAROPOLYSPORA* PHAGES -

AN EXAMPLE FOR STUDYING MODULAR ORGANIZATION OF ACTINOPHAGE GENOMES

Jörg Schneider

Angewandte Genetik der Mikroorganismen
FB Biologie/Chemie der Universität Osnabrück
D-4500 Osnabrück, FRG.

SUMMARY

27 actinophages of *Saccharopolyspora* (formerly *Faenia*) *rectivirgula*, *Sap. hirsuta*, *Sap. hordei* and *Sap. erythraea* were compared with respect to host range, genome length and G+C composition, virion morphology and DNA homology. According to the rules of the ICTV it was thus possible to allocate the phages to 8 new bacteriophage species. Interspecific DNA hybridization indicated a modular organization of the phage genomes: some DNA regions, presumably modules, are distributed independently among many of the phage species described. For further studies of such modules different approaches were chosen: (1) the A-"module" was used to integrate a thiostrepton resistance gene (*tsr*) by homologous recombination into the prophages of øFR113 and øFR755R to be able to monitor module exchange in future studies. (2) Several virulent mutants of the temperate øFRb-D have lost the same short DNA region; the corresponding fragments of the wild-type phage, a deletion mutant and a virulent mutant without deletions were cloned for further analysis. (3) The restriction maps of øFRb-D and øFRb-P are presented; both phages are closely related but øFRb-P has a broader host-range among *Saccharoploysspora* than øFRb-D.

INTRODUCTION

During the last 5 years, various authors (Brzezinski *et al.*, 1986; Donadio *et al.*, 1986; Grund & Hutchinson, 1987; Kempf *et al.*, 1987; Schneider *et al.*, 1987; Katz *et al.*, 1988; Smorawinska *et al.*, 1988) characterised actinophages of the *Pseudonocardiaceae* (Embley *et al.*, 1988) genus *Saccharopolyspora*, namely of *Sap. erythraea* (Labeda, 1987), *Sap. hirsuta* (Lacey & Goodfellow, 1975), *Sap. hordei* (Goodfellow *et al.*, 1989) and *Sap. rectivirgula* (Korn-Wendisch *et al.*, 1989). Recently, DNA hybridization experiments (Schneider & Kutzner, 1989a,b) revealed complex relationships on DNA level between most of these phages, indicating that the 27 phages studied should be divided into 8 homology groups: members of the same group exhibit strong homology (>92% stringency) between their genomes whereas phages from different groups might have in common short or extended parts of their DNA. Even homologies with phages from *Saccharomonospora* (Schneider *et al.*, 1989), another *Pseudonocardiaceae* genus, were demonstrated. Out of the 27 phages only the genome of øSE6 (Grund & Hutchinson, 1987) did not hybridize with any other phage DNA tested.

The distribution of homologies correlates very well with the concept of modular organization and evolution of bacteriophages (Botstein, 1980; Krylov *et al.*, 1985): bacteriophage genomes consist of independent genetic units (modules) which can be distributed among otherwise not necessarily closely related phages. So far, however, nothing is known about the genetic information of the presumed modules of the *Saccharopolyspora* phages. In this contribution the grouping of the phages into 8 species is demonstrated by various criteria, and first attempts are reported to recognize the genetic relevance of some of the potential modules.

MATERIALS AND METHODS

Bacterial strains and phages used in this work and their origin are listed in Tab. 1 and Tab. 2, respectively. All methods applied were described recently (Schneider & Kutzner, 1989a,b; Schneider *et al.*, 1989; Gayer-Herkert *et al.*, 1989).

Table 1 Bacterial strains used in this study

	strain	growth temperatures used for phage typing	provided by
Amycolatopsis orientalis	DSM 40040	37° C	F. KORN-WENDISCH[1]
Amycolatopsis mediterranei	DSM 40773	37° C	E. GRUND[2]
Pseudonocardia thermophila	G37	48° C	M. EMBLEY[3]
Saccharomonospora caesia	DSM 43044	48° C	DSM[4]
Saccharomonospora glauca	DSM 43771	48° C	DSM
Saccharomonospora viridis	DSM 43017	48° C	DSM
Saccharopolyspora rectivirgula	DSM 43747	48° C	DSM
Saccharopolyspora rectivirgula	DSM 43755	48° C	DSM
Saccharopolyspora erythraea	DSM 40517	37° C	DSM
Saccharopolyspora hirsuta	DSM 43463	37° C + 48° C	DSM
Saccharopolyspora hordei	TD8	37° C + 48° C	A. KEMPF[1]
Saccharopolyspora hordei	A65	37° C + 48° C	M. EMBLEY
Saccharopolyspora gregorii	A85	30° C	M. EMBLEY

[1]Darmstadt, FRG; [2]Bielefeld, FRG; [3]London, GB; [4]Braunschweig, FRG.

Table 2 Actinophages used in this study[1]

Actinophage	belonging to[2] phage species	kindly provided by	reference
øFR114, øFR113, øFR371	øFR114	A. Kempf[3]	Kempf *et al*, 1987
øFR747, øFRG9	øFR114	E. Greiner-Mai[3]	Kempf *et al*, 1987
øFR755R	øFR114		Schneider *et al*, 1987
Mp1	øFR114	H. Prauser[4]	Prauser & Momirova, 1970
øFRa-A, -C, -E	øFRa-C		Schneider & Kutzner, 1989a
øFRb-B, -D, -M, -P, -O	øFRb-D		Schneider & Kutzner, 1989a
øFRv-J, -N	øFRv-J		Schneider & Kutzner, 1989a
P113	øFRv-J	F. Korn-Wendisch[3]	Kempf *et al*, 1987
P517	P517	F. Korn-Wendisch	Kempf *et al*, 1987
121	121	R. Brzezinski[5]	Brzezinski *et al*, 1986
SE-3	121	R. Brzezinski	Smorawinska *et al*, 1988
øSE60	øSE60	C.R. Hutchinson[6]	Grund & Hutchinson,
øSE6	øSE6	C.R. Hutchinson	1987

[1]øFR13, øFR9 (øFR114) and øFRv-S, -T (øFRv-J) have not been described yet
[2]compare with Tab. 3
[3]Darmstadt, FRG; [4]Jena, GDR; [5]Sherbrooke, Canada; [6]Madison, USA.

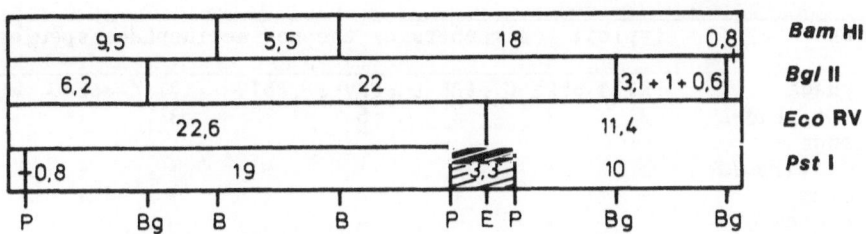

Fig. 1 Crude restriction map of P113. The hatching indicates the area of strong homology with øFR114, i.e. the J-module.

RESULTS AND DISCUSSION

Characterization of Eight Actinophage Species of *Saccharopolyspora*

As a major criterion for grouping the phages listed in Tab. 2, hybridization of the phage DNAs was used: in Southern analyses of *Bst*EII and *Pvu*II digested phage DNA (12-22 restriction fragments) strong homology (high stringency conditions; i.e. >92% homology) over 90-100 % of the genome length was demanded in order to allocate two phages to the same homology group. Dot blot analyses could not be applied as detailed homology studies (Schneider & Kutzner, 1989b) had shown that otherwise not hybridizing genomes might have a single short region (3-5kb) of very strong DNA homology, e.g. the J-module of the temperate øFR114 and the virulent øFRv-J and P113 (Schneider & Kutzner, 1989a; Fig. 1). In dot blot hybridizations, this would be enough to simulate a close relationship between the respective genomes.

The characterization of the phages with other criteria supported the grouping suggested by the DNA homology studies: phages within one homology group are very similar with respect to most features investigated. Thus it is proposed to regard the 8 groups of *Saccharopolyspora*-phages as 8 new actinophage species according to the rules (Matthews, 1982) of the International Committee on Taxonomy of Viruses (ICTV). The typical features as well as the typus phage of each species are summarized in Tab. 3.

Introduction of a Thiostrepton Resistance Gene (*tsr*) into the "A-Module"

In DNA hybridization studies the genomes of the øFR114, øFR113 (both temperate) and Mp1 (lytic on *Sap. rectivirgula* but temperate on *Sap. hordei*) exhibit a strong overall homology with the exception of two regions: øFR114 and øFR113/Mp1 differ by the alternative J/N-modules and Mp1 and øFR114/øFR113 by the alternative A/M-modules (Schneider & Kutzner, 1989a; Fig. 2). øFR371, a natural deletion derivative of øFR113 has lost the A-module (Fig. 2); as no phenotypic differences between øFR113 and øFR371 could be detected (Schneider *et al.*, 1987) it seems that the genetic information coded by the A-module is not necessary for the phage under laboratory conditions. It was thus chosen for the attempt to furnish øFR-prophages with a thiostrepton resistance gene (*tsr*) as shown in Fig. 2:

A 1.5 kb *Kpn*I fragment, covering most of the A-module from øFR114, was subcloned in two steps with the *E. coli* vector pUC18 (pJS55, pJS140). The 1.1 kb *Bcl*I fragment from pIJ702 containing the *tsr* was ligated into the *Bgl*II site of pJS140 (pJSøint), leaving 1.3 kb and 0.2kb of A-module DNA on both sites of the *tsr* available for homologous recombination.

Table 3   Characterization of the new actinophage species

features	typical for members of the new actinophage species							
typus phage	øFR114	øFRa-C	øFRb-D	øFRv-J	P517	121	øSE60	øSE6
number of phages	8	3	5	5	1	3[1]	1[2]	1[2]
host-range								
S. rectivirgula[3]	+	+	+	+	+	+	+	−
S. hordei[3]	+	+	+	+	+	+	+	−
above 43°C[3]	+	+	+	+	−	−	−	−
S. erythraea	−	+	+	−	+	+	+	+
S. hirsuta	−	+	+	−	−	−	−	+
other (Tab. 1)	−	−	−	−	−	−	−	−
life cycle[4]	t	t	t	v	l	l	t	v
genome size [kb]	41-43	43-45	41-42	32-34	42	44[1]	52[2]	42[2]
cohesive ends	+	+	+	+	+	+[1]	+[2]	−[2]
average[5] G+C [%]	58.4	58.5	60.0	66.0	58.9	59.0	nd	68.0
virion head [nm]	51x49	51x49	52x50	48x45	51x49	51x49	nd	54x50[2]
virion tail [nm]	115-120	135-140	110-115	110	120	115	nd	130[2]
remarks necessary	Mpl[6]		host range[7]					øSE[8]

[1] Brzezinski *et al.*, 1986; Smorawinska *et al.*, 1988
[2] Grund & Hutchinson, 1987
[3] note that the phages originally isolated for *Sap. erythraea* (P517, 121, øSE60) can be propagated only at temperatures below 43°C
[4] t: temperate; v: virulent; l: lytic, i.e. no lysogenization, but phage is repressed by prophages of other phages (Schneider & Kutzner, 1989a)
[5] ± 0.5% G+C; øFRb-B not included, it contains 62% G+C
[6] lytic on *Sap. rectivirgula* but temperate on *Sap. hordei*
[7] øFRb-B/D/M cannot be propagated on *Sap. hirsuta* or *Sap. erythraea* whereas øFRb-P/O can after overcoming considerable host restriction barriers
[8] prelim. description of 3 temperate relatives by Grund & Hutchinson (1987)

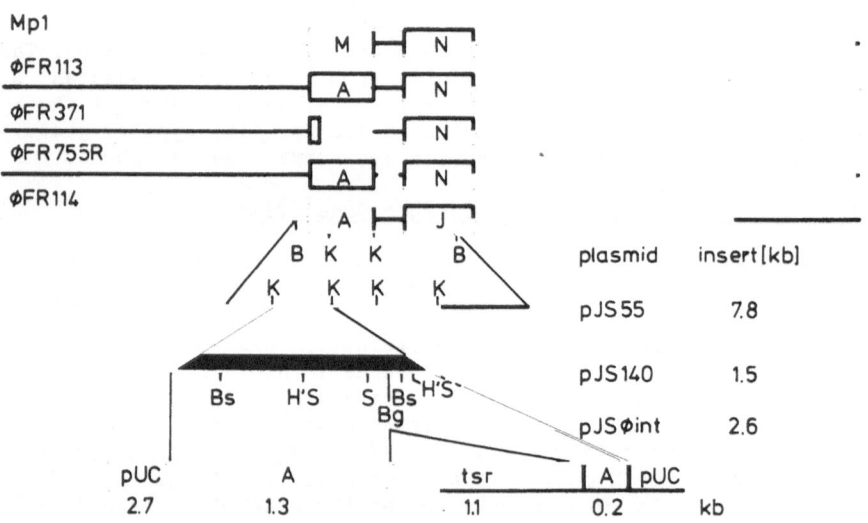

Fig. 2   Construction of a plasmid vector for integration of the thiostrepton resistance gene into the A-module of øFR-prophages by homologous recombination.
For explanation see text above. Abbreviations:
A/M/N/J: A/M/N/J-module; pUC: pUC18; tsr: thiostrepton resistance gene from pIJ702; B: *Bam*HI; Bg: *Bgl*II; Bs: *Bst*EII; H': *Hind*II; K: *Kpn*I; S: *Sac*I.

pJSøint, unable to replicate in *Sap. rectivirgula*, was used to transform lysogens of øFR113 or øFR755R. The latter (Fig. 2; Schneider *et al.*, 1987) is another deletion derivative of øFR113 which still has the A-module but lost 1kb of DNA adjacent to it. In several experiments transformation reproducibly resulted in approximately 50 thiostrepton resistant colonies per 1µg of DNA. Southern analysis of total DNA isolated from these transformants revealed that *tsr* was integrated into the phage DNA; pUC18 DNA could not be detected. All thiostrepton resistant lysogens tested so far, however, had lost the ability of spontaneous phage release: no plaques were obtained when culture supernatants were tested with suitable indicator strains whereas lysogenic cultures usually contain 10^4–10^7 pfu/ml. Thus it seems that the disruption of the A-module by *tsr* results in a defective prophage. It is not sure whether this allows to study module exchange, but the transformants might be suitable for mapping the attachment-site of the phage: this has not yet been possible as total DNA usually contains very high amounts of free (replicative?) phage DNA.

It is worth mentioning that the *tsr*-prophages seem to be stably maintained, even without pressure by thiostrepton. Recently Gayer-Herkert *et al.* (1989) showed that the *Streptomyces* plasmid vectors pIJ702 and pWOR120 are rapidly lost from *Sap. rectivirgula* transformants grown without thiostrepton, and similar difficulties were reported from other not-*Streptomyces* actinomycetes. Certainly not for shotgun cloning experiments, but perhaps for expression studies within the homologous host pJSøint-like constructs might become interesting alternatives to conventional vectors. Currently it is tested whether pJSøint can be integrated into the *Sap. erythraea* and *Sap. hirsuta* genomes by integration into a øFRa-C prophage.

Comparison of øFRb-D and øFRb-P

øFRb-D and øFRb-P belong to the same phage species but differ with respect to their host range: øFRb-D forms plaques on *Sap. hordei* and *Sap. rectivirgula* whereas øFRb-P will also lyse *Sap. erythraea* and *Sap. hirsuta* (Schneider & Kutzner, 1989a). The preliminary restriction maps presented in Fig. 3 (Tab. 4) reveal – despite obvious similarities – some variation in the restriction site patterns of their right half. Such observations preceded the detection of the J-and N-module in øFR114 and øFR113 by DNA hybridization (Schneider *et al.*, 1987), but in the case of øFRb-D and øFRb-P no pair of alternative modules was discovered. Provided that the resolution of hybridization limited by the rather crude restriction maps was high enough, the different host ranges of the two phages rather are due to short genomic variations.

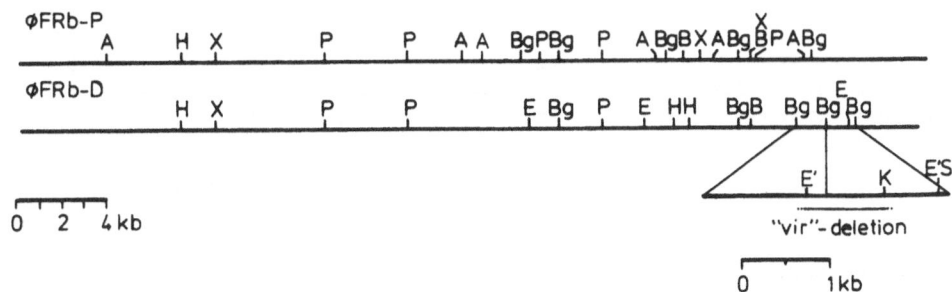

Fig. 3 Restriction maps of øFRb-P and øFRb-D. See Tab. 4 for abbreviation of enzymes; not all sites listed there are mapped yet. The missing *Bgl*II sites are located to the right of the most right *Bgl*II site shown in both cases. The "vir"-deletion (see text below) covers exactly the sites for *Eco*RI, *Bgl*II and *Kpn*I underlined.

Table 4   Restriction sites on the genomes of øFRb-D and øFRb-P

enzyme	abbreviation in Fig. 3	on the genome of øFRb-D	øFRb-P	further enzymes shown in Fig. 3	
*Apa*I	A	7	6		
*Bam*HI	B	1	2	*Eco*RI	E'
*Bgl*II	Bg	6	8	*Kpn*I	K
*Hind*III	H	3	1	*Sac*I	S
*Eco*RV	E	3	4		
*Pst*I	P	3	5		
*Xho*I	X	1	3		

Characterization of a Virulent Deletion Mutant of øFRb-D

Most of the *Saccharopolyspora* phages form huge plaques (up to 2cm in diameter) on indicator lawns of *Sap. rectivirgula* and *Sap. hordei*. As less than 40 pfu might cause a confluent lysis on conventional petri dishes, it is nearly impossible to screen large numbers of individual pfu for plaque mutants.   In the case of øFRb-D clear sectors sometimes appear within the turbid spots when 10^3-10^5 pfu are spotted onto the indicator lawn.   Clear plaques were isolated from 10 indepentently obtained sectors and the phage DNA was compared to the wild-type DNA by restriction analysis. 9 of them revealed the same short deletion which could easily be mapped on the øFRb-D restriction map (Fig. 3). One mutant obtained by D. Roberts (London) did not show any deletion. The mutants are able to lyse indicator strains which carry a øFRb-D prophage (D. Roberts, personal communication) and are thus regarded as virulent mutants. The deletion can therefore be assumed to have affected the prophage repression system of the temperate phage (Sinclair & Bibb, 1988). For more detailed investigations, the *Bgl*II frag- ments from the wild-type phage corresponding to this area of the øFRb-D genome as well as from one deletion mutant and the "point"-mutant were cloned with pUC18 (Roberts & Schneider, unpublished data).

Conclusion

The concept of modular evolution of bacteriophage genomes is well established and accepted for phages of enterobacteria and *Pseudomonas* (Campbell & Botstein, 1983; Kamp, 1987; Krylov *et al.*, 1985). Actinophages (Ackermann & Dubow, 1987), however, have scarcely been investigated for their evolution and genetic organization yet: Haroutunian *et al.* (1986) suggested modular organization of some *Brevibacterium flavum* phages, and the *Streptomyces* phages R4 and SH10 are known to be closely related though differing in the size of their genomes (Chater, 1986). The same was observed with the *Brevibacterium flavum* phage-like particle øGA1 and the virulent *Corynebacterium glutamicum* phage Cog (Sonnen *et al.*, 1990a,b). Recently, Schneider *et al.* (1990) demonstrated that the genome of the *Streptomyces* phage øSC623 consist of two halves, each with strong homology to the genomes of either øSC347 or øSC681.

These examples were chosen to underline the fact that the homology patterns described for *Saccharopolyspora* phages (Brzezinski *et al.*, 1986; Smorawinska *et al.*, 1988; Schneider & Kutzner, 1989a,b) are not exotic exceptions but might be common features of many actinophage groups, if not of most A,B or C morphotype (Ackermann & Dubow, 1987) bacteriophages (e.g. Braun *et al.*, 1989).

The distribution of DNA segments hybridizing with each other among the *Saccharopolyspora* phages is a hint but certainly not a proof of modular organization and evolution of these actinophages. Modules are defined as functional genetic units (Botstein, 1980; Campbell & Botstein, 1983), and hence some genetic information has to be collected for further studies, e.g. on the modules coding for prophage regulation, prophage integration into the host chromosome, or host range. First attempts on this have been reported in this contribution.

Modular evolution of bacteriophage genomes has important implications for interspecific genetic exchange – just consider the well accepted possibility of a coliphage lambda recombinant exchanging genetic information with the *Salmonella* phage P22 (Susskind & Botstein, 1978; Campbell & Botstein, 1983). Investigations on the fate of genetically marked phages in their natural habitats might thus be of great importance for risk assessment of genetic engineering. Derivatives of the best known *Streptomyces* phage øc31 (Chater, 1986) are currently used for such studies in soil (E. Wellington, Warwick; personal communication) and the *Saccharopolyspora* phages described above would be suitable for similar investigations, e.g. in the closed system of decomposing plant material.

Acknowledgment. The author expresses his thanks for the generous support by Prof. H.J. Kutzner (Darmstadt) and Prof. H. Schrempf (Osnabrück), in whose laboratories this work was carried out. The kind gift of strains and phages by the researchers mentioned in Tab. 1 and 2 is gratefully appreciated. J.v. Twickel is thanked for her help with the pJSøint transformations and D. Roberts (London) for his cooperation on øFRb-D, and G. Schneider for her help with the manuscript.

REFERENCES

Ackermann H-W, Dubow MS (1987). Viruses of Procaryotes. Boca Raton, Florida, CRC Press.

Botstein D (1980). A theory of modular evolution for bacteriophages. *Ann NY Acad Sci* 354, 484-491.

Braun V, Hertwig S, Neve H, Geis A, Teuber M (1989). Taxonomic differentiation of bacteriophages of *Lactococcus lactis* by electron microscopy, DNA-DNA hybridization, and protein profiles. *J Gen Microbiol* 135, 2551-2560.

Brzezinski R, Surmacz E, Kutner M, Piekarowicz A (1986). Restriction mapping and close relationship of the DNA of Streptomyces erythraeus phages 121 and SE-5. *J Gen Microbiol* 132, 2937-2943.

Campbell A, Botstein D (1983). Evolution of lambdoid phages. *In* Lambda II. RW Hendrix, JW Roberts, FW Stahl, RA Weisberg, eds., Cold Spring Harbor, New York.

Chater KF (1986). *Streptomyces* phages and their application to *Streptomyces* genetics. *In* Antibiotic-producing Streptomyces. SW Queener, LE Day, eds., Academic Press, London.

Donadio S, Paladino R, Costanzi I, Sparapani P, Schreil W, Iaccarino M (1986). Characterization of bacteriophages infecting *Streptomyces erythreus* and properties of phage-resistant mutants. *J Bacteriol* 166, 1055-1060.

Embley TM, Smida J, Stackebrandt E (1888). The phylogeny of mycolateless wall chemotype IV actinomycetes and description of Pseudonocardiaceae fam. nov. *J Syst Appl Microbiol* 11, 44-52.

Gayer-Herkert G, Schneider J, Kutzner HJ (1989). Transfection and transformation of protoplasts of the thermophilic actinomycete Faenia rectivirgula. *Appl Microbiol Biotechnol* 31, 371-375.

Goodfellow M, Lacey J, Athalye M, Embley TM, Bowen T (1989). *Saccharopolyspora gregorii* and *Sap. hordei*: two new actinomycete species from fodder. *J Gen Microbiol* 135, 2125-2139.

Grund AD, Hutchinson CR (1987). Bacteriophages of *Saccharopolyspora erythraea. J Bacteriol* 169, 3013-3022.

Haroutunian SJ, Kajoian SV, Karabekov BP, Reulets MA, Khrenova EA, Akhverdian VZ, Krylov VN (1986). Comparative studies of bacteriophages specific for *Brevibacterium flavum*. *Biotechnologia* 6, 21-27.

Kamp D (1987). The evolution of Mu. *In* Phage Mu. N Symonds, A Toussaint, P Van De Putte, MM Howe, eds., Cold Spring Harbor, New York.

Katz L, Chiang S-JD, Tuan JS, Zablen LB (1988). Characterization of bacteriophage øC69 of *Saccharopolyspora erythraea* and demonstration of heterologous actinophage propagation by transfection of *Streptomyces* and *Saccharopolyspora*. *J Gen Microbiol* 134, 1765-1771.

Kempf A, Greiner-Mai E, Schneider J, Korn-Wendisch F, Kutzner HJ (1987). A group of actinophages of *Faenia rect. Curr Microbiol* 15, 283-285.

Korn-Wendisch F, Kempf A, Grund E, Kroppenstedt RM, Kutzner HJ (1989). Transfer of *Faenia rectivirgula* Kurup and Agre 1983 to the genus *Saccharopolyspora* Lacey and Goodfellow 1975, elevation of *Sap. hirsuta* subsp. *taberi* Labeda 1987 to species level, and emended description of the genus *Saccharopolyspora*. *Int J Syst Bacteriol* 39, 430-441.

Krylov VN, Akhverdyan VZ, Bogush VG, Khrenova EA, Reulets MA (1985). The modular structure of transposable phage genomes of *Pseudomonas aeruginosa*. *Genetika* 21, 724-734.

Labeda DP (1987). Transfer of the type strain of *Streptomyces erythraeus* (Waksman 1923) Waksman and Henrici 1948 to the genus *Saccharopolyspora* Lacey and Goodfellow 1975 as *Sap. erythraea* sp. nov. and designation of a neotype strain for *Streptomyces erythraeus*. *Int J Syst Bacteriol* 37, 19-22.

Lacey G, Goodfellow M (1975). A novel actinomycete from sugar cane bagasse *Saccharopolyspora hirsuta* gen. et sp. nov. *J Gen Microbiol* 88, 75-85.

Matthews REF (1982). Classification and nomenclature of viruses. Fourth Report of the International Committee on Taxonomy of Viruses. *Intervirol* 17, 1-200.

Prauser H, Momirova S (1970). Phagensensitivität, Zellwandzusammensetzung und Taxonomie einiger thermophiler Actinomyceten. *Zentralbl Allg Mikrobiol* 10, 219-222.

Schneider J, Aguilera-Garcia I, Kutzner HJ (1987). Characterization of a family of temperate actinophages of Faenia rectivirgula. *J Gen Microbiol* 133, 2263-2268.

Schneider J, Kutzner HJ (1989a). Distribution of modules among the central regions of the genomes of several actinophages of *Faenia* and *Saccharopolyspora*. *J Gen Microbiol* 135, 1671-1678.

Schneider J, Kutzner HJ (1989b). Distribution of homologies among the genomes of several actinophages of *Faenia* and *Saccharopolyspora* as determined by DNA hybridization. *Intervirol* 30, 237-240.

Schneider J, Kramer D, Grund E, Kutzner HJ (1989). Preliminary characterization of a group of actinophages of the thermophilic actinomycete genus *Saccharomonospora*. *Intervirol* 30, 323-329.

Schneider J, Korn-Wendisch F, Kutzner HJ (1990). øSC623, a temperate actinophage of *Streptomyces coelicolor* Müller, and its relatives øSC347 and øSC681. *J Gen Microbiol* 136, in press.

Sinclair RB, Bibb MJ (1988). The repressor gene (*c*) of the *Streptomyces* temperate phage øc31: Nucleotide sequence, analysis and functional cloning. *Mol Gen Genet* 213, 269-277.

Smorawinska M, Denis F, Dery C, Magny P, Brzezinski R (1988). Characterization of SE-3, a virulent bacteriophage of *Saccharopolyspora erythraea*. *J Gen Microbiol* 134, 1773-1778.

Sonnen H, Schneider J, Kutzner HJ (1990a). Characterization of øGA1, an inducible phage particle from *Brevibacterium flavum*. *J Gen Microbiol* 136, 567-571.

Sonnen H, Schneider J, Kutzner, HJ (1990b). Corynephage Cog, a virulent bacteriophage of *Corynebacterium glutamicum*, and its relation to øGA1. *J Gen Virol* 71, in press.

Susskind MM, Botstein D (1978). Molecular genetics of bacteriophage P22. *Microbiol Rev* 42, 385-413.

ANTIBIOTIC BIOSYNTHESIS

Iain S. Hunter

Institute of Genetics, University of Glasgow, Church
Street, Glasgow, G11 5JS, UK

Our knowledge and understanding of antibiotic biosynthesis has
advanced significantly in recent years. The initial cloning of
resistance genes in the early 1980's (e.g. 1,2) was followed by the
cloning of complete pathways (e.g. 3,4). Without exception, antibiotic
production genes in streptomycetes are clustered on the chromosome.
Often a resistance genes (or genes) is also located within the cluster
(e.g. 5,6) which provides a facile strategy for cloning the others.
Many different strategies nave been used for cloning production genes
(reviewed 7).

Now that they have been cloned for some time, DNA sequence
information is becoming available for pathway genes which specify
antibiotic production. It is clear (see Piepersberg et al., this volume)
that they have evolutionary relationships with other 'house-keeping'
genes, and genes from other species.

Polyketides are a major class of antibiotics, and each of the
contributions in this section relates to polyketide structures.
However, significant progress has also been made on antibiotics derived
from amino acids, e.g. bialaphos (8) and on streptomycin (9). For B-
lactams (which are also derived from amino acids), recent advances in
the fungi (10) are being mirrored in the streptomycetes (see Demain &
Piret, this volume).

A major thrust in research on antibiotics is to define the reaction
carried out by each protein. This may be achieved by deriving mutants
blocked in the pathway, identifying the step at which each is blocked
(usually by a structural analysis of the metabolite which is built up)
and then defining the gene involvlved by complementation studies. An
alternative (and less rigorous) approach is to infer the function of the
protein from the similarity of the DNA sequence of its gene to another
gene of known function. In this way, the 'architecture' of gene
clusters is beginning to emerge.

A great deal of work needs to be done on the transcription patterns
of the gene clusters. Only a few pilot studies have been done (e.g. 9,

11). They show that (not unexpectedly) transcription of clusters is complex, involving polycistronic mRNA's and divergent, overlapping transcription. For any one gene, there may be several possibilities for transcriptional start or stop sites, but it is not yet clear what the significance of these is. It is perhaps surprising that the early studies on RNA polymerase heterogeneity in streptomycetes (12) have not been applied to antibiotic pathways in a similar way to those on catabolic genes (13,14). There is no information on the type(s) of RNA polymerase used to transcribe pathway genes, which are expressed temporally, after growth has ceased. What IS clear is that the cell takes great care to control the expression of antibiotic pathways and often uses 'activator'(positive regulator) genes, such as actII to ensure that the regulation is tight.

Antibiotic production and morphological differentiation have always been considered as different facets of the life cycle of streptomycetes. However, the discovery (see Chater et al., this volume) that the whiE gene product (whiE being a morphology gene involved in spore formation) encodes a polyketide biosynthetic gene array has modified our thinking on this. Perhaps antibiotics exist as signals or effectors of steps in differentiation, rather than a consequence of it.

Studies on polyketide synthesis are now a preoccupation of many biologists and chemists interested in antibiotic production. The discovery that the biosynthetic genes of different polyketides shared homology with each other (15) signalled the beginning of a massive effort in structure/function analysis of the enzymes responsible for polyketide assembly. This is being approached first at the level of DNA sequence (16,17). The 'simple' polyketides, such as actinorhodin and tetracycline) are derived solely from acetate precursors. The gene clusters for their biosyntheses are small and compact - of the order of 30kb- and contain single copies of the genes for polyketide biosynthesis. By contrast, the macrolide pathways are much larger - up to 100kb. They use a variety of different precursors (both straight and branched chain). It now appears (18) that these more complicated antibiotic structures have 'modules' of polyketide biosynthetic genes. Each module contains a complete set of biosynthetic genes and is predicted to be responsible for only one step in the elongation of the backbone of the molecule. It is likely that the polyethers will also use the module principle to carry out their biosynthesis. As the DNA sequence analyses progress, attention is now being turned to the expression and purification of the proteins involved and their eventual reconstitution into a cell-free system in which to study the reactions of antibiotic biosynthesis.

Although our knowledge has advanced significantly, it should be clear that it will require a concerted effort from molecular biologists, biochemists and microbial physiologists to approach a fuller understanding of antibiotic biosynthesis. It is particularly pleasing to see these aspects grouped round the tetracycline family of antibiotics for the contributions to this symposium and book.

REFERENCES

(1) Bibb, M.J., Schottel J.L. & Cohen, S.N. (1980) Nature 284, 284-286.

(2)   Thompson, C.J., Kieser, T., Ward, J.M. & Hopwood, D.A. (1982) Gene
        20, 51-58.
(3)   Malpartida, F.M. & Hopwood, D.A. (1984) Nature 309, 462-4.
(4)   Motamedi, H & Hutchinson C.R. (1987) Proc. Natl. Acad. Sci. USA.
        84, 4445-9.
(5)   Rhodes, P.M, Hunter, I.S., Friend, E.J. & Warren, M. (1984)
        Biochem. Soc. trans. 12, 586-7.
(6)   Chater, K.F. & Bruton, C.J. (1985) EMBO Journal 4, 1893-7.
(7)   Hunter,I.S. & Baumberg, S. (1989) SGM Symposia 44, 121-162.
(8)   Hara, O., Anzai, H., Imai, S., Kumoda, Y., Murakami, T., Itoh, K.,
        Tanako, E., Satoh, A. & Nagoaka, K. (1988) J. Antibiotics 41, 538-
        47.
(9)   Mansouri, K., Pissowotzki, K., Distler, J., Mayer, G., Heinzel,
        P., Braun, C., Ebert, A. & Pipersberg, W. (1989) in Genetics and
        Molecular Biology of Industrial Microorganisms. eds C.L.
        Hershberger, S.W. Queener & G. Hegeman. ASM.
(10)   McCabe, A.P., Riach, M.B.R., Unkles, S.E. & Kinghorn, J.R. (1990)
        EMBO Journal 9, 279-287.
11)   Malpartida, F.M., & Hopwood, D.A. (1986) Molec. Gen. Genetics 205,
        66-71.
(12)   Westpheling, J.E., Raines, M. & Losick, R. (1985) Nature 31322-7.
(13)   Buttner, M.J., Smith, A.M. & Bibb, M.J. (1988) Cell 52, 599-607.
(14)   Westpheling, J., Brawner, M., Fornwald, D., Huang, D-Y., Ingram,
        C. & Mattern, S. (1990) J.Cellular Biochem. S14, 98.
(15)   Malpartida, F.M., Hallam, S.E. Kieser, H.M., Motamedi, H.,
        Hutchinson, C.R., Butler, M.J., Sugden, D.A., Warren, M.,
        McKillop, C., Bailey, C.R., Humphreys, G.O. & Hopwood, D.A. (1987)
        Nature, 325 818-20.
(16)   Bibb, M. J., Biro, S., Motamedi, H., Collins, J.F. & Hutchinson,
        C.R. (1989) EMBO Journal 8, 2727-2736.
(17)   Sherman, D.H., Malpartida, F.M., Kieser, H.M., Bibb, M.J. &
        Hopwood, D.A. (1989) EMBO Journal 8, 2717-2726.
(18)   Katz, L., Staver, M., Tuan, J., Brown, D. & Donadio, S. (1990) J.
        Cellular Biochem. S14, 90.

CEPHAMYCIN PRODUCTION BY STREPTOMYCES CLAVULIGERUS

A. L. Demain[1] and J.M. Piret[2]

[1]Department of Biology, Massachusetts Institute of
Technology, Cambridge, MA 02139, USA
[2]Northeastern University, Boston, MA 02115, USA

INTRODUCTION

The classical ß-lactam antibiotics can be divided into
hydrophobic and hydrophilic fermentation products. The
hydrophobic members, e.g. benzylpenicillin (penicillin G)
and phenoxymethylpenicillin (penicillin V), contain
non-polar side chains, e.g. phenylacetate and
phenoxyacetate, respectively, and are made only by
filamentous fungi; the best known of these is Penicillium
chrysogenum. The hydrophilic types are penicillin N,
cephalosporins and 7-a-methoxycephalosporins (cephamycins)
which are made by fungi, actinomycetes and unicellular
bacteria. They all contain the polar side chain,
D-α-aminoadipate.

We can draw a sequence of reactions which describes the
biosynthesis of penicillins and cephalosporins (Fig. 1),
however the total sequence exists in no one microorganism.
All penicillin and cephalosporin biosynthetic pathways
possess the first three steps in common and all
cephalosporin pathways go through deacetylcephalosporin C.
However, there are many subsequent biosynthetic reactions
which vary in the different producing organisms.

BIOSYNTHESIS OF CEPHALOSPORINS

1. The early pathway leading to isopenicillin N

The initial reaction of the common pathway is the
condensation of L-cysteine (L-cys) and L-a-aminoadipic acid
(L-aaa) to form L-a-aminoadipyl-L-cysteine (AC). L-valine

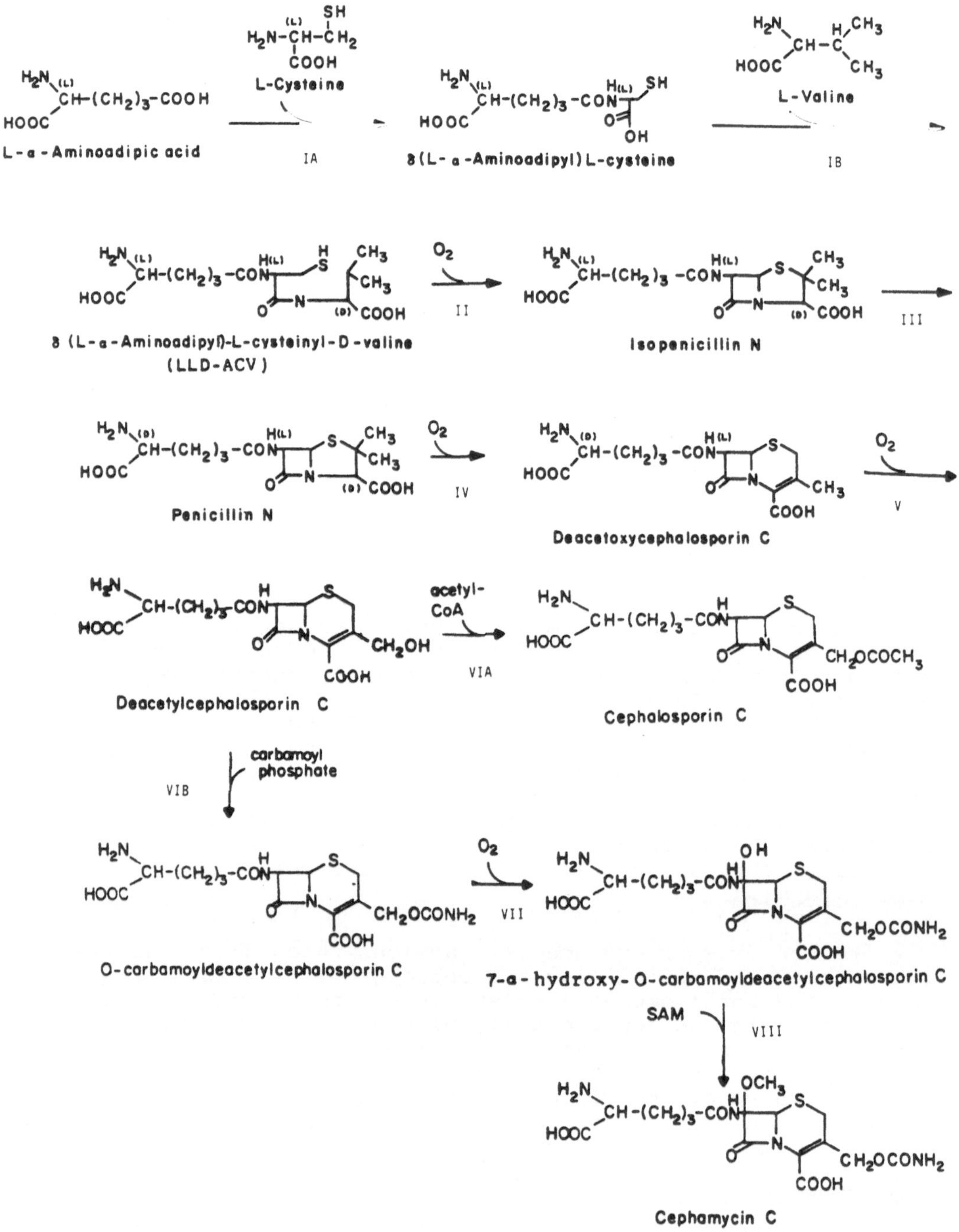

Fig. 1.   Pathway of Cephamycin C Biosynthesis

(L-val) is epimerized to the D-form during activation and addition to form the LLD-tripeptide, i.e. d-(L-a-aminoadipyl)-L-cysteinyl-D-valine (ACV). ACV is the key intermediate in the formation of all penicillins and cephalosporins by eukaryotic and prokaryotic microorganisms.

Banko et al. (1986) were the first workers able to reproducibly obtain net production of AC by cell-free extracts of C. acremonium. Activity was extremely unstable unless glycerol was added before preparation and storage of the extracts. This was the key finding which led to the development of a successful system. The activity was dependent on L-aaa, L-cys, ATP and Mg^{2+} or Mn^{2+}. The very low levels of AC observed appeared to be due to its role as an enzyme-bound intermediate of a multifunctional ACV synthetase. Banko et al. (1987) found that ACV formation from L-aaa + L-cys + L-val was much faster than from AC + L-val. When L-val was added to an AC-forming reaction at 3 h, the ensuing rate of ACV formation was much higher than that previously observed for AC formation, and although no more AC was accumulated, previously formed AC did not decrease. Thus the AC detected in the AC-forming system was probably that small amount which fell off the enzyme. The ACV formed could be converted to isopenicillin N by isopenicillin N synthase (cyclase).

The conversion of ACV to isopenicillin N by isopenicillin N synthase ("cyclase") was first demonstrated in cell-free systems from C. acremonium (Fawcett et al., 1976; Konomi et al., 1979; O'Sullivan et al, 1979a). Since then, it has been demonstrated in cell-free systems from virtually all producers of penicillins and cephalosporins. Mutants of S. clavuligerus which completely lack or have a low level of cyclase are either non-producers or low producers of cephamycin (Mahro and Demain, 1987; Romero et al., 1988).

The gene encoding cyclase has been the object of interest to recombinant DNA technologists in the past few years. The first cyclase gene to be cloned was that from C. acremonium and the recipient was Escherichia coli (Samson et al., 1985). A plasmid vector was used and cyclase was produced in E. coli at a level of 20% of cell protein. Cyclase genes have also been cloned and sequenced from P. chrysogenum (Carr et al., 1986)., Aspergillus nidulans (Ramon et al., 1987; Weigel et al., 1988), S. clavuligerus (Leskiw et al., 1988), Streptomyces lipmanii and Streptomyces jumoninensis (Weigel et al., 1988; Shiffman et al., 1988). It appears that the cyclase genes from prokaryotes and eukaryotes are related and probably evolved from a common ancestral gene. The three fungal genes show 74-80% similarity in nucleotide sequence and their proteins 73-81% similarity in amino acid sequence. The corresponding figures for the actinomycetes are 70-81% and over 70%. The relatedness between fungal and actinomycete genes are 56-62% and the enzymes 54-56%. Ramon et al. (1987) suggested that the cyclase gene was transferred from the streptomycetes to a common fungal ancestor of the fungi. Cephalosporium was thought to branch off this common ancestral evolutionary pathway first, and then P. chrysogenum at a later date. Weigel et al. (1988) concluded that there was a horizontal transfer of the cyclase gene from a prokaryote to a eukaryote about 370 million years ago. No introns have been found in any cyclase gene.

Cyclase contains two cysteine residues at positions 106 and 255. By site-directed mutagenesis, three mutant cyclases were prepared with serines replacing one or both of the cysteines (Samson et al., 1987). Substitution at cys 255 reduced activity by 50%. Substitution at cys 106 or at both cysteine sites, reduced activity by 95%. Thus the cysteine residues are important but not essential for cyclase activity. Cyclase from a number of Streptomyces species contain the same cysteine residues (Shiffman et al., 1988).

2. Conversion of isopenicillin N to cephalosporins

The next step is catalyzed by an extremely labile enzyme which epimerizes the L-aaa side chain of isopenicillin N to the D-configuration of penicillin N. This enzyme activity (previously called "racemase") was discovered in C. acremonium by Konomi et al. (1979). The epimerase in S. clavuligerus appears to be somewhat more stable than that in C. acremonium. In both S. clavuligerus (Jensen et al., 1983) and C. acremonium (Lübbe et al., 1986), pyridoxal phosphate appeared to stabilize the enzyme. The enzyme from S. clavuligerus has recently been purified to electrophoretic homogeneity from S. clavuligerus (Usui and Yu, 1989). It is a monomer of molecular weight 47,000 to 50,000, containing 1 mole of pyridoxal-5'-phosphate per mole of enzyme. The enzyme acts on both isopenicillin N and penicillin N producing an equimolar mixture of the two diastereoisomers. The epimerase gene has been cloned and expressed in E. coli (Ingolia and Queener, 1989).

Although for many years it was thought that penicillin N and cephalosporin C were products of different biosynthetic branches, ring-expansion of penicillin N to deacetoxycephalosporin C was discovered in 1976 by Kohsaka and Demain (1976) The ring-expansion reaction catalyzed by deacetoxycephalosporin synthase ("expandase") requires Fe^{++}, ascorbate (Hook et al., 1979; Sawada et al., 1979), oxygen and α-ketoglutarate (Felix et al., 1981; Kupka et al., 1983a, b); CO_2 is liberated from the α-ketoglutarate during the reaction.

The ring-expansion enzyme has many of the characteristics of an α-ketoglutarate-linked dioxygenase (Shen et al., 1984) although it does not technically fit the definition since the two atoms of oxygen do not end up in the products (De Jong et al., 1982). One atom of oxygen is incorporated into succinate during the oxidative decarboxylation of the cosubstrate, α-ketoglutarate. The

other oxygen atom probably goes to an intermediate   which
is converted to deacetoxycephalosporin C and water (Pang et
al., 1984). The second oxygen atom would thus end up in
H_2O.

Although expandase and the next enzyme in the sequence,
deacetoxycephalosporin C hydroxylase, are separate
dioxygenases in S. clavuligerus (Jensen et al., 1985;
Rollins et al., 1988), the two enzyme activities are present
on a single protein in C. acremonium. (Scheidegger et al.,
1984; Dotzlaf and Yeh, 1987).

The S. clavuligerus expandase is a monomer with a
molecular weight of about 36,000 and a K_m for penicillin N
of 50μM. It requires α-ketoglutarate, Fe^{2+}, oxygen and is
specifically stimulated by ascorbate and DTT (Rollins et
al., 1988). The enzyme has been purified almost to
homogeneity.

Cloning and expression of the expandase gene from S.
clavuligerus in E. coli revealed an enzyme molecular weight
of 34,519. The expandase genes from S. clavuligerus and C.
acremonium (Kovacevic et al., 1989) are very similar. The
DNA sequences are 65% identical and at the predicted amino
acid level, the identity is 57% (Miller and Ingolia, 1989).
The 3' end of the streptomyces cefE gene is shy of the 20
codons of the fungal cefEF gene which codes for
deacetoxycephalosporin C hydroxylase as well. In S.
clavuligerus, the cyclase gene and the expandase gene may be
clustered together in that a single cosmid contains both
genes.

Deacetoxycephalosporin C is hydroxylated to
deacetylcephalosporin C by an α-ketoglutarate-linked
dioxygenase (Fujisawa et al., 1977; Liersch et al., 1976;
Brewer et al., 1977). The hydroxylase is stimulated by
α-ketoglutarate, ascorbate, DTT and Fe^{++}, incorporates
oxygen from molecular oxygen (O'Sullivan et al., 1979b) and
is specific for deacetoxycephalosporin C (Turner et al.,
1978).

Once the deacetylcephalosporin C stage is reached,
there is a branch in the pathway. C. acremonium acetylates
deacetylcephalosporin C to cephalosporin C whereas
actinomycetes carbamoylate the intermediate. S.
clavuligerus converts deacetylcephalosporin C to
O-carbamoyldeacetylcephalosporin C using carbamyl phosphate
as the carbamoyl donor. The enzyme, carbamoyl
phosphate-3-hydroxymethylcephem-O-carbamoyl transferase,
carries out the ATP-dependent reaction (Whitney et al.,
1972; Brewer et al., 1980).

Carbamoyldeacetylcephalosporin C is hydroxylated to
7-α-hydroxycarbamoyldeacetylcephalosporin C by another

α-ketoglutarate-linked dioxygenase (Hood et al., 1983);
molecular oxygen provides the oxygen atom (O'Sullivan et al,
1979b). This dioxygenase has been purified to near
homogeneity recently (Xiao and Demain, unpublished data).
This intermediate is methylated to cephamycin C by a
methyltransferase using S-adenosylmethionine (O'Sullivan and
Abraham, 1980). Semi-synthetic cephamycins, e.g. cefoxitin,
are very useful in the clinic against β-lactam-resistant
bacteria. Cell-free extracts of S. clavuligerus carry out
the two-step methoxylation of carbamoyldeacetylcephalosporin
C in the presence of S-adenosylmethionine, α-ketoglutarate,
Fe^{2+} and a reducing agent such as ascorbate. Methoxylation
works only feebly on deacetoxycephalosporin C and not at all
on deacetylcephalosporin C. Surprisingly, cephalosporin C
can be methoxylated by the cell-free S. clavuligerus enzyme
to 7a-methoxycephalosporin C (O'Sullivan and Abraham, 1980)
although this is not an in vivo reaction since S.
clavuligerus does not produce cephalosporin C.

REGULATION OF CEPHALOSPORIN BIOSYNTHESIS

      Production of ß-lactam antibiotics occurs best under
conditions of nutrient imbalance and at low growth rates.
Nutrient imbalance can be brought about by limitation of the
carbon, nitrogen or phosphorus source. In addition to these
factors, lysine exerts a positive effect on production of
cephamycins.

1.   Carbon source regulation

      Carbon sources regulate ß-lactam antibiotic formation
in actinomycetes. Aharonowitz and Demain (1978) found that
glycerol and maltose support extensive growth of S.
clavuligerus but the specific production of cephamycin
decreases as carbon source concentration is increased. The
organism does not use glucose for growth, but does use
glycerol, maltose and starch.

      Glycerol suppresses cephamycin biosynthesis by S.
clavuligerus while starch is less suppressive (Aharonowitz
and Demain, 1978; Hu et al., 1984). Suppression of
cephamycin synthesis is accompanied by repression of
expandase (Lebrihi et al., 1988b). When added to a starch
fermentation, glycerol is more repressive the earlier it is
added. Neither glycerol nor starch inhibit expandase
activity however glucose-6-phosphate and
fructose-1,6-diphosphate (but not fructose-6-phosphate) were
strong inhibitors in the experiments of Lebrihi et al.
(1988b). Zhang and Demain (unpublished data) found
glyceraldehyde-3-phosphate to be a strong inhibitor of ACV
synthetase in S. clavuligerus; no evidence of repression of
ACV synthetase was obtained with glycerol.

2. Nitrogen source regulation

The ammonium ion exerts a negative effect on ß-lactam production in S. clavuligerus (Aharonowitz and Demain, 1979; Aharonowitz, 1980). The effect is reversed by the ammonium-trapping agent, magnesium phosphate. Better nitrogen sources for cephamycin production are asparagine, aspartate, urea and glutamine. In addition to NH_4^+, alanine and histidine are very suppressive nitrogen sources. In this organism, cyclase appearance preceeds expandase in the time-course of events. The cyclase is more severely repressed than the expandase by NH_4^+.

Braña et al. (1985) compared three cephalosporin synthases in S. clavuligerus NRRL 3585 and found cyclase to be the most sensitive to ammonium repression, expandase moderately sensitive and epimerase not significantly affected; none of these synthases was inhibited in their action by NH_4^+. Since cyclase is not the first enzyme in the pathway, the existence of a control mechanism affecting the synthesis of the ACV tripeptide was postulated by Braña et al. (1985). Recent studies (Zhang et al., 1989a) revealed that ACV synthetase was the most repressible enzyme among the four tested (75% decrease by 120 mM NH_4^+ added to the medium), followed by cyclase (70% decrease) and expandase (50% decrease); epimerase was only slightly affected. Little to no inhibition of ACV synthetase action by NH_4^+ was observed. Thus repression of ACV synthetase, cyclase and expandase appears to be the major factor contributing to the negative effect of ammonium on cephamycin biosynthesis in S. clavuligerus. The ammonium effect is exerted to a greater extent in S. clavuligerus on cephamycin formation than on production of the co-product, clavulanic acid (Romero et al., 1984).

In an attempt to understand the mechanism of the ammonium effect, nitrogen assimilation in S. clavuligerus was studied (Braña et al., 1986a). Significant levels of glutamine synthetase (GS), glutamate synthase (GOGAT) and alanine dehydrogenase (ADH) were found in crude extracts after growth in media containing different defined nitrogen sources. GS activity varied markedly depending on the nitrogen source although depressed levels were always found in the presence of ammonium. GOGAT activities were rather constant and ADH was induced when high ammonium or high alanine were used. Glutamate dehydrogenase and alanine aminotransferase were not detected. After mutagenesis, mutants lacking GS, GOGAT or ADH were obtained. The data obtained by examining the utilizable nitrogen sources for wild-type and mutant cultures showed that the GS-GOGAT pathway is the only means of ammonium assimilation in S. clavuligerus.

Data on production of cephamycin and its synthases during growth of the wild-type culture on different nitrogen sources showed no correlation between these parameters and intracellular levels of GS, GOGAT and ADH (Braña et al., 1986b). The lack of correlation was further supported by the behavior of the mutants blocked in each enzyme, i.e. the ammonium effect on cephamycin and synthase levels was still evident in the mutants. Thus active GS, GOGAT or ADH is not necessary for ammonium regulation of cephamycin synthesis in S. clavuligerus.

Looking for a hypothetical intracellular mediator of ammonium repression of cephamycin production, Braña et al. (1986b) analyzed the intracellular pool of 21 free amino acids and NH_4^+ in cells growing with different nitrogen sources. Only glutamine, alanine and ammonium showed marked changes. Although no correlation was found between the intracellular level of these compounds and the titers of cephamycin and cyclase, the amino acids alanine, glutamine and glutamate did inhibit cephamycin synthesis by resting S. clavuligerus cells in the absence of protein synthesis.

3. Phosphorus source regulation

It has been reported that the biosynthesis of ß-lactams by S. clavuligerus (Aharonowitz and Demain, 1977; Lübbe et. al., 1985a; Lebrihi et al., 1987) is subject to phosphate control. In this species, Lübbe et al., (1984, 1985) found that expandase is subject to phosphate repression and both expandase and cyclase are inhibited by phosphate. Lebrihi et al. (1987) confirmed the phosphate repression of expandase. Zhang et al. (1989b) recently found that high phosphate concentrations strongly reduce cephamycin production, and represses all the four synthases examined (ACV synthetase, cyclase, epimerase and expandase). ACV synthetase is the main repression target and expandase the next. Phosphate inhibition of ACV synthetase action is relatively slight, i.e. 30% inhibition at 80mM phosphate (Zhang et al., 1989b), compared to almost 100% inhibition of expandase and 50% inhibition of cyclase by 20mM phosphate reported by Lübbe et al. (1985a).

In S. clavuligerus, the negative effect of phosphate appears to be much greater on clavulanic acid formation than on production of cephamycin (Romero et al., 1984).

4. Regulation by lysine

The relationship between lysine addition and cephalosporin synthesis in the actinomycetes is completely different from that in the fungi. Since L-aaa is not an intermediate of lysine biosynthesis in actinomycetes as it is in the fungi, the cephalosporin side chain is not derived from lysine biosynthesis, but instead from lysine degradation via 1-piperideine-6-carboxylate. A L-lysine epsilon-aminotransferase was detected in Nocardia lactamdurans (Kern et al., 1980). Cephamycin-producing actinomycetes (S. clavuligerus, S. griseus, N. lactamdurans) metabolize lysine by two different pathways (Madduri et al. 1989). For growth, they use the cadaverine pathway:

Lysine $\rightarrow$ cadaverine $\rightarrow$ [1-piperideine] $\rightarrow$

δ-aminovalerate

    lysine   cadaverine   1-piperideine $\downarrow$
      decarboxylase   aminotransferase   dehydrogenase $\downarrow$
                                                        CO_2 +

$$H_2O + NH_2$$

For cephamycing biosynthesis, they use the lysine aminotransferase pathway to α-aminoadipate, the precursor of cephamycin:

Lysine $\rightarrow$ 1-piperideine-6 carboxylate $\rightarrow$ α-aminoadipate

     lysine   1-piperideine-
     ε-aminotransferase   6-carboxylate
                                   dehydrogenase

The addition of lysine, as well as its precursor, diaminopimelate, or its product, L-aaa, stimulates the production of cephalosporins in S. clavuligerus (Medelovitz and Aharonowitz, 1982). The positive lysine effect in S. clavuligerus has been further exploited (Mendelovitz and Aharonowitz, 1983; Aharonowitz et al., 1984). These investigators studied the prokaryotic pathway to lysine from aspartic acid. The first biosynthetic enzyme of the aspartic acid family of amino acids, aspartokinase, is the rate-limiting step of cephamycin biosynthesis. Seventy percent of thialysine-resistant, aspartokinase-deregulated strains of S. clavuligerus overproduced cephamycin C. These mutants accumulate a high concentration of diaminopimelate in their intracellular pool. As much as 30% of the intracellular amino acids is diaminopimelate; in the parent, the value is only 0.5%. Growth of the mutants in defined medium was somewhat slower than that of the parent but specific cephamycin production was 2- to 5-fold higher. It thus appears that after aspartokinase, diaminopimelate decarboxylase is the next limiting enzyme in S. clavuligerus.

5. Growth rate regulation

With S. clavuligerus, growth rate may be a key factor in controlling cephamycin production (Lebrihi et al., 1988a). In both batch and chemostat cultures, there was an inverse relationship between growth rate on the one hand, and cephamycin and expandase production on the other, irrespective of whether glycerol, ammonium or phosphate was the limiting substrate. On the other hand, Braña et al. (1986b) failed to observed a correlation between growth rate and cephamycin production in batch cultures of S. clavuligerus growing on different nitrogen sources.

GENETIC STUDIES ON STREPTOMYCES CLAVULIGERUS

1. Cloning of a DNA region involved in cephalosporin biosynthesis

Mahro and Demain (1987) set up a mutant complementation strategy for S. clavuligerus NRRL 3585 to clone DNA sequences involved in early steps of cephalosporin biosynthesis. A screening assay was developed to isolate cephamycin-deficient mutants following nitrosoguanidine treatment. The assay could detect mutants blocked in steps prior to penicillin N formation, i.e. deficiencies in the ACV synthetase, cyclase or epimerase activities. One mutant, called NP1, which produced an extremely low level of cephamycin was found among ~5000 colonies screened (Mahro and Demain, 1987). The mutant exhibited close to normal expandase and epimerase activities but only about 35% of normal levels of cyclase and 25% of normal ACV synthetase activity.

The Streptomyces multicopy plasmid vector pIJ702 (Katz et al., 1983) was used to shotgun clone wild type DNA fragments into mutant NP1 (Piret et al., 1990). Size-fractionated wild type chromosomal DNA fragments of 4-10 kb, obtained by partial digestion with MboI, were ligated into the multicopy plasmid vector pIJ702 propagated in S. clavuligerus. pIJ702 had been digested with BglII, inactivating the tyrosinase gene which imparts the ability of plasmid-bearing transformants to produce the pigment melanin (Katz et al., 1983). The frequency of inserts is normally assessed by insertional inactivation of the tyrosinase, however, this gene is poorly expressed in S. clavuligerus (Katz et al., 1983; Bailey et al., 1984; our observations). Therefore a portion of the ligation mixture was transformed into Streptomyces lividans 1326, where the tyrosinase gene is expressed, and the insert frequency determined to be about 75%. Transformation of mutant NP1 yielded ~4000 transformants which were screened for cephamycin production in the plate bioassay. One transformant produced significantly more cephamycin than did the mutant NP1, although the amount was only about 10% of that of the wild type strain. This clone was named S. clavuligerus MR1.

2. Plasmid copy number effects in S. clavuligerus and isolation of pNBR1

Attempts to isolate a plasmid from strain MR1 were unsuccessful by routine plasmid isolation procedures, from cultures grown on plates or liquid medium. We attributed this to low plasmid copy number or preferential integration of the plasmid into the S. clavuligerus chromosome. In order to "rescue" the plasmid, the concentrated plasmid DNA fraction prepared from strain MR1 (no plasmid band being visible by gel electrophoresis) was transformed into S. lividans 1326. Eight thiostrepton-resistant transformants were obtained and all carried a plasmid of about 13 kb, implying a cloned insert size in pIJ702 of about 7 kb. The plasmid copy number in S. lividans was typical for pIJ702 derivatives in this strain. The plasmid was named pNBR1.

Retransformation of mutant NP1 with pNBR1 showed that the plasmid conferred the restored cephalosporin production phenotype of S. clavuligerus MR1. The transformants obtained produced cephamycin. Both the copy number of pNBR1 and the level of cephamycin produced by these secondary transformants varied (from 10-70% of wild type levels) between transformants. In hybridizations of nick-translated pNBR1 to Southern blots of genomic DNA digests prepared from these pNBR1-bearing strains, the copy number of pNBR1 varied between transformants by at least 20-fold. Our experience suggests that these copy number variations are dependent upon the method used to prepare and transform protoplasts. We believe that the physiological conditions under which mycelium is produced for the preparation of protoplasts strongly affects not only the efficiency of protoplast transformation, but the subsequent distribution or maintenance of the plasmid in the growing colony. Thus, the early transformation protocol used in the shotgun cloning work (Bailey et al., 1984) in our hands resulted in low transformation frequencies and low plasmid copy number. The secondary transformants were obtained using a protocol modified from Bailey and Winstanley (1986) and Garcia-Dominguez et al.(1987) which yields higher transformation frequencies. These strains grew rapidly and confluently on selective medium and carried plasmid DNA in quantities sufficient to be easily visualized in agarose gels.

Southern blotting of genomic DNA from the wild type and mutant
NP1 and probing with pNBR1 confirmed that the cloned insert originated from S. clavuligerus. The hybridization banding pattern was as predicted from the restriction map of pNBR1 indicating that the cloned DNA had not been visibly rearranged during the cloning experiment. This observation turned out to be important later (this section, part 6).

3. Presence of homologous DNA regions in other ß-lactam producing species

To gain supporting evidence that the cloned DNA was directly involved in ß-lactam biosynthesis, total genomic DNA isolated from a collection of ß-lactam-producing and non-producing species were probed with pNBR1. The ß-lactam producers were Streptomyces griseus MA 2837, S. jumonjinensis MA 4646, N. lactamdurans MA 2908, S. lipmanii MA 4250 and S. lipmanii ATCC 27357. The non-producing species were Streptomyces glaucescens ETH 22794, Streptomyces peucetius ETH 3180, S. lividans 1326, and Streptomyces coelicolor M145. The DNA was digested with BglII, blotted and probed with pNBR1 purified from S. lividans. Under stringent conditions, specific hybridization was obtained to each producing strain but not to the non-producing strains.

4. ß-lactam enzyme activities in the complemented strain

The secondary transformant MR1.3, which produced the highest level of cephalosporin (70% of wild type levels), was used to obtain time courses of growth, antibiotic production and ß-lactam enzyme activities in liquid culture. Both ACV synthetase and cyclase activities were restored to the mutant (about 60% and 75%, respectively) (Piret et al., 1990). This suggested several possibilities: (i) the cloned region might carry the structural genes for both the ACV synthetase and the cyclase; (ii) it might carry either the ACV synthetase or the cyclase structural gene and the expression of the two genes or the activities of their products might be interdependent; (iii) the cloned region might carry a regulatory gene or element which controls the expression of both ACV synthetase and cyclase.

Table 1. Phenotypes of S. clavuligerus NRRL3585 mutants deficient in ß-lactam production (Romero et al.,1988; Piret et al.,1990)

Strain	Cephamycin	ACV Synthetase	Cyclase	Epimerase	Expandase
nce1	-	-	low	low	low
nce2	-	-	-	wt	wt
ncc4	low	-	low	wt	wt
NP1	low	low	low	wt	wt

-, not detectable; wt, wild-type levels.

5. Complementation of other blocked mutants

We introduced pNBR1 into several other S. clavuligerus mutants blocked in cephalosporin production to test for phenotypic complementation. Romero et al. (1988) isolated and characterized several such mutants from S. clavuligerus NRRL 3585. As in the isolation of mutant NP1 (this section, part 1), Romero et al. (1988) screened for mutants blocked in early steps of the ß-lactam pathway. Relative to the parental strain, the mutants are essentially unaffected in their growth kinetics and ability to utilize various amino acids (Romero et al., 1988), suggesting that they are specifically impaired in the antibiotic pathway. The following table summarizes the phenotypes of three mutant strains which we have tested for complementation by pNBR1. The same information for mutant strain NP1 is also given.

The four mutants are clearly deficient in ACV synthetase and cyclase activities. In three strains, epimerase and expandase activities remain essentially normal. Only in the case of mutant nce1 are the two latter activities also strongly affected. This pattern suggests that nce1 may be blocked in a regulatory element controlling all four enzymes and which is present on the cloned DNA in pNBR1. Alternatively, nce1 may bear more than one mutation in the biosynthetic steps of the pathway and which are present in the cloned region.

Upon transformation with pNBR1, all four mutants regained cephamycin production. Transformation with the cloning vector, pIJ702, without insert had no effect on the mutant phenotypes. Thus the 7 kb insert in pNBR1, barring suppression effects, carries information sufficient to complement all four mutants which we have in hand.

6. Nature of the cloned DNA sequences

In cooperation with the laboratory of Dr. Susan Jensen (University of Alberta, Edmonton, Canada), we obtained strong evidence that the cloned sequences in pNBR1 originate from a chromosomal region which lies in close proximity to the S. clavuligerus cyclase gene. Since in Streptomyces it has been found that antibiotic biosynthetic genes of a given pathway are clustered (Hopwood, 1988), this argued for a direct involvement of these sequences in the cephalosporin pathway. Probing of digests of pNBR1 with an internal fragment of the S. clavuligerus cyclase gene (Leskiw et al., 1988) gave no hybridization signal (S. Jensen and B. Leskiw, pers. comm.), thus excluding the presence of the cyclase gene in pNBR1. However, probing of the same blot with pBL1 which contains the 9.2 kb DNA insert carrying the cyclase gene and surrounding regions (Leskiw et al., 1988) gave hybridization signals which map to the "left" end of the cloned region. The reciprocal experiment of probing pBL1 digests with pNBR1 confirmed the homology. However, a simple overlap of the pNBR1 and pBL1 inserts cannot be drawn. pNBR1 hybridizes to a region in the center of the pBL1 insert, near the IPNS gene. This may be due (S. Jensen, personal communication) to the fact that pBL1 carries a discontinuous region of the S. clavuligerus genome which was probably formed by coligation of two unrelated DNA fragments during the cloning of the cyclase gene.

Our hybridization experiments (this section, part 2) showed that the cloned insert in pNBR1 is probably not rearranged and originates from a contiguous chromosomal region.

Leskiw et al. (1988) reported that DNA sequencing of the S. clavuligerus cyclase gene region revealed the presence of an open reading frame (ORF) terminating just 16-31 bp upstream of the cyclase ORF which would be transcribed in the same direction as the cyclase gene. The sequence for a likely ribosomal binding site (RBS) lies in the region between these two ORF's. This suggests that the cyclase gene is co-transcribed with the upstream ORF and that the sequences controlling the transcription of this operon lie 5' to the upstream ORF. Although speculative at this point, this arrangement could provide the means by which the cyclase and the ACV synthetase genes are co-regulated. The upstream ORF might encode the ACV synthetase itself, or a protein which controls ACV synthetase formation or activity.

We recently have obtained evidence that our isolated DNA sequence does contain the ACV synthetase gene or at least part of it. Hybridization experiments done between the fragment and the ACV synthetase gene from P. chrysogenum were positive (unpublished data).

Additional recent data indicate that the fragment is capable of improving cephamycin production. In recent work in our laboratories, we have found that introduction of the fragment into certain blocked mutants of S. clavuligerus resulted in cephamycin production considerably higher than that of the wild-type parent (Serpe, 1990; Agayn, 1990).

SUMMARY

Streptomyces clavuligerus produces cephamycin C in an 8-step pathway from the amino acids L-a-aminoadipic acid, L-cysteine and L-valine. Our groups have been involved in elucidation of the biosynthetic pathway as well as its regulation and genetics. The path is the same as that occurring in the eukaryotic Cephalosporium acremonium up to the formation of deacetylcephalosporin C. Then the pathway diverges with the prokaryotic S. clavuligerus carrying out sequential carbamoylation, oxidation and methylation reactions to form cephamycin C. The main regulatory controls exerted by nutrients appear to be repression of ACV synthetase by ammonium and phosphate and inhibition by glyceraldehyde-3-phosphate; repression of cyclase by ammonium and phosphate and inhibition by phosphate; and repression of expandase by carbon sources, ammonium and phosphate and inhibition by phosphate.

We have characterized mutant NP1 of S. clavuligerus NRRL 3585 which is almost completely blocked in cephalosporin biosynthesis and exhibits depressed activities of both the ACV synthetase and cyclase enzymes of the biosynthetic pathway. A wild type DNA region was cloned which partially restores antibiotic production, ACV synthetase and cyclase activities to this mutant. The recombinant plasmid exhibits a variable copy number in different transformants. Hybridization experiments indicate that sequences homologous to the cloned region are present in various ß-lactam producing Streptomyces strains but absent in strains which are not known to produce this class of antibiotics. Furthermore, the chromosomal copy of the cloned region lies in close proximity to a gene coding for the isopenicillin N synthase gene of the cephalosporin pathway. We recently observed hybridization between our DNA fragment and the ACV synthetase gene from Penicillium chrysogenum indicating that the S. clavuligerus fragment contains all or part of the ACV synthetase gene. In some completely blocked mutants, cloning of the fragment results in hyperproduction of cephamycin C.

REFERENCES

Agayn, V. (1990). Master of Science Thesis, Massachusetts Institute of Technology, Cambridge, Massachusetts.

Aharonowitz, Y. (1980). Ann. Rev. Microbiol. 34, 209.

Aharonowitz, Y., and Demain, A. L. (1977). Arch. Microbiol. 115, 169.

Aharonowitz, Y., and Demain, A. L. (1978). Antimicrob. Agents Chemother. 14, 159.

Aharonowitz, Y., and Demain, A. L. (1979). Can. J. Microbiol. 25, 61.

Aharonowitz, Y., Mendelovitz, S., Kirenberg, F., and Kufer, V. (1984). J. Bacteriol. 157, 337.

Bailey, C.R., Butler, M.J., Normansell, I.D., Rowlands, R.T. and Winstanley, D.J. (1984). Bio/Technology 2, 808.

Bailey, C.R. and Winstanley, D.J. (1986). J. Gen. Microbiol. 132, 2945.

Banko, G., Wolfe, S., and Demain, A. L. (1986). Biochem. Biophys. Res. Commun. 137, 528.

Banko, G., Demain, A. L., and Wolfe, S. (1987). J. Amer. Chem. Soc. 109, 2858.

Braña, A. F., Wolfe, S., and Demain, A. L. (1985). Can. J. Microbiol. 31, 736.

Braña, A. F., Paiva, N. P., and Demain, A. L. (1986a). J. Gen. Microbiol. 132, 1305.

Braña, A. F., Wolfe, S., and Demain, A. L. (1986b). Arch. Microbiol. 146, 46.

Brewer, S. J., Farthing, J. E., and Turner, M. K. (1977). Biochem. Soc. Trans. 5, 1024.

Brewer, S. J., Taylor, P. M., and Turner, M. K. (1980). Biochem. J. 185, 555.

Carr, L. G., Skatrud, P. L., Scheetz, M. E. II, Queener, S. W., and Ingolia, T. D.(1986). Gene 48, 257.

De Jong, L., Albracht, S. P. J., and Kemp, A. (1982). Biochim. Biophys. Acta 704, 326.

Dotzlaf, J. E., and Yeh, W.-K. (1987). J. Bacteriol. 169, 1611.

Fawcett, P. A., Usher, J. J., and Abraham, E. P. (1976). In "Second International Symposium on the Genetics of Industrial Microorganisms" (Macdonald, K. D.,ed.), p. 129. Academic Press, New York.

Felix, H. R., Peter, H. H., and Treichler, H. J. (1981). J. Antibiot. 34, 567.

Fujisawa, Y., Kikuchi, M., and Kanzaki, T. (1977). J. Antibiot. 30, 775.

Garcia-Domingues, M., Martin, J.F., Mahro, B., Demain, A.L. and Liras, P. (1987). Appl. Environ. Microbiol. 53, 1376.

Hood, J. D., Elson, A., Gilpin, M. L., and Brown, A. G. (1983). J. Chem. Soc. Chem. Commun. 1187.

Hook, D. J., Chang, L. T., Elander, R. P., and Morin, R. B. (1979). Biochem. Biophys. Res. Commun. 87, 258.

Hopwood, D.A. (1988). In "Biology of the Actinomycetes '88" (Okami, Y., Beppu, T., Ogawara, H.,eds.), p. 3, Japan Scientific Press, Tokyo.

Hu, W. S., Brana, A. F., and Demain, A. L. (1984). Enzyme Microb. Technol. 6, 155.

Ingolia, T.D. and Queener, S.W. (1989). Med. Chem. Revs. 9, 245.

Jensen, S. E., Westlake, D. W. S., and Wolfe, S. (1983). Can. J. Microbiol. 29, 1526.

Jensen, S. E., Westlake, D. W. S., and Wolfe, S. (1985). J. Antibiot. 38, 263.

Katz, E., Thompson, C.J. and Hopwood, D.A. (1983). J. Gen. Microbiol. 129, 2703.

Kern, B. A., Hendlin, D., and Inamine, E. (1980). Antimicrob. Agents Chemother. 17, 679.

Kohsaka, M., Demain, A. L. (1976). Biochem. Biophys. Res. Commun. 70, 465.

Konomi, T., Herchen, S., Baldwin, J. E., Yoshida, M., Hunt, N. A., and Demain A. L. (1979). Biochem. J. 184, 427.

Kovacevic, S., Weigel, B.J., Tobin, M.B., Ingolia, T.D. and Miller, R.J. (1989). J. Bacteriol. 171, 754.

Kupka, J., Shen, Y.-Q., Wolfe, S., and Demain, A. L. (1983a). Can. J. Microbiol. 29, 488.

Lebrihi, A., Germain, P., and Lefebvre, G. (1987). Appl. Microbiol. Biotechnol. 26, 130.

Lebrihi, A., Lefebvre, G., and Germain, P. (1988a). Appl. Microbiol. Biotechnol. 28, 39.

Lebrihi, A., Lefebvre, G., and Germain, P. (1988b). Appl. Microbiol. Biotechnol. 28, 44.

Leskiw, B. W., Aharonowitz, Y., Mevarech, M., Wolfe, S., Vining, L. C., Westlake, D. W. S., and Jensen, S. E. (1988). Gene 62, 187.

Liersch, M., Nuesch, J., and Treichler, H. J. (1976). In " Second International Symposium on the Genetics of Industrial Microorganisms " (K. D. Macdonald, ed.). p. 179. Academic Press, New York.

Lubbe, C., Jensen, S. E., and Demain, A. L. (1984). FEMS Microbiol. Lett. 25, 75.

Lubbe, C., Wolfe, S., and Demain, A. L. (1985). Arch. Microbiol. 140, 317.

Lubbe, C., Wolfe, S., and Demain, A. L. (1986). Appl. Microbiol. Biotechnol. 23, 367.

Madduri, K., Studdard, C. and Vining, L.C. (1989). J. Bacteriol. 171, 299.

Mahro, B., and Demain, A. L. (1987). Appl. Microbiol. Biotechnol. 27, 272.

Mendelovitz, S., and Aharonowitz, Y. (1982). Antimicrob. Agents Chemother. 21, 74.

Mendelovitz, S., and Aharonowitz, Y. (1983). J. Gen. Microbiol. 129, 2603.

Miller, J.R. and Ingolia, T.D. (1989). Molec. Microbiol. 3, 689.

O'Sullivan, J., and Abraham, E. P. (1980). Biochem. J. 186, 613.

O'Sullivan, J., Bleaney, R. C., Huddleston J. A., and Abraham E.P. (1979a). Biochem. J. 184, 421.

O'Sullivan, J., Aplin, R. T., Stevens, C. M., and Abraham, E. P. (1979b) Biochem. J. 179, 47.

Pang, C. P., White, R. L., Abraham, E. P., Crout, D. H. G., Lutstorf, M., Morgan, P. J., and Derome, A. E. (1984). Biochem. J. 222, 777.

Piret, J., Resendiz, B., Mahro, B., Zhang, J., Serpe, E., Romero, J., Connors, N. and Demain, A.L. (1990). Appl. Microbiol. Biotechnol. 32, 560.

Ramon, D. Carramolino, L., Patino, C., Sanchez, F., and Penalva, M. A. (1987). Gene 57, 171.

Rollins, M. J., Westlake, D. W. S., Wolfe, S., and Jensen, S. E. (1988). Can. J. Microbiol. 34, 1196.

Romero, J., Liras, P., and Martin, J. F. (1984). Appl. Microbiol. Biotechnol. 20, 318.

Romero, J., Liras, P., and Martin, J. F. (1988). Appl. Microbiol. Biotechnol. 27, 510.

Samson, S. M., Belagaje, R., Blankenship, D. T., Chapman, J. L., Perry, D., Skatrud, P. L., Frank, R. M., Abraham, E. P., Baldwin, J. E., Queener, S. W., and Ingolia, T. D. (1985). Nature 318, 191.

Samson, S. M., Chapman, J. L., Belgaje, R., Queener, S. W., and Ingolia, T. D. (1987b). Proc. Natl. Acad. Sci. USA 84, 5705.

Sawada, Y., Hunt, N. A., and Demain, A. L. (1979). J. Antibiot. 32, 1303.

Scheidegger, A., Kuenzi, M. T., and Nuesch, J. (1984). J. Antibiot. 37, 522.

Serpe, E. (1990). Master of Science Thesis. Northeastern University, Boston, Massachusetts.

Shen, Y.-Q., Wolfe, S., and Demain, A. L. (1984a). Enzyme Microb. Technol. 6, 402.

Shiffman, D., Mevarech, M., Jensen, S. E., Cohen, G., and Aharonowitz, Y. (1988). Mol. Gen. Genet. 214, 562.

Turner, M. K., Farthing, J. E., and Brewer, S. J. (1978). Biochem. J. 173, 839.

Usui, S. and Yu, C.-A. (1989). Biochim. Biophys. Acta 999, 78.

Weigel, B. J., Burgett, S. G., Chen, V. J., Skatrud, P. L., Frolik, C. A., Queener, S. W., and Ingolia, T. D. (1988). J. Bacteriol. 170, 3817.

Whitney, J. D., Brannon, D. R., Mabe, J. A., and Wicker, K. J. (1972). Antimicrob. Agents Chemother. 1, 247.

Zhang, J., Wolfe, S., and Demain, A. L. (1989a). Can. J. Microbiol. 35, 399.

Zhang, J., Wolfe, S., and Demain, A. L. (1989b). FEMS Microbiol. Lett. 57,145.

MOLECULAR GENETICS OF OXYTETRACYCLINE PRODUCTION

BY *Streptomyces rimosus*

Kenneth J. McDowall, Deirdre Doyle, Michael J. Butler*, Craig Binnie*, Melvyn Warren*, and Iain S. Hunter

Institute of Genetics, University of Glasgow
Church Street, Glasgow
G11 5JS, U. K.

*Pfizer Limited, IPDG
Sandwich, Kent
CT13, 9NJ, U. K.

Oxytetracycline (OTC) is a member of the tetracycline family of broad spectrum antibiotics produced by the filamentous bacteria. These antibiotics have maintained their importance in aspects of human health care (especially the semi-synthetic, doxycycline), but they are also used increasingly in animal healthcare and animal feed supplements.

Oxytetracycline (5-hydroxytetracycline) is produced by *Streptomyces rimosus* (Fig.1), while *Streptomyces aureofaciens* is used in the commercial production of 7-chlorotetracycline. Mutants for the chlorination step have been isolated, and they produce tetracycline. Some natural soil isolates make both, in which case the culture conditions must be modified to enhance the ratio of chlorotetracycline:tetracycline. A large number of natural tetracyclines have been reported, some of which are made by microorganisms other than the streptomycetes, e.g. Nocardia, (Sinclair *et al*, 1962).

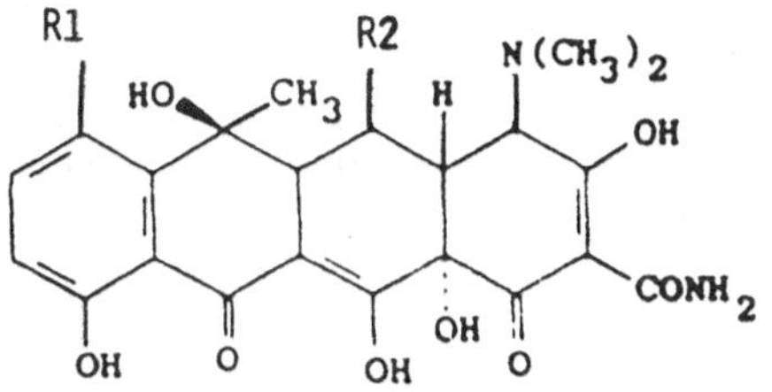

Fig. 1

Tetracycline structure.

tetracycline	:	R1 = H	R2 = H
oxytetracycline	:	R1 = H	R2 = OH
chlorotetracycline	:	R1 = Cl	R2 = H

Genetics and Product Formation in Streptomyces
Edited by S. Baumberg *et al.*, Plenum Press, New York, 1991

The plausible biochemical pathway for tetracycline production was first studied by the Lederle Group (reviewed McCormick & Jensen, 1969) by careful biorganic chemistry with mutants blocked in steps of the pathway. Tetracyclines are polyketides: condensation products of a starter unit (most likely malonamyl-CoA) with eight sequential additions of acetate units (from malonyl-CoA). After formation of the tetracycline nucleus (the first stable intermediate being 6-methylpretetramid), the various substituent function groups are added. The question of whether the tetracycline nucleus was synthesised 'clockwise' or 'anticlockwise' was answered (Fig. 2) by ^{13}C-labelling studies (Thomas & Williams, 1983).

Fig. 2

Result of ^{13}Cnmr labelling of tetracycline nucleus (after Thames & Williams, 1982). ●—— represents ^{13}CH$_3$COOH labelling.

GENETIC MAPPING IN S. rimosus

The pioneering work of Hopwood and co-workers on S. coelicolor (reviewed Hopwood, 1967) was followed closely by the deduction of a genetic map for S. rimosus (Friend and Hopwood, 1971). The map was quite clearly similar to that of S. coelicolor. The positions of some OTC mutations were mapped (Boronin & Mindlin, 1971; Pigac & Alacevic, 1974). But the first integrated study of the genetics and biochemistry of OTC production was undertaken at Pfizer (Rhodes et al, 1981; Fig.3). Mutants blocked in OTC production were isolated, characterised biochemically and their map positions were determined. The production genes for OTC were apparently clustered on the chromosome. At that time, the interpretation of the data implied two clusters of production genes. The first (named the '4 o'clock cluster') was responsible for conversion of malonyl-CoA to anhydrotetracycline (ATC) while the second ('10 o'clock') cluster contained genes whose products converted ATC to OTC. Although this formed the working model of OTC genetics for several years, the completion of the cloning of the pathway (see later) required the model to be revised.

A resistance gene for OTC production was mapped to the 'early' cluster (Rhodes et al., 1984). The proximity of resistance genes to production clusters has been observed in many other systems (e.g. methylenomycin (Chater & Bruton, 1985); streptomycin (Ohnuki et al., 1985b); erythromycin (Stanzak et al,1986) and bialaphos (Murakami et al, 1986). Resistance genes provide a convenient selectable marker to begin to clone adjacent production genes.

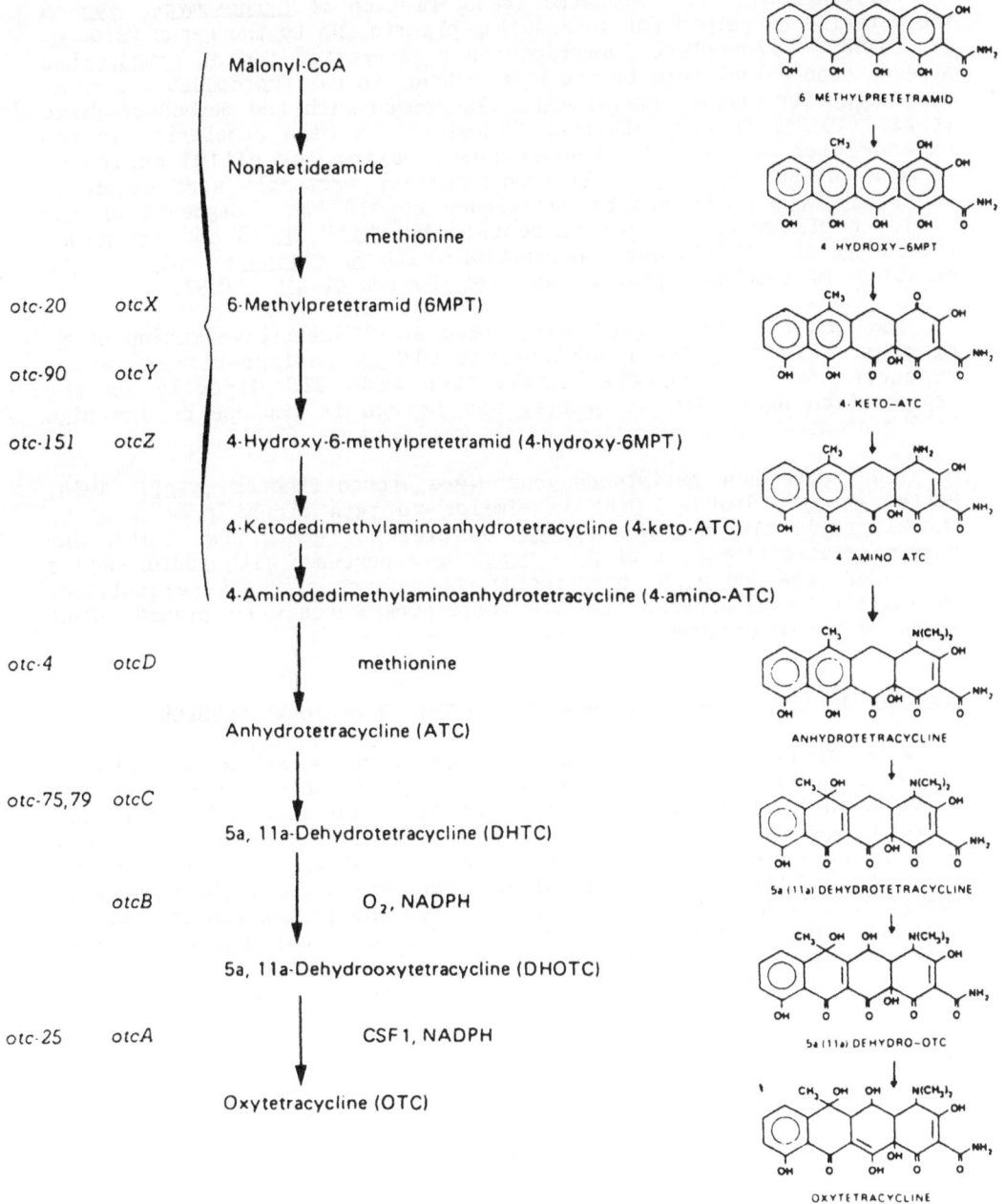

Malonyl-CoA

Nonaketideamide

methionine

otc-20   otcX   6-Methylpretetramid (6MPT)

otc-90   otcY

otc-151   otcZ   4-Hydroxy-6-methylpretetramid (4-hydroxy-6MPT)

4-Ketodedimethylaminoanhydrotetracycline (4-keto-ATC)

4-Aminodedimethylaminoanhydrotetracycline (4-amino-ATC)

otc-4   otcD   methionine

Anhydrotetracycline (ATC)

otc-75,79 otcC   5a, 11a-Dehydrotetracycline (DHTC)

          otcB   O$_2$, NADPH

5a, 11a-Dehydrooxytetracycline (DHOTC)

otc-25   otcA   CSF1, NADPH

Oxytetracycline (OTC)

Fig.3

The OTC pathway and assignment of blocked mutants (data of Rhodes et al., 1981).

CLONING OF TWO RESISTANCE GENES

Polyethylene glycol-mediated transformation of Streptomyces spp is the established method for introducing plasmid DNA to the genus (Bibb et al., 1978). Protoplast formation and regeneration for S. rimosus had already been established by the Pfizer Group to use protoplast fusion as a strategy for strain improvement. In common with the method of Pigac et al. (1982), it was important to maintain a high osmolarity in the regeneration medium with 0.6M sucrose. Using the pIJ101 series of vectors (Kieser et al., 1982) transformation protocols were developed for S. rimosus which had an efficiency of $>10^6$/ug. Segments of the pIJ101 replicon [now known to contain the kilB, korA and tra genes (Stein et al., 1989)] were incompatible with S. rimosus; only in vivo deletions of the basic plasmid survived (Butler et al., 1989).

The initial cloning experiments used an OTC-sensitive mutant of S. rimosus, selecting for resistance to OTC by replica-plating after regeneration was complete. Selection with OTC directly on the regeneration agar with an overlay was impossible because of the high Ca^{++} concentration in the agar, which complexed the OTC.

Two independent resistance genes were cloned (Rhodes et al., 1984; Butler et al., 1989). OtrA is similar to tetA cloned previously by Ohnuki et al. (1985a), whereas otrB relates to tetB. When lambda and cosmid genomic libraries of S. rimosus were screened with radiolabelled probes of otrA and otrB, overlapping recombinant clones were isolated. By restriction analysis, the two resistance genes were placed 25kb apart on the chromosome.

CLONING AND COMPLEMENTATION ANALYSIS OF THE '4 o'clock' CLUSTER

Although mutants blocked in the genes which mapped to this cluster could only be transformed poorly, it was possible to obtain recombinants if substantial quantities of plasmid DNA were used. DNA segments between, and including, the two resistance genes were subcloned. With the high copy-number pIJ101-derived vectors, analysis of complementation of blocked mutants was complicated. Many of these high copy-number subclones switched off production of OTC in the production strain. If this was also occurring in blocked mutants, the failure to see complementation with a subclone had to be interpreted carefully. Failure to make OTC by a blocked mutant containing a recombinant plasmid implied that the subclone did not contain the corresponding production gene. However, it was also possible that it did contain the production gene - but failure to see synthesis of OTC was due to this 'switch off' phenomenon. Only when a subclone had been shown to have no effect on the production strain could it be evaluated for complementation of the blocked mutants (Butler et al., 1989).

An alternative approach was to subclone segments of the cluster onto low copy-number vectors. The SCP2-derivatives (Hopwood et al., 1985) could be used to transform S. rimosus [albeit inefficiently (10^4/ug)], and subclones of the cluster with these vectors had no adverse effect on production of OTC. Fragments which complemented each of the 4 classes of mutant blocked at this locus were defined (Fig. 4). In some cases complementation was directly by expression of the plasmid-encoded gene, but in others it was due to repair of the chromosomal lesion by recombination with homologous sequences on the plasmid, as reported for other systems (Seno _et al._, 1984; Murakami _et al._, 1986). Thus the block diagram of Fig.4 represents only the approximate locations of the genes. The otcY locus was intriguing because a large fragment was necessary to effect complementation. OtcY was the most frequent class of blocked mutants obtained in the isolation of blocked mutants (Rhodes _et al._, 1981). Since these mutants were pale and nonpigmented, and many of the later biochemical intermediates and shunt metabolites of the pathway are highly coloured, the mutants were likely blocked in the early biosynthetic step(s). The otcY locus is extremely complex (see later).

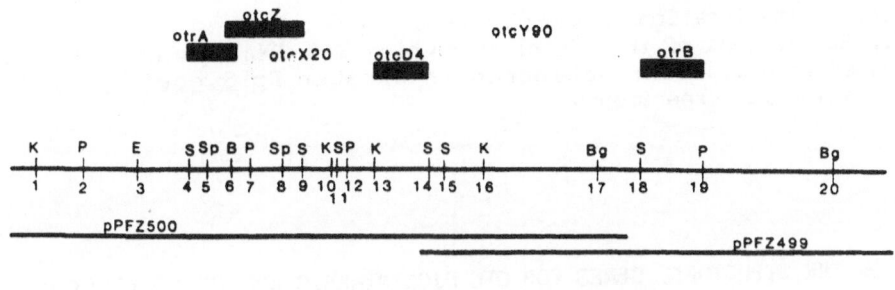

Fig.4

Locations of genes otrA, otrB, otcD, otcX, otcY, otcZ within the '4 o'clock' cluster. For clarity, only the relevant restriction sites are shown. B, BamHI: Bg, BglII: E, EcoRI: K, KpnI: P, PstI: S, SacI: Sp, SphI.

CLONING OF THE otcC GENE

With the completion of the complementation analysis of the '4 o'clock' cluster, attention was turned to the genes involved in the later stages of the biosynthetic pathway (ostensibly located within the '10 o'clock ' cluster.

Anhydrotetracycline (ATC) is converted to 5a, 11a-dehydro-tetracycline by ATC oxygenase. The availability of ATC and a published enzyme assay (Behal _et al._, 1979) provided the opportunity to purify the protein. Simultaneously the Prague group had purified the homologous protein from _Streptomyces aureofaciens_ (Vancurova _et al._, 1988).

The gene encoding ATC oxygenase was cloned by the 'reverse genetics' approach (Binnie _et al._, 1989). The purified protein was subjected to amino terminal amino acid analysis. Forty-five residues were sequenced. From the available codon usage pattern for streptomycete genes, two synthetic 42-mer oligonucleotides were synthesised and a genomic fragment cloned which hybridised particularly well to one of the oligonucleotides. DNA sequencing of the fragment confirmed that it did indeed encode ATC oxygenase. However, the location of the gene came as a surprise: it mapped to within the early cluster, between otcZ and otcX (Fig. 5a).

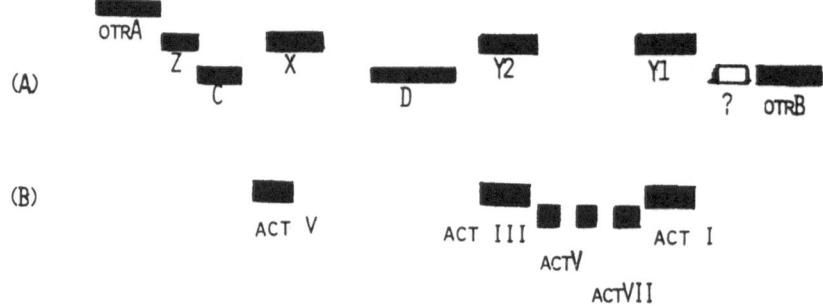

Fig. 5.

Fig.5a The location of the otcC gene.
Fig.5b Regions of the OTC cluster which show DNA homology with the named genes of the actinorhodin cluster fo S.coelicolor in hybridisation experiments.

ALL OF THE STRUCTURAL GENES FOR OTC BIOSYNTHESIS ARE IN ONE CLUSTER

The definitive localisation of the otcC gene, formerly thought to map in the second cluster, within the 4 o'clock cluster prompted the question, 'Were all of the structural genes located within one cluster?' There are EcoRI sites flanking the '4 o'clock cluster' including both resistance genes which enabled the entire 34kb segment of DNA to be subcloned onto the SCP2 derivative, pIJ916. When this plasmid construct was transferred into S. lividans, it made an antibiotic which was chemically indistinguishable from OTC (Binnie et al., 1989). Thus, all of the structural genes were located within this single cluster. In OTC-sensitive strains, the entire cluster and at least 40kb of flanking DNA is deleted (Butler et al., 1989).

WHAT COULD MAP AT THE OTHER CLUSTER?

The genetic map position assigned previously to otcC within the '10 o'clock' cluster depended principally on data derived from the otcC75 mutant (Rhodes et al., 1981). This mutant quite clearly was unable to convert ATC to dehydrotetracycline. But as the cloning of the otcC structural gene had located it to the '4 o'clock' cluster (Binnie et al., 1989), the otcC75 mutant had to be defective in some ancillary enzymic function (or co-factor) necessary for ATC oxygenase enzyme activity.

Cosynthetic factor 1 (CSF1) has been shown to be necessary for the last step in tetracycline biosynthesis (Miller et al., 1960), and indeed CSF1⁻ mutants were obtained during the isolation of the mutants blocked in OTC production (Rhodes et al., 1981). They mapped to the second cluster, which is now assumed to be responsible for the biosynthesis of CSF1 - identified chemically as 8-hydroxy-5-deazaflavin (McCormick & Morton, 1982).

Confirmation that the mutation in the otcC75 strain was definitely outwith the OTC biosynthetic cluster (at '4 O'clock') was obtained when the otcC75 (CSF1$^+$) mutant cosynthesised with an OTC-sensitive (i.e. deleted for all OTC structural genes) CSF1$^-$ mutant to make OTC (Binnie et al, 1989).

Thus the CSF1 biosynthetic genes and the otc75 lesion affecting ATC oxygenase activity both map to the '10 o'clock' cluster. It is tempting to speculate that the flavin requirement for ATC oxygenase may be somehow provided by genes associated with the 'CSF1 cluster', though not necessarily CSF1. Donation of electrons from CSF1 to the tetracycline biosynthetic enzyme which converts dehydro-OTC to OTC is mediated by NADPH:CSF1 oxidoreductase - a F420 enzyme (J. Novotna, personal communication). It will also be important to establish the map position and clone this gene.

The demonstration that S. lividans can synthesize OTC when provided with only the OTC structural gene cluster implies that it must also produce CSF1. Indeed a recent survey of streptomycetes showed that this unusual flavin is made by many members of the genus (Kuo et al., 1988).

DNA SEQUENCING AND CHARACTERISATION OF THE OTC CLUSTER

A number of the genes of the OTC cluster have now been sequenced. These DNA sequences give some clues to the biochemical functions of the gene products.

otrA

OtrA encodes a polypeptide of Mr 71,329, which is involved in non-covalent modification of the ribosomes (Ohnuki et al.' 1985a). Ribosomes which are prepared whilst otrA is being expressed are resistant to the arrest of translation by tetracycline in an in vitro assay. However, washing the ribosomes in 1M salt renders them sensitive to tetracycline once again.

The otrA gene product has a high degree of sequence identity with the tetM (Sanchez-Pescador et al., 1988) and tetO (Manavathu et al., 1988) genes associated with gram-positive tetracycline-resistant transposons. It is tempting to speculate that the transposons have acquired the resistance factors through genetic transfer from an ancestor of S.rimosus/aureofaciens. The otrA gene product also shares sequence identity with various elongation factors, but this may be no more than conservation around a GTP/GDP binding domain (Salyers et al., 1990).

otrB

The otrB gene product (Mr 35,818) was sequenced by Reynes et al. (1988), who named it tet347. This very hydrophobic protein has some sequence identity with the enteric tetracycline resistance factors. When expressed in E.coli, an additional protein was identified in the membrane fraction. A homologue of otrB, called tetB, was cloned from a soil isolate of S.rimosus (Ohnuki et al., 1985a). When expressed in S.griseus, tetB resulted in reduced accumulation of tetracycline within the cell. Taken together, these data imply that the otrB gene product is responsible for the export of (oxy)tetracycline during the production phase.

otcC

The aminoterminal sequence of the purified protein from S.rimosus had already indicated that the otcC gene product has a high sequence identity with the amino-terminus of p-hydroxybenzoate hydrolase of Pseodomonas fluorescens (Wierenga et al., 1979), whose crystal structure is known. The DNA sequence of the S.rimosus gene has yet to be completed. However, when the preliminary sequence is translated, it has significant identity with the p-hydroxybenzoate hydroxylase of Ps. aeruginosa (Entsch et al., 1988). This is another example of an enzyme of an antibiotic production pathway sharing commonality with a catabolic or anabolic enzyme from another species - indicating that there is nothing mysterious or novel about the evolution of antibiotic production genes.

otcZ

otcZ encodes a polypeptide with significant sequence identity to hydroxyindole O-methyltransferase from bovine pineal gland (Ishada et al.,1988). There are three methylation steps in OTC biosynthesis. The addition of 2 methyl groups to 4-amino-ATC has been assigned to otcD (Rhodes et al., 1981). If otcZ indeed encodes a methylase function, it would be implicated in 6-methylation of the pretetramid structure. Careful disruption of the otcZ gene, wihthout destroying the complex transcriptional pattern around this locus, is in progress to establish whether 6-demethyl-OTC will be made in the recombinant. That 6-demethyl tetracyclines have been reported before (McCormick, 1969), implies that the methylation is not a prerequisite for folding and cyclisation of the tetracycline ring structure.

otcZ also shares significant sequence identity with tcm orf4. The tcm gene is implicated in both cyclisation and O-methylation of tetracenomycin (C.R. Hutchinson, pers comm and cited in Hopwood & Sherman, 1990). It is thought to be a bifunctional protein, but the sequence identity with otcZ only occurs at only the C-terminal domain of tcm.

otcY

Results with blocked mutants (see above) implied that otcY was involved early in the OTC pathway. The hypothesis was substantiated when the actI sequence from the actinorhodin cluster of S.coelicolor was shown to hydridise to the otcY locus (Malpartida et al., 1987). The actI locus is involved in polyketide assembly of actinorhodin. The actIII gene (thought to involved in a reductive step during actinorhodin polyketide assembly (Hallam et al., 1988)) also hybridsed to the otcY locus. However, unlike in S.coelicolor where act I & III lie adjacent on the chromosome, the homologues in S.rimosus are around 8kb apart (Fig.5b) - indicating a different topological arrangement for the polyketide synthase genes in S.rimosus.

The actI-like genes for tetracenomycin and granatacin biosyntheses have now been sequenced (Bibb et al., 1989; Sherman et al., 1989). Analyses of these sequences has answered a long-standing question about polyketide synthesis by the streptomycetes.

Polyketide synthesis is chemically analagous to the synthesis of fatty acids. Different species have different forms of fatty acid synthase (FAS). The bacteria have a 'type-II FAS', where the different enzymic steps of fatty acid biosynthesis are encoded by different proteins which form a multienzyme (dissociable) complex. By contrast, yeast has a 'type-I' FAS, where all of the components are on 2 multi-domain proteins. Other eukaryotes have 'type-I' FAS systems, although the order and contents of the multidomain components varies.

The tetracenomycin and granatacin loci encode separate proteins of 'type-II', i.e. the polyketide assembly enzymes are dissociable complexes. Our own preliminary DNA sequence of this part of the otcY locus (and more extensive work carried out by D.H. Sherman at the John Innes Institute, Norwich (cited in Hopwood & Sherman, 1990)) show that S.rimosus has a 'type-II' polyketide synthase.

TRANSCRIPTION STUDIES

As a prerequisite to undertaking transcription studies of the OTC cluster, it was necessary to establish fermentation condidtions in defined media where growth and subsequent production of the antibiotic would take place, thus permitting reproducible fermentations and facilitating studies on the repression of antibiotic synthesis by carbon, nitrogen and phosphorous. Conditions have now been established which fulfil this requirement (Fig.6).

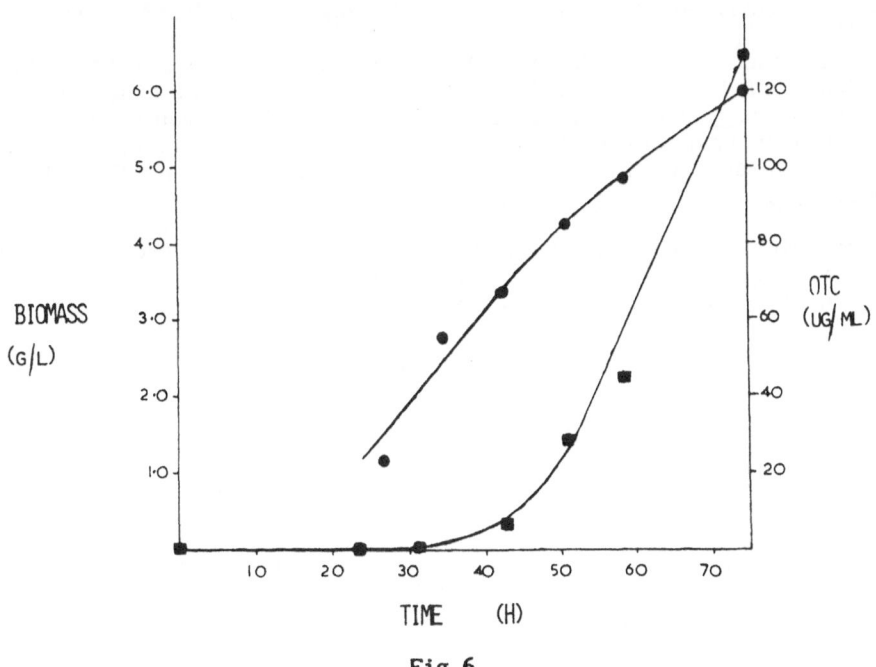

Fig.6.

Growth (●) and OTC production (■) by .rimosus.

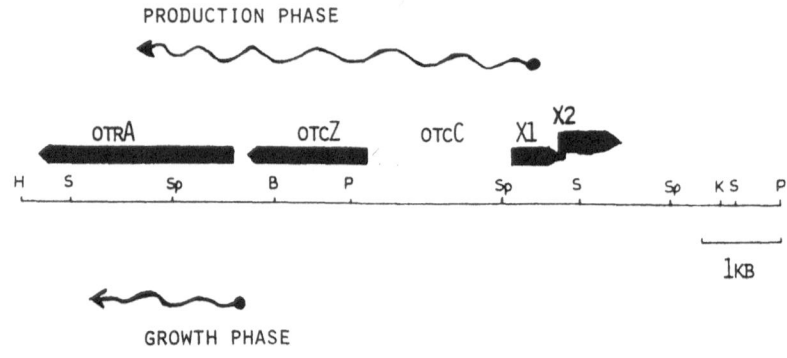

Fig.7.

Transcription pattern of otcC,Z andotrA during growth and OTC production phases.

At present only the 'left' end of the OTC cluster has been studied in detail. During vegetative growth, the otrA gene is transcribed from its own promoter (otrAp1; Doyle et al., 1988), with the transcription start 127 nucleotides upstream of the translation start (Fig.7). The sequence of the otrAp1 promoter element is very similar to the consensus sequence identified for many E.coli genes (Hawley & McClure, 1983).

During antibiotic production, the otrA gene is no longer transcribed from the otrAp1 promoter, but as the 3' gene of a polycistronic mRNA containing the genes otcC-otcZ-otrA (Fig.7). It is probable that cessation of transcription from otrAp1 is due to loss of the sigma factor necessary to initiate transcription at that point (Westpheling et al., 1985).

Preliminary DNA sequence upstream of the otcC gene shows divergent open reading frames (Fig.5) indicating that this locus may be an important point of control of expression of the pathway. Likewise, the otcY polyketide synthase genes are transcribed divergently from otrB, which could be another site of control for the pathway.

PROSPECTS

With a significant part of the OTC cluster now sequenced, computer-assisted analysis has revealed important clues to the functions of the genes contained within it. This analysis will continue for some time, until the entire cluster is sequenced. The next steps will be to understand how expression of the pathway is regulated, and how the enzymes specified interact to produce the tetracycline structure, rather than that of some other polyketide.

REFERENCES

Behal, V., Hostalek, Z. and Vanek, Z. (1979). Biotechnol. Lett. 1, 177-182.

Bibb, M.J., Ward, J.M. and Hopwood, D.A. (1978). Nature 274, 398-400.

Binnie, C., Warren, M. and Butler, M.J. (1989). J. Bacteriol. 171 (2), 887-895.

Boronin, A.M. & Mindlin, S.Z. (1977). Genetika 8, 125-131.

Butler, M.J., Friend, E.J., Hunter, I.S., Kaczmarek, F.S., Sugden, D.A. and Warren, M. (1989). Mol. Gen. Genet. 215, 231-238.

Chater, K.F. and Bruton, C.J. (1985). EMBO J. 4, 1893-1897.

Distler, J., Brown, C., Ebert, A. and Pipersberg, W. (1987). Molec. & Gen. Genetics 208, 204-210.

Doyle, D., Butler, M.J. and Hunter, I.S. (1988). Heredity 61, 305.

Entsch, B., Nan, Y., Weaich, K. and Scott, K.F. (1988). Gene 71, 279-291.

Friend, E.J. & Hopwood, D.A. (1971). J. Gen. Microbiol. 68, 1187-1197.

Hallam, S.E., Malpartida, F.M. and Hopwood, D.A. (1988). Gene 74, 305-320.

Hanley, D.K. and McClure, W.R. (1983). Nuc. Acid Res. 11, 2237-2255.

Hopwood, D.A. (1967). Bacteriological Reviews 31, 373-403.

Hopwood, D.A., Bibb, M.J., Chater, K.F., Kieser, T., Bruton, C.J., Kieser, H.M., Lydiate, D.J., Smith, C.P., Ward, J.M. and Schrempf, M. (1985). Genetic Manipulation of Streptomyces; a laboratory manual. John Innes Foundation, Norwich.

Hopwood, D.A. and Sherman, D.H. (1990). Ann. Rev. Genetics 24 (in press).

Ishada, I., Obinata, M. and Deguchi, T. (1988). J. Biol. Chem. 262, 2895-2899.

Kieser, T., Hopwood, D.A., Wright, H.M. and Thompson, C.J. (1982). Mol. Gen. Genet. 185, 223-238.

Kuo, M-S.T., Yurek, D.A., Coats, J.H. and Li, G.P. (1989). J. Antibiotics 42 (3), 475-478.

McCormick, J.R.D. and Jensen, E.R. (1969). J. American Chem. Soc. 91, 206-221

McCormick, J.R.D. (1969). In: Genetics and Breeding of Streptomyces, Sennart, G. and Alacevic, M., Yugoslav Academy of Sciences and Arts, Zagreb. p. 163.

Malpartida, F., Hallam, S.E., Kieser, H.M., Motamedi, H., Hutchinson, C.R., Butler, M.J., Sugden, D.A., Warren, M., McKillop, C., Bailey, C.R., Humphreys, G.O. and Hopwood, D.A. (1987). Nature 325, 818-820.

Manavathu, E.K., Hiratsuko, K. and Taylor, D.E. (1988). Gene 62, 17-26.

Miller, P.A., Sjolander, N.O., Nalesnyk, S., Arnold, N., Johnson,S., Doerschuk, A.D. and McCormick, J.R.D. (1960). J. Am. Chem. Soc. 82, 5002-5003.

Murakami, T., Anzai, H., Imari, S., Satch, A., Nagaoka, K. and Thompson, C.J. (1986). Mol. Gen. Genetics 205, 42-50.

Ohnuki, T., Imanaka, T. and Aiba, S. (1985b). J. Bacteriol. 164, 185-194.

Ohnuki, T., Katoh, T., Imanaka, T. and Aiba, S. (1985a). J. Bacteriol. 161, 1010-1016.

Pigac, J. and Alacevic, M. (1979). Period. Biol. 81, 575-82.

Pigac, J., Hranueli, D., Smokvina, T. and Alacevic, M. (1982). Appl. Environ. Mircobiol. 44, 1178-1186.

Reynes, J-P., Calmeb, T., Drocourt, D., and Tiraby, G. (1988). J.Gen. Microbiol. 134, 585-598.

Rhodes, P.M., Winskill, N., Friend, E.J. and Warren, M. (1981). J. Gen. Microbiol. 124, 329-398.

Rhodes, P.M., Hunter, I.S., Friend, E.J. and Warren, M. (1984). Biochem. Soc. Trans. 12, 586-587.

Salyers, A.A., Speer, B.S. and Shoemaker, N.B. (1990). Mol. Microbiol. 4 (1), 151-156.

Seno, E.T. and Baltz, R.H. (1982). Antimicrobiol Agents and Chemotherapy 21, 758-763.

Sinclair, A.C., Schenk, J.R., Post, G.G., Cardinal, E.V., Burakas, S. and Frida, H.H. (1962). Antimicrobial Agents & Chemotherapy 1961, 892-895.

Sherman, D.H., Malpartida, F., Kieser, H.M., Bibb, M.J. and Hopwood, D.A. (1989). EMBO J. 8, 2717-2726.

Sanchez-Pescader, R., Brown, J.T., Roberts, M. and Urdia, M.S. (1988). Nuc. Acid Res. 16, 1216-1217.

Stanzak, R., Matsushima, P., Baltz, R.H. and Rao, R.N. (1986). Biotechnology 4, 229-232.

Stein, D., Kendall, K.J. and Cohen, S.N. (1989). J. Bacteriol. 171, 5768-5775.

Thomas, R. and Williams, D.J. (1983). J. Chem. Soc. Chem. Comm. 677-679.

Vancurova, I., Volc, J., Flieger, M., Neuzil, J., Novotna, J., Vlach, J. and Behal, V. (1988). Biochem. J. 253, 263-267.

Westpheling, J.E., Raines, M. and Losick, R. (1985). Nature 313, 22-27.

Wierenga, R.K., de Jong, R.J., Kalk, K.H., Hol, W.G. and Dreuth, R. (1979). J. Mol. Biol. 131, 55-73.

GENETIC CHARACTERIZATION AND CLONING OF GENES FOR CHLORTETRACYCLINE RESISTANCE IN STREPTOMYCES AUREOFACIENS STRAINS

Natalie Lomovskaya, Gennady Sezonov,
Lyubov Isayeva and Tatyana Chinenova

Institute of Genetics and Selection of Industrial
Microorganisms, I Dorozhny Pr. 1
Moscow 113545, U.S.S.R.

Introduction

Tetracycline antibiotics and their semisynthetic derivatives are now being widely used clinically and veterinarily. Industrial strains — the producers of oxy- and chlortetracycline (Otc and Ctc) are the objects of long-term conventional selection and possess high level of antibiotic activity. In recent years considerable success has been met in research on genetic control of Otc biosynthesis in _Streptomyces rimosus_ strains. All genes for Otc biosynthesis are localized on one long DNA fragment of about 34 kb[1,2,8]. The genes for biosynthesis are flanked by genes responsible for resistance to Otc ,otrA and otrB, "early" biosynthesis genes being linked to otrB and "late" genes to otrA. Coordinated transcription of resistance and biosynthesis genes has been demonstrated .

Significant progress has been also made in the studies of S. aureofaciens , the producer of Ctc. Stages of the Ctc biosynthesis pathway were defined, their similarity to those of Otc biosynthesis shown,and enzymes catalyzing the stages of Ctc biosynthesis characterized[7,11]. Recently,a resistance gene has been cloned from S.aureofaciens and strong similarity of restriction maps demonstrated for this and the otrB gene[13]. In addition, it appeared that DNA fragments from S. rimosus and S. aureofaciens, equal in size, hybridize with the actI gene[3] responsible for synthesis of polyketide synthase . These data suggest that genes for Otc and Ctc biosynthesis have similar organization.

In this work, main attention is paid to genetic identification and cloning of genes for Ctc resistance of an S. aureofaciens strains to comparative analysis of the structure of these genes in producers of Ctc and Otc, and to obtaining mutations leading to increase in resistance, revealing the impact of these mutations on biosynthesis.

Materials and Methods

Standard strain[6] S.lividans TK64 and two divergent lines of selection of S.aureofaciens strain (under VNIIgenetika collection numbers S755 and S799) were used.

Complete media CM-6[14] and minimal medium MM[5] were used for S.lividans TK64, and complete medium CM-12[3] and minimal medium MMA[3] were used for S. aureofaciens strains.

Isolation, regeneration and transformation of protoplasts of S.lividans strain TK64[5] were performed according to the method described elsewhere. Preparative isolation of plasmid DNA was made by the alkaline denaturation method[10]. Chromosomal DNA was extracted from actinomycete strains as described[4].

DNA nick-translation was performed as reported elsewhere using the mixture of DNA polymerase I and DNAse ("Amersham", U.K.).[10] DNA-DNA hybridization was carried out by the standard method.[10]

Cloning of genes for resistance to Ctc was made as follows: chromosomal DNA from S. aureofaciens S755 and S799 strains was digested with Bgl II and BamHI restriction endonucleases, respectively. The vector DNAs(plasmids pIJ702 and pIJ699) were cut with Bgl II enzyme and the former was ligated with Bgl II fragments and the latter with BamHI fragments. After transformation of S. lividans TK64 the clones resistant to thiostrepton (TsnR) obtained were replica plated to minimal medium MM containing 25μ/ml Ctc. Recombinant DNA was isolated from TsnR TetR clones and used for repeated transformation of S. lividans TK64 protoplasts.

The level of resistance of strains carrying recombinant plasmids to antibiotics was assayed by efficiency of plating (e.o.p.) of the culture - the ratio of spore titer on selective MM containing antibiotics of different concentrations to that on MM without antibiotics.

In oder to detect inducible resistance, spore suspensions were spread in dilutions onto Petri plates with MMA containing sublethal concentrations of antibiotics to be used as inducers. After 24 h semiliquid MM with the antibiotic of concentrations needed for ensuring selective conditions was layered onto the agar plate. Increase in culture e.o.p. in the presence of the inducer was taken as indicative of induction.

Results

Resistance of S. aureofaciens S755 to a number of antibiotics and Ctc

The study of the action of exogenous antibiotics and Ctc on S. aureofaciens S755 strain demonstrated that the strain was initially sensitive to low concentrations of macrolide antibiotics erythromycin (Erm), oleandomycin (Oln), tylosin (Tyl), turimycin (Tur), to thiostrepton (Tsn) and rifampicin (Rfn), less sensitive to streptomycin (Smn), kanamycin (Kmn), chloramphenicol (Cml), resistant to lincomycin (Lmn), neomycin (Neo), viomycin (Vio), nourseothricin (Ntn) (Fig.1).

In Fig.1 concentrations are indicated (dark columns) which were used for scoring AntR mutants - ErmR, OlnR, TylR, SmnR, KmnR, CmlR. Efficiency of strain mutagenic treatment is assessed by increase in frequency of these induced forward AntR mutations.

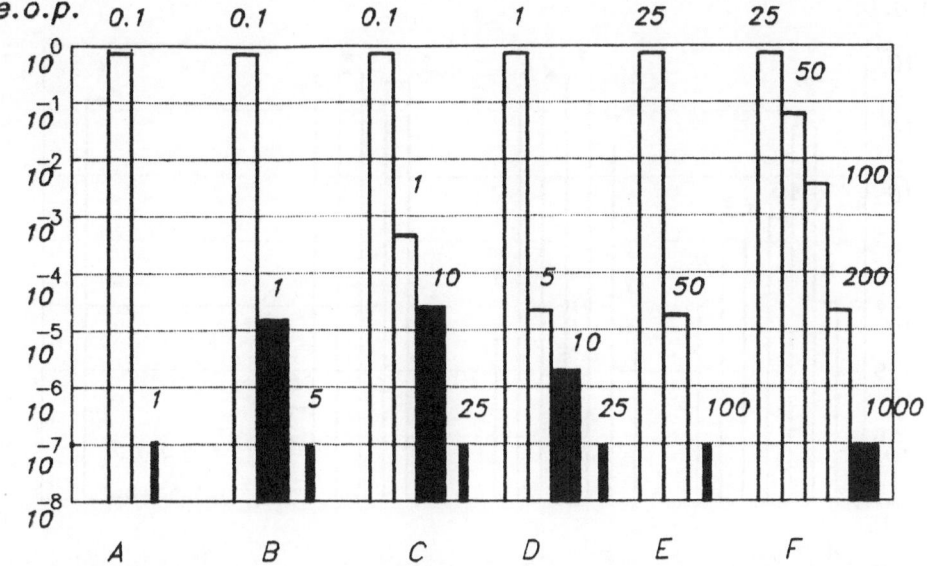

Fig.1. Resistance of S. aureofaciens S755 to a number of antibiotics.
Numbers over columns indicate concentrations of antibiotics (μg/ml) in selective medium: A - Tsn, Tur; B - Erm, Rfn, Tyl; C - Oln; D - Smn, Kmn, Cml; E - Lmn, Neo, Vio, Ntn; F - Ctc.
Numbers over dark thin columns indicate the concentrations of antibiotics at which e.o.p. $< 10^{-7}$
Dark heavy columns indicate concentrations which are used for scoring <u>AntR</u> mutants - ErmR, OlnR, TylR, SmnR, KnmR, CmlR, CtcR

The highest level of resistance is observed to autogenous Ctc antibiotic. The strain is constitutively resistant to 25 μg/ml of Ctc. Endogenous induction of resistance is expressed by a 10-fold e.o.p. elevation 10-fold after 24 h of culture incubation. Effective exogenous inducers of resistance are Ctc at 25 μg/ml and ethidium bromide (EB) at 30 μg/ml, increasing e.o.p. 200 and 1000 times, respectively.

On the medium CM12 containing Ctc at 1000 μg/ml, <u>CtcR</u> mutants isolated, then distributed into four groups, depending on the level of resistance to Ctc.

<u>Genetic characteristics of resistance determinants in S. aureofaciens</u>

Choosing among three alternative explanations for strain sensitivity to macrolides and thiostrepton (the absence of determinants, "silent" genes, inducible genes), the levels of resistance to these antibiotics were analyzed. As seen in Fig.2, expression of resistance determinant TsnR as well as ErmR and

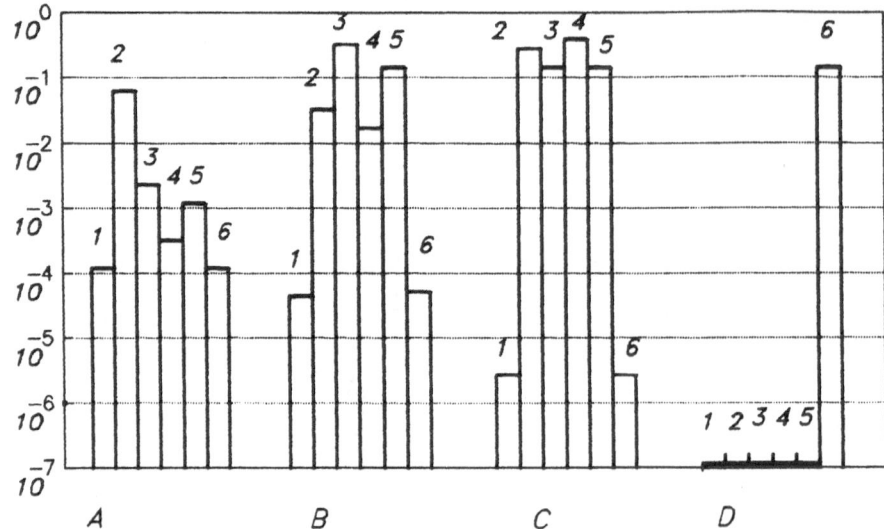

Fig. 2. Exogeneous induction of resistance to macrolides antibiotics Erm, 1 μg/ml (A) and Oln, 10 μg/ml (B); Ctc, 200 μg/ml (C); Tsn, 1 μg/ml (D); Inducers (μg /ml): 1 – inducer omitted; 2 – Erm (0,1); 3 – Oln (1,0), 4 – Tyl (0,1); 5 – Tur (0,1); 6 – Tsn (0,1).

OlnR determinants (resistance to erythromycin and oleandomycin, respectively) were inducible. Most efficient inducers of ErmR and OlnR determinants expression are Erm and Oln, respectively, i.e. antibiotics, towards which resistance is induced. Expression of TsnR determinant is only induced by thiostrepton. These data point to the presence in S. aureofaciens of inducible genes responsible for resistance to macrolides, and of inducible genes determining resistance to Tsn.
Very important in these series of experiments is that macrolides are not only inducers of resistance to exogenous macrolides (Fig. 2) but they efficiently induce increase in resistance of the strain to endogeneous antibiotic Ctc. No resistance to Tsn is induced . This suggests the presence in the chromosome of inducible mtr gene(s) responsible both for resistance to macrolides and Ctc. Functioning of the latter leads to simultaneous elevation of strain resistance to macrolides and autogenous Ctc antibiotic (MtcR phenotype), as shown by the data indicating that when Ctc is used as exogeneous inducer, efficient induction of resistance (10^3–10^4 times) to macrolides Erm, Oln, Tyl and Tur takes place, in addition to induction of resistance to Ctc. As seen in Fig.3, there is clear-cut dependence of induction level on the dose of Ctc. Thus, functioning of these inducible gene(s) depends on the presence in the medium of exogeneous inductors, both macrolides and Ctc. Also, MtcR mutants with OlnR ErmR TylR CtcR and ErmR TylR TurR

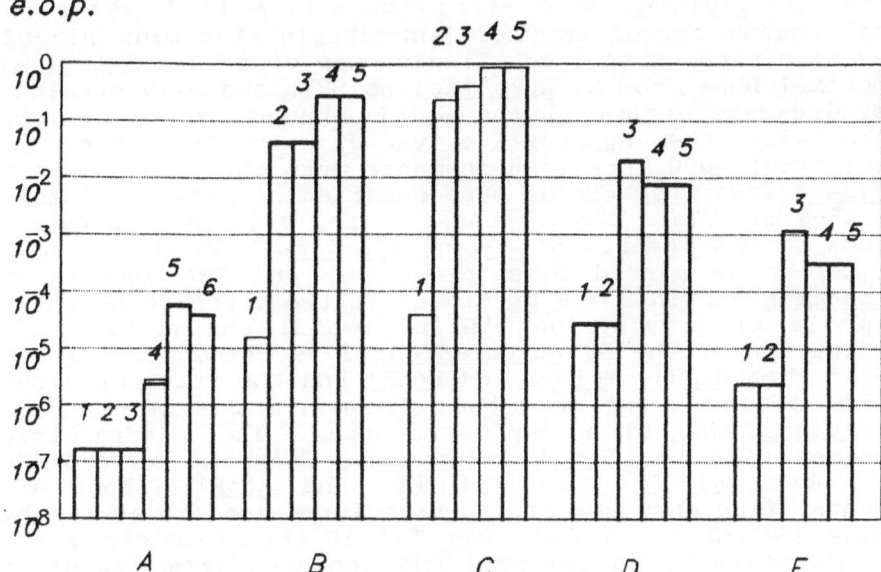

Fig. 3. Exogeneous induction of resistance to Ctc, 1000 μg/ml
(A), and macrolides: Erm, 1 μg/ml (B); Oln,10 μg/ml (C);
Tyl, 1 μg/ml (D); Tur, 1 μg/ml (E) using Ctc as inducer at
doses (μg/ml): 1. – inductor omitted, 2. – 10, 3. – 25,
4. – 50, 5. – 100

CtcR phenotypes have been isolated which are resistant both to
Ctc and macrolides. The frequency of occurrence of these mutants
is characteristic of that for single mutations. The mutants
possess constitutive resistance to concentrations of macrolide
antibiotics lethal for the initial strains and to Ctc. So,
appearance of MtcR single mutants implies that mutations occur
in the mtr gene(s).

The effect of CtcR mutations on Ctc biosynthesis

Among CtcR mutants of groups I–IV and MtcR mutants, those
having increased level of Ctc biosynthesis are found at
significantly higher frequency than among total population of
the strain studied. Induction of expression of inducible Ctc
resistance genes by Ctc and EB also results in elevation of Ctc
biosynthesis level by 30–35%. These data suggest that inducible
Ctc resistance genes participate in determination of Ctc
biosynthesis level.

Cloning of genes for resistance to Ctc from S. aureofaciens
strains S755 and S799

Shot-gun experiments were performed to clone genes for
resistance to Ctc from the chromosomes of S. aureofaciens

strains. S. lividans TK64 strain sensitive to 5 µg/ml Ctc in minimal medium served as recipient strain. Plasmids pIJ702 and pIJ699 were used as vectors. Cloning was done into Bgl II sites. Chromosomal DNAs from S. aureofaciens S755 and S799 strains were digested excess of Bgl II and BamHI enzymes, respectively, and ligated with the appropriate vector. In each experimental series, about 4000 TsnR transformants were obtained, among which S.lividans TK64 CtcR clones were detected on selective medium MM with 25µg/ml Ctc. Electrophoretic analysis demonstrated the presence of DNA inserts of S. aureofaciens S755 strain equal to about 20 bp in hybrid plasmid pVG530, and DNA inserts of S. aureofaciens strain S 799 of 6.6 kb in two opposite orientations within the three plasmids pVG520, pVG521 and pVG522. DNA-DNA hybridization of the S. aureofaciens S799 chromosomal fragments digested with BamHI, Bgl II and EcoRI and the fragment cloned on plasmid pVG520 confirms the fragment to belong to S. aureofaciens DNA (data not presented). The cloned fragment hybridizes with the BamHI chromosomal fragment of 6.6 kb and with longer Bgl II (about 20 kb) and EcoRI (about 18 kb) fragments.Efficiency of transformation by hybrid plasmids pVG530 and pVG520 was 2-5.10^{4}transformants per 1 µg DNA, respectively. The plasmid DNA isolated from transformants obtained after repeated transformation with pVG520 retained the initial structure. After repeated transformation with the pVG530 all transformants kept the CtcR phenotype. However, deletions occurred inside the cloned fragment in plasmids isolated from all the transformants analysed. Quantitative comparison of resistance levels for the strains bearing pVG530 and pVG520 plasmids, and also pIJ699, exhibits (Fig.4 A) the highest resistance in S. lividans TK64 to be mediated by the pVG530. These data confirm the presence on the cloned fragments of determinants of resistance to Ctc.

It seemed of interest to bring about comparative study of restriction maps for the cloned fragments of S.aureofaciens chromosome carrying CtcR determinants as well as otrA [1,2,12] and otrB [13] resistance genes of S. rimosus. Fig.5 presents restriction maps of cloned fragments of 6.4, 2.1 and 6.6 kb carrying CtcR determinants, plasmids pVG531, pVG533 and pVG520, respectively. A portion of the BglII cloned fragment is deleted in pVG531 and pVG533 as compared with pVG530. Comparison of restriction maps of these fragments cloned on pVG531 and pVG533 plasmids with the otrA gene of S. rimosus reveals resemblance in the area of overlap of cloned fragments and otrA. As seen from Fig.5, this similarity is shown by the presence of similar restriction sites for BamHI , SmaI , MluI , BstEII in the regions of otrA and cloned fragments and by the similar order of their location. The cloned resistance gene is designated ctrA in this work.The presence of PvuII site in a supposed region of ctrA in strain S755 and PvuII, SacI and PstI sites in the same region in strain S799 in contrast to otrA is observed.

As mentioned above, the PvG520 carries a cloned fragment with the resistance gene from the other divergent line of selection of S. aureofaciens strain S799 in contrast to plasmid pVG530 carrying a fragment from S. aureofaciens S755 chromosome. Similarity of restriction maps of the BglII-SmaI subfragments localized downstream of ctrA and otrA genes of S. aureofaciens and S. rimosus is established. However,rather than

122

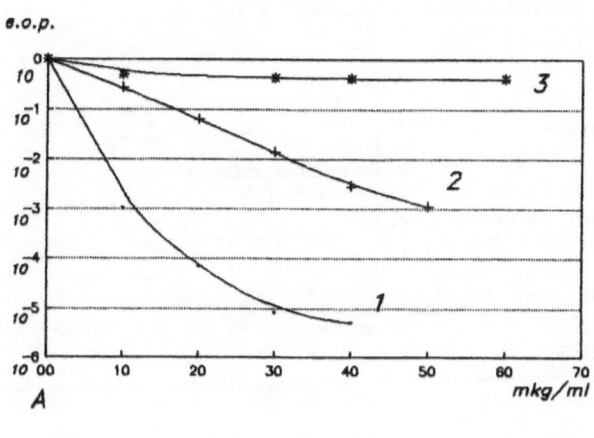

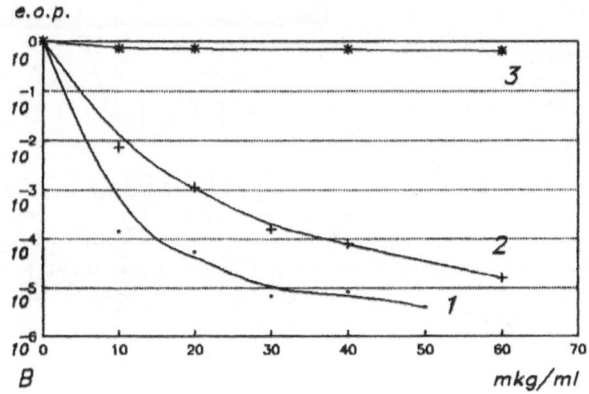

Fig. 4. Resistance of S. lividans TK64 strains carrying pIJ699
plasmid (1) and recombinant plasmids pVG520 (2) and
pVG530 (3) with ctrA gene to Ctc(A) and Erm(B)

resembling restriction maps of S. rimosus and the plasmids of pVG530 series, this BglII-SmaI fragment in the plasmid pVG520 has an inverted order of BglII, BstE2,HindIII, and SmaI sites. In the cloned fragment of plasmid pVG520 the subfragment BglII-SmaI is separated from the ctrA gene by a DNA sequence of 3.2 kb. The order of a number of restriction sites in this region is inverted in relation to the region situated downstream of the BglII-SmaI fragment of the otrA gene of S. rimosus . These data permit us to assume the presence in the area downstream of the ctrA gene of S. aureofaciens S799, of an inverted region which equals about 4 kb.

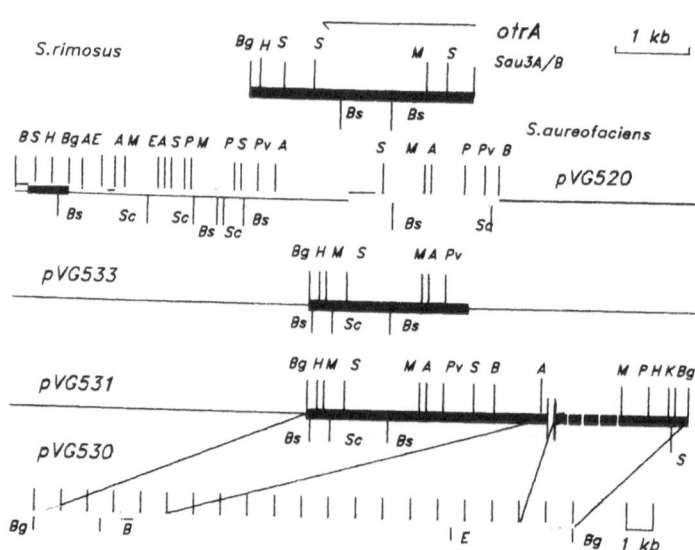

Fig.5. Restriction maps of the otrA[1, 2, 12] gene of S. rimosus and the ctrA gene on the cloned BglII fragment of S. aureofaciens strain S755 (pVG531 and pVG533 being the deletion variants of pVG530) and the BamHI fragment of S. aureofaciens S799 (pVG520) B — BamHI, Bg — BglII, A — ApaI, Bs —BstEII, Sc — SacI, S — SmaI, P — PstI, Pv — PvuII, H — HindIII, M — MluI, E — EcoRI

We were interested in seeing, whether the cloned ctrA can function as mtr gene. As may be seen in Fig.4B, the TK64 strains bearing recombinant pVG530 and pVG520 plasmids are characterized by both high resistance to Ctc and Erm as compared to S. lividans TK64(pIJ699) and S. lividans TK64(pIJ702) strains. Moreover, the strains S. lividans TK64(pVG530) and TK64(pVG520) are of higher resistance to other macrolides (data not presented). So, there is good reason to believe that the cloned ctrA gene functions as an mtr gene. Constitutive expression of the cloned gene in the absence of inducers might be connected with the absence of all indispensable components of the regulation system.

Discussion

Data are presented on resistance of S. aureofaciens strain S755 , the producer of chlortetracycline, to a number of exogenous antibiotics and the autogenous antibiotic Ctc. It was shown that genes responsible for strain resistance to a number of macrolide antibiotics and to thiostrepton are inducible and their functioning only takes place in the presence in the medium of appropriate antibiotics. It was also demonstrated that resistance to Ctc is mediated by genes with constitutive and inducible expression, Ctc and EB playing role as efficient inducers of inducible genes.

Functions of inducible mtr gene(s) which may simultaneously increase resistance to a number of macrolides and Ctc were characterized in detail. Functioning of the mtr genes only occurs in the presence of exogenous inducers, such as both macrolides and Ctc. Mutants were isolated with constitutive resistance to macrolides and Ctc, some of them having increased biosynthesis of Ctc. The mtr gene(s) may be assumed to take part in acquisition by the strain of inducible resistance to macrolides in the environment. At the same time, mtr gene(s) ensure elevation of the level of resistance to self-produced antibiotic Ctc. This, in turn, may provide initiation of Ctc biosynthesis aimed at active protection of the strain against competitor strains.

The fact described earlier of high frequency loss, in many Streptomyces strains, of resistance to several antibiotics simultaneously, including the autogenous antibiotic, may be explained by deletions in genes resembling mtr gene(s). Intrinsic high resistance of a number of initial Streptomyces strains to a great number of antibiotics may well be explained by selection of strains with mutated inducible genes, which results in constitutive expression of these genes. It also seems that mechanisms of natural resistance of Streptomyces strains may be similar to those mediated by appearance of induced resistance mutations.

In the present work, the ctrA gene seemingly to be linked to the genes for Ctc biosynthessis was cloned from two divergent lines of selection of S. aureofaciens strains S755 and S799. Restriction maps of ctrA genes are similar and resemble that of the otrA gene of S. rimosus. The resemblance was also observed in the region downstream of the ctrA and otrA genes. In this region of S. aureofaciens strain S799 an inversion has been revealed. It seems likely that the S. aureofaciens S799 chromosomal region covering the area from the BglII site to the ctrA gene cloned on the pVG520 plasmid has regulatory functions, since it was noted that incorporation into S. lividans TK64 of recombinant pVG520 plasmid leads to significant increase in actinorhodin biosynthesis. The data on simultaneous acquisition of resistance to Ctc and macrolides in the presence of recombinant plasmids carrying the ctrA gene suggest that ctrA has the functions of the mtr gene. If so, then using the cloned mtr gene as a labelled probe may reveal the extent of distribution of genes functioning as mtr gene(s) in Streptomyces strains.

Recently, H. Shrempf et al[9] reported the cloning from S. lividans of a natural tetR determinant which displays weak homology with the otrA gene for resistance to Otc. In this light, it was of interest to define whether determinants responsible for simulataneous resistance to Ctc and macrolides function in S. lividans strain TK64 . If so, a gene homologous to otrA might possess the functions of mtr in S. lividans. Based on our data, it appeared that the level of resistance to Ctc in S. lividans TK64 increases 100 and more times in the presence of Ctc as exogenous inducer. In addition, the level of resistance to Ctc was shown to significantly increase when Erm and Oln were used as inducers. In turn, Ctc acts as inducer of resistance to Erm and Oln. These data might point to the presence in S. lividans TK64 of inducible determinant(s) mediating resistance both to macrolides and Ctc. Based on these data, it may also be assumed that streptomycetes have genes, whose functioning ensures strain resistance to a number of unrelated antibiotics. The products of these genes can simultaneously protect the strain against its own toxic antibiotics. These genes with resembling functions may be wide- spread among Streptomyces strains.

References

1. Binnie C., Warren M., Butler M. (1989) Cloning and heterologous expression in Streptomyces lividans of Streptomyces rimosus genes involved in oxytetracycline biosynthesis. J. Bacteriol. 171: 887–895.
2. Butler M.J., Friend E.J., Hunter I.S., Kaczmarek F.S., Sugden D.A., Warren M. (1989) Molecular cloning of resistance genes and architecture of the linked gene cluster involved in biosynthesis of oxytetracycline by Streptomyces rimosus. Mol.Gen.Genet. 215: 231–238.
3. Chinenova T.A., Biryukova I.V., Voeykova T.A., Emelyanova L.K. Klochkova O.A., Sezonov G.V., Lomovskaya N.D. (1989) Identification of determinants for resistance to chlortetracycline and other antibiotics in S. aureofaciencs strain – the producer of chlortetracycline. Genetika (USSR) 26: 00.
4. Hinterman G., Grameri R., Kieser T., Hutter R. (1981) Restriction analysis of the Streptomyces glaucescens genome by agarose gel electrophoresis. Arch. Microbiol. 131: 218–222.
5. Hopwood D.A., Bibb M.J., Chater K.F.,Kieser T., Bruton C.J., Kieser H.M., Lydiate D.J., Smith C.P., Ward J.M., Schrempf H. (1985) Genetic manipulation of Streptomyces: a laboratory manual. John Innes Foundation, Norwich, 355 p.
6. Hopwood D.A., Kieser T., Wright H.M., Bibb M.J. (1983) Plasmids, recombination and chromosome mapping in Streptomyces lividans 66. J. Gen. Microbiol. 129: 2257–2269.
7. Hostalek Z., Vanek Z. (1985) Biosynthesis of the tetracyclines. In: Handbook Experim. Pharmacol. 78: P.137.
8. Hunter I.S., Deidre D., McDowall K.J. (1990) Molecular genetics of oxytetracycline production in Streptomyces rimosus J. Cell. Biochem., Abstracts 19th Annual Meetings UCLA Symposia on Mol. Cell. Biol. CC012, p.91
9. Kessler A., Dittrich W., Betzler M., Schrempf H. (1989) Cloning and analysis of a deletable tetracycline–resistance

        determinant of <u>Streptomyces</u> <u>lividans</u> 1326. <u>Molecular</u>
        <u>Microbiology</u>. 3(8): 1103-1109.
10. Maniatis T., Fritsch E.E., Sambrook J. (1982) Molecular
        cloninng. A laboratory manual.Cold Spring Harbor Lab. P.
        480
11. Mc.Cormick I.R.D. (1969) Point-blocked mutants and, the
        biogenesis f tetracyclines. In: Sermonti G., Alacevic M.
        (eds) . Genetics and breeding of Streptomyces. Yugoslavian
        Academy of Sciences and Arts. Zagreb, P.163.
12. Ohnuki T., Katoh T., Imanaka T., Aiba S. (1985) Molecular
        cloning of tetracycline resistance genes from Streptomyces
        rimosus into S.griseus and characterization of the cloned
        genes. J.Bacteriol. 161: 1010-1016.
13. Reunes J.P., Calmels T., Drocourt D., Tiraby G. (1988)
        Cloning,expression in <u>Escherichia</u> <u>coli</u> and nucleotid
        sequense of a tetracyclin-resistance gene from <u>Streptomyces</u>
        <u>rimosus. J.Gen.Microbiol.</u> 134: 585-598
14. Zvenigorodsky V.I., Lomovskaya N.D. (1972) Induction of
        prophage by a virulent mutant of actinophage ΦC31
        <u>Streptomyces</u> <u>coelicolor</u> A3(2). <u>Genetika</u> (USSR) 8: 94-100.

BROMOPEROXIDASE FROM <u>STREPTOMYCES AUREOFACIENS</u> TÜ24 AND ITS

ROLE IN 7-CHLOROTETRACYCLINE BIOSYNTHESIS

Helmut Häcker, Otto Pfeifer, Frauke Thiermann,
Maria Weng, Franz Lingens, and Karl-Heinz van Pée

Institut für Mikrobiologie
Universität Hohenheim
Garbenstr. 30
D-7000 Stuttgart 70
FRG

INTRODUCTION

<u>Streptomyces</u> <u>aureofaciens</u> Tü24 and <u>S.</u> <u>aureofaciens</u> ATCC 10762 both produce the antibiotics tetracycline and 7-chlorotetracycline. The biosynthetic pathway for tetracycline and 7-chlorotetracycline is identical with the exception of the chlorination step (1, Fig. 1).

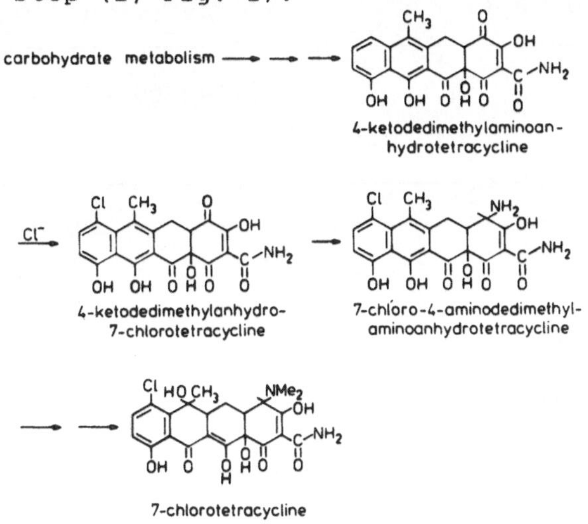

Fig. 1. Tetracycline and 7-chlorotetracycline biosynthetic
pathway.

In this step chlorine is introduced into the aromatic part of the molecule. This reaction is catalysed by a chlorinating enzyme. The substrate for the chlorinating enzyme is 4-ketodedimethylaminoanhydrotetracyline. After transamination chlorination is not possible anymore. Obviously the 4-keto group is necessary for the chlorinating enzyme to recognize its substrate.

As 4-ketodedimethylaminoanhydrotetracycline, the "natural" substrate for the chlorinating enzyme involved in 7-chlorotetracycline biosynthesis was not available, monochlorodimedone, the substrate normally used for the detection of halogenating enzymes (2), had to be used for the isolation of a halogenating enzyme from the 7-chlorotetracycline-producing S. aureofaciens strains.

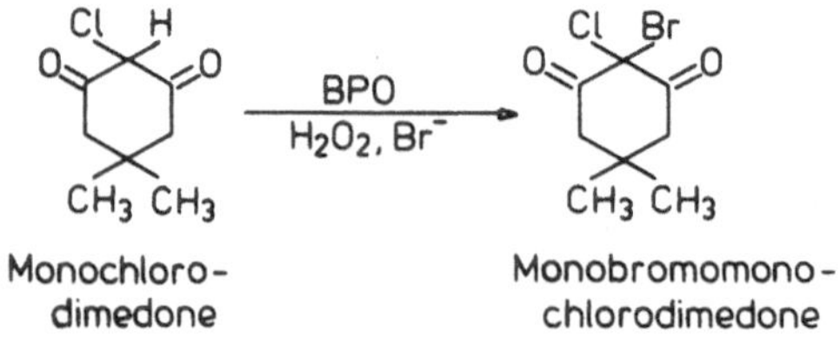

Fig. 2. Monochlorodimedone assay used for the detection of halogenating enzymes.

Using the monochlorodimedone assay, a bromoperoxidase was isolated from S. aureofaciens Tü24 (3). This bromoperoxidase was not able to catalyse the chlorination of monochlorodimedone, but only its bromination.

Thus the question arose, whether this enzyme is actually the one involved in 7-chlorotetracycline biosynthesis.

Table 1. Comparison of bromoperoxidase from S. aureofaciens Tü24 (3) and bromoperoxidase from S. aureofaciens ATCC 10762 (4).

Properties	S. aureofaciens	
	ATCC 10762	Tü24
Molecular weight	65,000	90,000-95,000
Subunit size	32,000	31,000
Number of subunits	2	3
Specific activity (units/mg)	65	7
Inhibition by fluoride	non-competitive	uncompetitive

Some time later, another bromoperoxidase was isolated from S. aureofaciens ATCC 10762 by Krenn et al. (4). This enzyme showed some important differences to the bromoperoxidase from S. aureofaciens Tü24 (Table 1).

ENZYMATIC AND IMMUNOLOGICAL INVESTIGATIONS

When antibodies raised against the bromoperoxidase from

S. aureofaciens Tü24 were used, a bromoperoxidase could be detected in extracts from S. aureofaciens ATCC 10762, which was immunologically identical to the Tü24-bromoperoxidase (Fig. 3).

However, this enzyme was not identical with the one isolated by Krenn et al. (4), which was also present in the crude extract. This could be demonstrated by polyacrylamide gel electrophoresis followed by activity staining (Fig. 4).

The two brominating activities could be separated by ion exchange chromatography (Fig. 5).

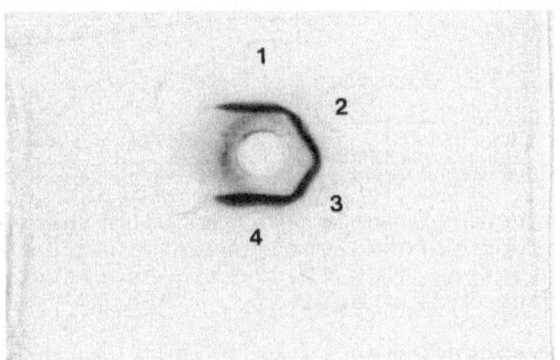

Fig. 3. Ouchterlony agar diffusion assay. The central well contained antiserum raised against bromoperoxidase from S. aureofaciens Tü24. Wells 1 and 4 contained S. aureofaciens Tü24 bromoperoxidase; wells 2 and 3 contained partially purified extracts from S. aureofaciens ATCC 10762.

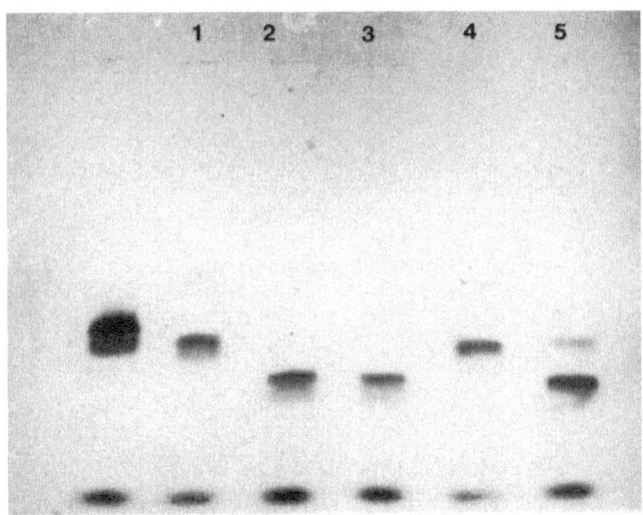

Fig. 4. Native polyacrylamide gel electrophoresis of extracts from S. aureofaciens ATCC 10762 (lanes 1 and 2), S. aureofaciens ATCC 13910 (lane 3), and S. aureofaciens ATCC 13911 (lanes 4 and 5). The gel was stained with phenol red for brominating activity.

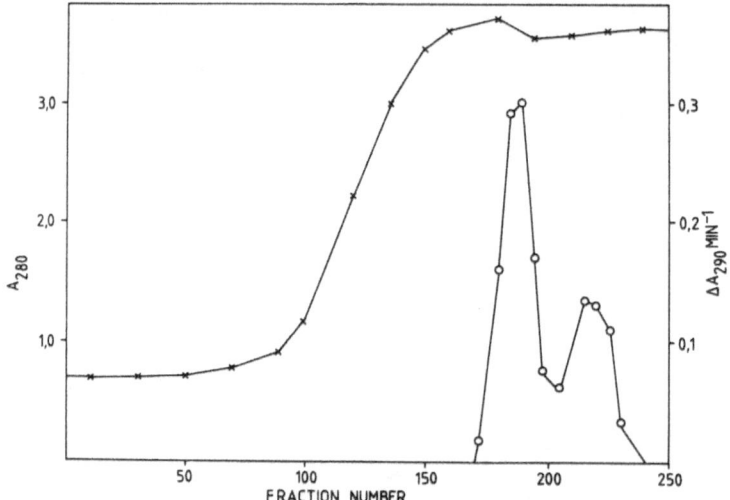

Fig. 5. Ion exchange chromatography showing the separation of the two bromoperoxidases present in crude extracts of S. aureofaciens ATCC 10762. x, Absorbance at 280 nm; o, bromoperoxidase activity.

Immunoblotting experiments revealed that only one of the two bromoperoxidases from S. aureofaciens ATCC 10762 reacted with antibodies raised against the Tü24-bromoperoxidase (Fig. 6).

As S. aureofaciens Tü24 only produces one bromoperoxidase, it was very likely that this enzyme is the one, involved in 7-chlorotetracycline biosynthesis. However, doubt was cast on this, when the strains S. aureofaciens, ATCC 13908, 13910,

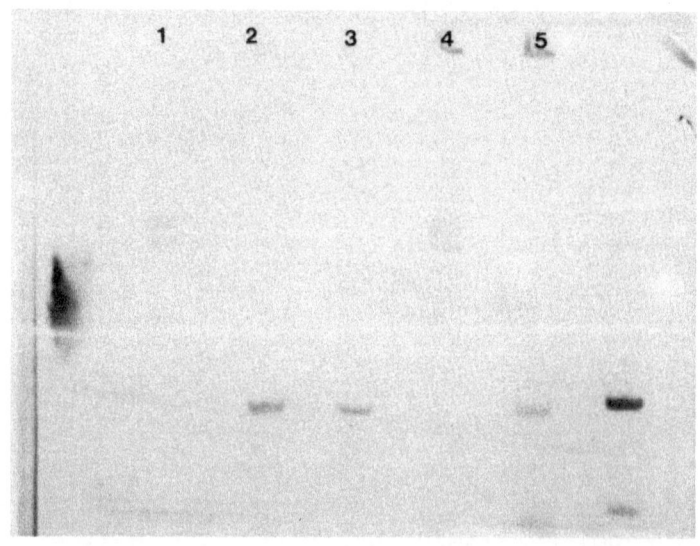

Fig. 6. Western blot of extracts from S. aureofaciens ATCC 10762 (lanes 1 and 2), S. aureofaciens ATCC 13910 (lane 3) and S. aureofaciens ATCC 13911 (lanes 4 and 5) after ion exchange chromatography.

and 13911, three mutants of S. aureofaciens ATCC 10762, which
are blocked in the halogenating step, were investigated. These
three mutants still produce tetracycline, but can not synthesize 7-chlorotetracycline anymore.

It was found that the mutant ATCC 13911 produces both
bromoperoxidases and the mutants ATCC 13908 and 13910 only
produce the one cross-reacting with antibodies raised against
the Tü24-enzyme. This result suggested that none of the
isolated bromoperoxidases is involved in 7-chlorotetracycline
biosynthesis, as they were still present in the mutants
blocked in the chlorination step.

GENETICAL ANALYSIS

Some more information was obtained, by genetical analysis. The gene of the bromoperoxidase from S. aureofaciens Tü24
was cloned and partially mapped (3, Fig. 7).

When the 4.3-kb SstI fragment of pHM621 was hybridized
with total DNA from S. aureofaciens ATCC 10762, digested with
SstI, only a single hybridizing band was obtained. Either
there is only one gene for both bromoperoxidases produced by
S. aureofaciens ATCC 10762 or the two genes have less than 63%
of homology.

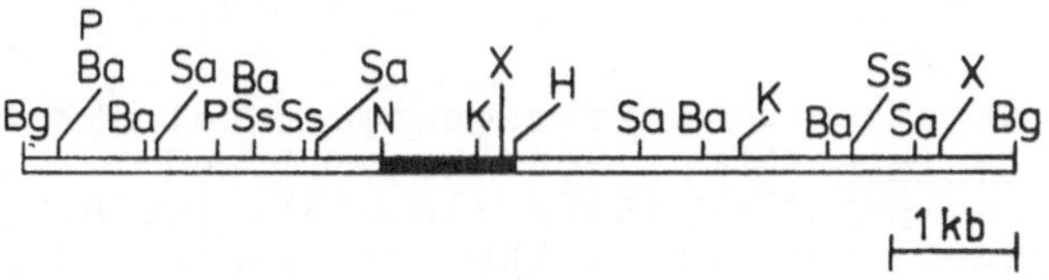

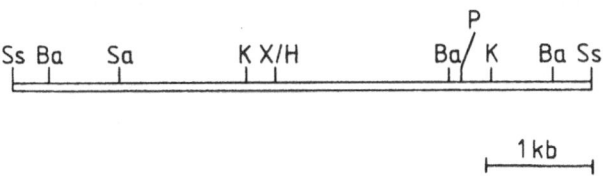

Fig. 7. Partial restriction map of the 8.0-kb insert of
pHM621 containing the bromoperoxidase gene from S.
aureofaciens Tü24. The solid line indicates the
approximate location of the bromoperoxidase gene. The
abbreviations Ba, Bg, H, K, N, P, Sa, Ss, and X indicate BamHI, BglII, HindIII, KpnI, NcoI, PstI, SalI,
SstI, and XhoI, respectively.

Using the 4.3-kb SstI fragment of pHM621 as a probe, the
corresponding bromoperoxidase gene from S. aureofaciens ATCC
10762 was cloned, too. Mapping of the cloned insert showed
that it was very similar to the SstI fragment isolated from S.
aureofaciens Tü24 (Fig. 8).

Fig. 8. Partial restriction map of the cloned 5.4-kb SstI
fragment of S. aureofaciens ATCC 10762 containing the
bromoperoxidase gene. The abbreviations Ba, H, K, P,
Sa, Ss, and X indicate BamHI, HindIII, KpnI, PstI,
SalI, SstI, and XhoI, respectively.

The 4.3-kb SstI fragment of pHM621 containing the bromo-peroxidase gene of S. aureofaciens Tü24 was hybridized with SstI-digested total DNA from the parent strain S. aureofaciens ATCC 10762 and the mutant strain ATCC 13908. The hybridizing bands were different in the two strains. Whereas the size of the hybridizing band of the parent strain was 5.4 kb, its size was only 4.4 kb in the mutant strain (Fig. 9).

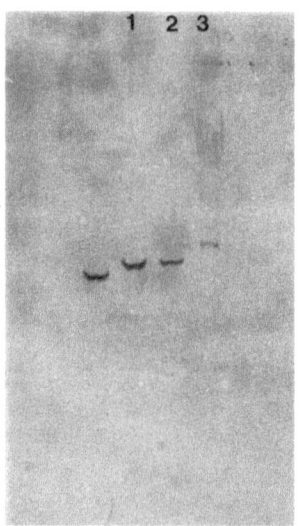

Fig. 9. Hybridization of SstI-digested total DNA from S. aureofaciens ATCC 13908 (lanes 1 and 2) and S. aureofaciens ATCC 10762 (lane 3). The hybridization probe used was the 4.3-kb SstI insert of pHM621.

As the mutant still produces the bromoperoxidase, which is immunologically identical to the Tü24-enzyme, this gene is obviously still present in this strain. However, there is a deletion close to the bromoperoxidase gene. This suggests that either the isolated bromoperoxidase is the wrong enzyme or a yet unknown compound, necessary for halogenating activity in vivo is missing in the mutants and that the gene encoding for this unknown compound is located on the deleted DNA fragment.

To shed some light on this, the bromoperoxidase gene from S. aureofaciens Tü24 was inactivated by digestion with KpnI/XhoI. The inactivated bromoperoxidase gene was ligated to the tyrosinase gene isolated from pIJ702. This construct was ligated into the temperature-sensitive vector pGM160 (5). The resulting plasmid pGRB1 (Fig. 10) conferred kanamycine and gentamycin resistance to E. coli cells and thiostrepton and gentamicin resistance to Streptomyces cells. S. lividans TK64 transformed with pGRB1 produced melanin but no bromoperoxidase.

The plasmid pGRB1 was then used to transform S. aureofaciens Tü24. However, the plasmid was unstable in this strain and only one clone, S. aureofaciens Tü2488, harbouring a plasmid that hybridized to the vector pGM160 could be obtained. After the second transfer the plasmid was lost completely and thus the resistance to thiostrepton and gentamicin. Hybridi-

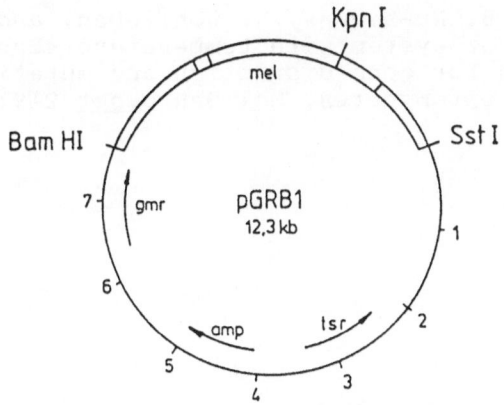

Fig. 10. pGRB1 containing the inactivated bromoperoxidase gene from S. aureofaciens Tü24.

zation experiments showed that the obtained clone had lost the chromosomal bromoperoxidase gene together with a DNA fragment of unknown size and did not produce any bromoperoxidase anymore.

HPLC analysis of the culture medium of S. aureofaciens Tü2488 showed that this clone still produced tetracycline but was not able to produce 7-chlorotetracycline. When S. aureofaciens Tü2488 was transformed with pHM621, which contained the bromoperoxidase gene from S. aureofaciens Tü24, the clones obtained produced bromoperoxidase in large amounts, but still did not synthesize 7-chlorotetracycline. The reason for this is probably that on the deleted DNA fragment the gene for a compound necessary for halogenating activity in vivo is located.

This result very strongly suggests that the bromoperoxidase isolated from S. aureofaciens Tü24 is actually the halogenating enzyme catalysing the chlorination step in 7-chlorotetracycline biosynthesis, and that a yet unknown compound is needed for the in vivo activity of this enzyme.

REFERENCES

1. J. D. R. McCormick, Tetracyclines, in: "Antibiotics Vol II-Biosynthesis," D. Gottlieb and P. D. Shaw, ed., Springer, Berlin-Heidelberg-New York (1967).
2. L. P. Hager, D. R. Morris F. S. Brown, and H. Eberwein, Chloroperoxidase II. Utilization of halogen anions, J. Biol. Chem. 241:1769 (1966).
3. K.-H. van Pée, Molecular cloning and high-level expression of a bromoperoxidase gene from Streptomyces aureofaciens Tü24, J. Bacteriol. 170:5890 (1988).
4. B. E. Krenn, H. Plat, and R. Wever, Purification and some characteristics of a non-heme bromoperoxidase from Streptomyces aureofaciens, Biochim. Biophys. Acta 952:255 (1988).

5. G. Muth, B. Nußbaumer, W. Wohlleben, and A. Pühler, A vector system with temperature-sensitive replication for gene disruption and mutational cloning in streptomycetes, <u>Mol Gen Genet</u> 219:341 (1989).

ENZYMES AND COENZYMES OF THE TERMINAL PART OF THE ANTIBIOTIC BIOSYNTHETIC PATHWAY IN STREPTOMYCETES PRODUCING TETRACYCLINES

J. Novotná, J. Neužil, I. Vančurová, V. Běhal, Z. Hošťálek

Institute of Microbiology, Czechoslovak Academy of Sciences
142 20 Prague 4, Czechoslovakia

INTRODUCTION

In spite of more than four decades of research on microbial antibiotic biosynthesis, little appears to be known about the enzymology of antibiotic production. Details about most of individual enzymes involved in the definitive biosynthetic pathway of tetracyclines, in which the sequence of essential reaction steps was ascertained more than 20 years ago, have also remained unknown.

However, research has recently been aimed at the enzymes participating in the terminal part of the biosynthetic pathway. This contribution, focused mainly on the role of flavins in the terminal reaction, summarizes the available data about anhydrotetracycline oxygenase (ATCox) and tetracycline dehydrogenase (TCdh) (Fig. 1) in *Streptomyces aureofaciens*, the producer of chlortetracycline and tetracycline. Some additional information concerning comparable reaction steps in *Streptomyces rimosus*, the producer of oxytetracycline, in which a three-step terminal part of the pathway was established[1], is also presented.

ANHYDROTETRACYCLINE OXYGENASE

The conversion of anhydrotetracycline to dehydrotetracycline is catalyzed in *S. aureofaciens* solely by ATCox (M_r 115,000), a homodimeric enzyme[2,3], in the presence of NADPH and O_2. The results of localization and purification experiments suggest that ATCox could be loosely bound to the membrane and become dislodged during purification[2-4]. The same enzyme catalyzes the reaction if the substrate is anhydrochlortetracycline[5] as can be expected for enzyme isolated from *S. aureofaciens* that produces more than 80 % chlortetracycline together with tetracycline. The broad substrate specificity of the purified enzyme is in agreement with the cell-free extract experiments[6]. However, the *S. rimosus* extracts catalyze the reaction only with anhydrotetracycline as a substrate, not with anhydrochlortetracycline which is not a natural intermediate in the biosynthetic pathway of oxytetracycline.

Prosthetic group of ATCox. Hydrophobic interaction chromatography on Phenyl-Sepharose during ATCox purification caused a dramatic drop of its activity which could be fully restored by the addition of a low-molar mass fraction of heat-denatured cell-free extract[3,7]. Absorption spectrum of the purified enzyme exhibiting maxima at 276, 376, and 452 nm and a shoulder at about 486 nm (Fig. 2) and the

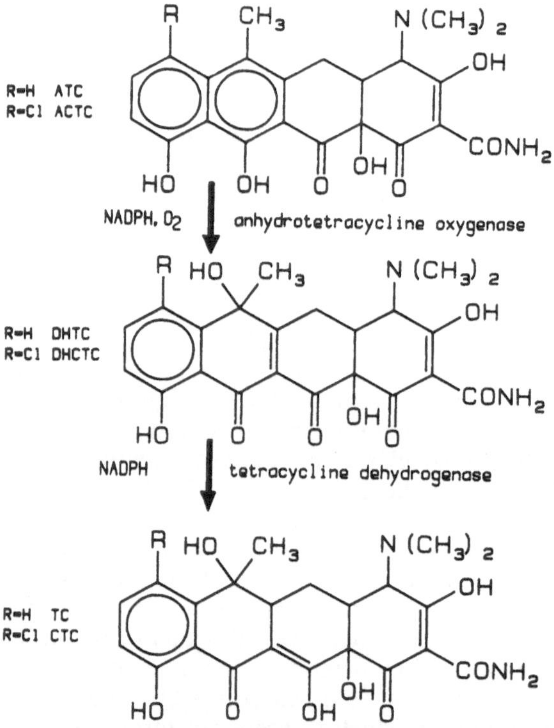

Fig. 1. Terminal part of biosynthetic pathway of tetracyclines in *Streptomyces aureofaciens*. ATC, Anhydrotetracycline; ACTC, anhydrochlortetracycline; TC, tetracycline; CTC, chlortetracycline.

absorption spectrum of the chromophore released from purified ATCox by heat denaturation (data not shown) indicated the involvement of flavin-like compounds. Hence riboflavin and its 5'-phosphate (FMN) and 5'-ADP (FAD) derivatives were tested for restoration of the decreased activity of purified ATCox in a way similar as in the case of the low-molar mass cell-free extract fraction[7]. FAD, but not FMN or riboflavin, reactivates significantly the activity of purified ATCox after pre-incubation with the enzyme suggesting FAD to be its essential component.

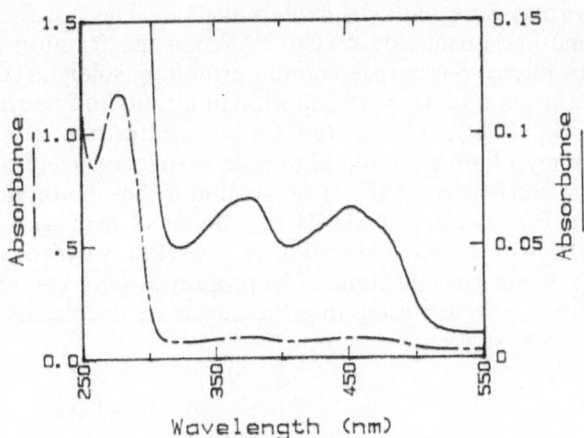

Fig. 2. Absoption spectrum of purified anhydrotetracycline oxygenase (1 mg/ml) in 0.02 M Bistris/HCl, 0.25 M NaCl, pH 6.5.

Calculation of the molar ratio of flavin to ATCox (dimeric) being 0.68 (before addition of exogenous FAD) was based on the absorption spectrum of purified ATCox (Fig. 2), molar absorptivity of free flavin of 1.13×10^4 and on the supposition that an enzyme-bound flavin has the same molar absorptivity as free flavin[8]. Lower chromofore contents indicated by molar ratio under 1 (or even 2) and the reversible drop in the activity of the enzyme was probably caused by its partial release during chromatography on Phenyl-Sepharose.

Easy release of flavin suggests that it is tightly, but not covalently, bound to the protein. Likewise, the absorption spectrum of the purified protein (Fig. 2) not exhibiting hypochromic shift of the maximum at 376 nm, typical for a covalent bond of flavin to protein, indicates a non-covalently bound flavin[9].

An important evidence supporting the involvement of FAD in the ATCox reaction is the presence of an FAD-binding fold in the N-terminal sequence of *S. rimosus* ATCox[10] homologous with other flavoproteins[11,12].

TETRACYCLINE DEHYDROGENASE

The last reaction of the biosynthetic tetracycline pathway reducing 5a,11a-dehydro(chlor)tetracycline [DH(C)TC] to the final antibiotic product (chlor)tetracycline (Fig. 1) proceeds in cell-free extracts of *S. aureofaciens* supplemented with NADPH and can be followed spectrophotometrically or by HPLC[5,13,14]. The instability of DHTC is avoided by using ATCox and anhydrotetracycline as a starting substrate in a coupled preceding reaction. Interestingly, the reduction of DHTC takes place only after complete conversion of anhydrotetracycline to DHTC. To characterize the enzyme(s) participating in the last biosynthetic reaction, it was necessary to purify the protein(s) responsible for the final reduction.

NADPH:8-hydroxy-5-deazaflavin oxidoreductase. The purification procedure was based on the methods used for ATCox[2,3]. When the fraction of hydrophobic proteins eluted from Phenyl-Sepharose column exhibiting solely ATCox activity was supplemented with a protein of $M_r = 40,000$ without affinity to Phenyl-Sepharose, the ATC-to-TC conversion via DHTC occurred. Cell-free extracts of *S. aureofaciens* and *S. rimosus* vegetative mycelium were found to reduce streptomycete-origin 8-hydroxy-5-deazaisoalloxazine derivatives (SF_{420}) or synthetic 7,8-didemethyl-8-hydroxy-5-deazariboflavin (F_0) (Fig. 3) using NADPH as a donor of hydrogen and electrons[15]. The purified 40-kDa protein was identified as NADPH:8-hydroxy-5-deazaflavin oxidoreductase (F_{420} reductase) and some of its properties were compared with a very similar enzyme from *S. griseus*[16] including the substrate specificity of the enzyme revealing SF_{420} as a better substrate than F_0.

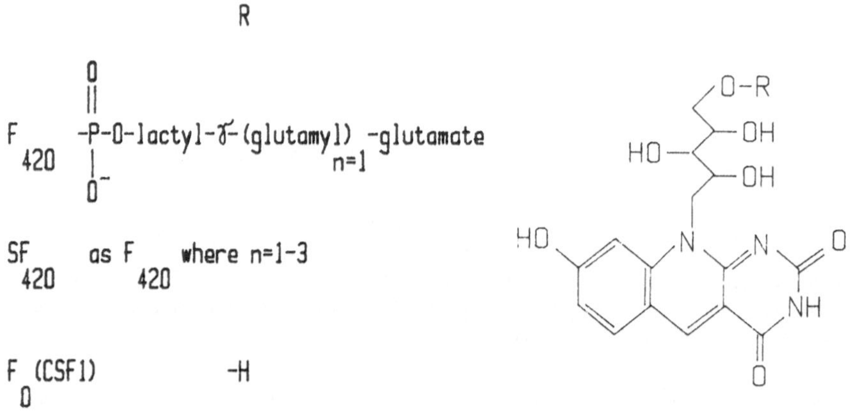

Fig. 3. Structure of 8-hydroxy-5-deazaflavins.

Homogeneous ATCox did not catalyze the terminal DHTC reduction even after the addition of F_{420} reductase; on the other hand, this does not rule out its participation in the reaction. TCdh activity was not observed after mixing F_{420} reductase and individual protein fractions separated from ATCox, which suggests the involvement of more than two components and/or the loss of the activity during the FPLC purification step. However, the redox role of 8-hydroxy-5-deazaflavin(s) in the conversion of DH(C)TC as shown by McCormick *et al.*[17,18], was confirmed by the fact that F_{420} reductase was an essential component of the TCdh system.

8-hydroxy-5-deazaflavins. In addition to F_0 in *S. aureofaciens*[17,18], 8-hydroxy-5-deazaflavin derivatives differing in the length of the lactyl-oligoglutamyl side chain (Fig. 3) were found in *S. griseus* where an unidentified derivative acting as a photon acceptor in a DNA-photoreactivating enzyme was also documented[19]. A widespread distribution of F_0 and/or other 8-hydroxy-5-deazaflavin derivatives in streptomycetes and related microorganisms was recently demonstrated[20,21] and the participation of F_0 in one of early steps of lincomycin biosynthetic pathway was proved[22].

To elucidate the role of 8-hydroxy-5-deazaflavins in the terminal reaction of tetracycline biosynthesis and in secondary metabolism in general, a detailed qualitative and quantitative analysis of the individual derivatives will be needed. The methods of separation and quantification of 8-hydroxy-5-deazaflavins from methanogenic bacteria[23,24] can be applied to streptomycetes, although their level in streptomycetes is more than 100-fold lower than in methanogens[20]. Typical HPLC analyses of 8–hydroxy-5-deazaflavins extracted from *S. aureofaciens* are shown in Fig. 4.

Individual peaks in the chromatograms were correlated with those achieved by the chromatographic analysis of *S. griseus* SF_{420} used as standard. The analyses show

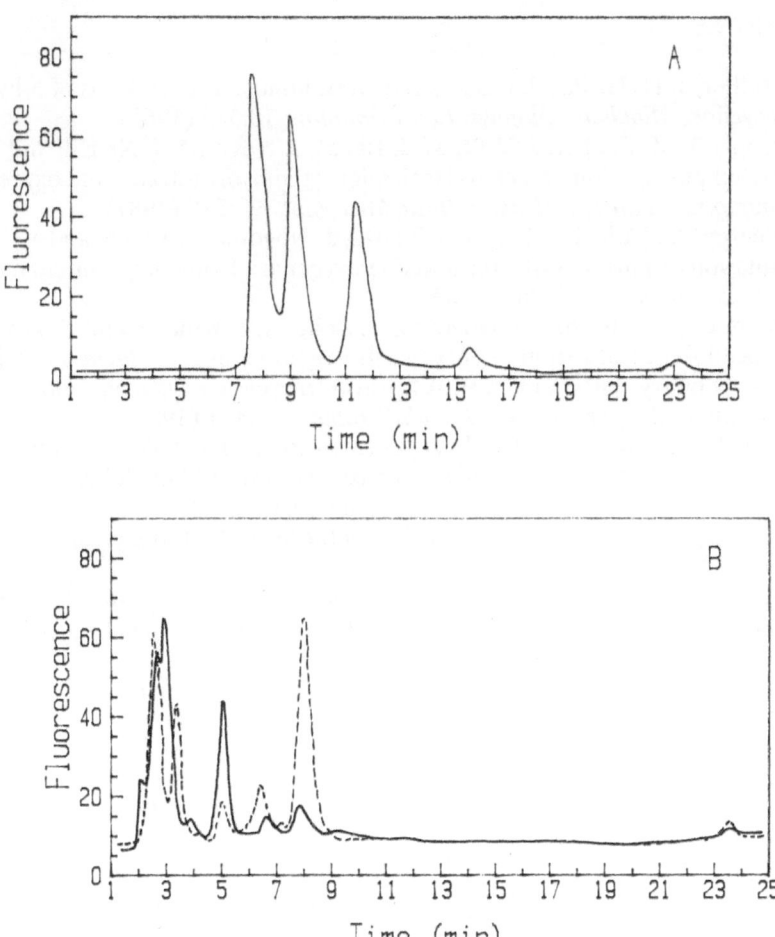

Fig. 4. HPLC analysis[23] of 8-hydroxy-5-deazaflavins. A, SF_{420} standard from *Streptomyces griseus* isolated by Eker *et al.*[19]; B, MeOH extract from 24-h (full line) and 72-h (dashed line) *S. aureofaciens* mycelium.

that 8-hydroxy-5-deazaflavins isolated from *S. aureofaciens* mycelium differ in the side chain as was demonstrated for *S. griseus*[19]. Although changes in the levels of individual derivatives with the age of the culture were observed, further experimental evidence is necessary for elucidation of the importance of total 8-hydroxy-5-deazaflavins and their individual derivatives for antibiotic biosynthesis.

ACKNOWLEDGEMENT

The authors wish to express their thanks to Dr. A. P. M. Eker of Erasmus University, Rotterdam, for the generous gift of SF_{420} and F_0 standards.

REFERENCES

1. P. A. Miller, J. H. Hash, M. Lincks, and N. Bohonos, Biosynthesis of 5-hydroxytetracycline, *Biochem. Biophys. Res. Commun.* 18:325 (1965).
2. I. Vančurová, M. Flieger, J. Volc, M. J. Beneš, J. Novotná, J. Neužil, and V. Běhal, Partial purification and characterization of anhydrotetracycline oxygenase of *Streptomyces aureofaciens*, *J. Basic Microbiol.* 27:529 (1987).
3. I. Vančurová, J. Volc, M. Flieger, J. Neužil, J. Novotná, J. Vlach, and V. Běhal, Isolation of pure anhydrotetracycline oxygenase from *Streptomyces aureofaciens*, *Biochem. J.* 253:263 (1988).
4. V. Erban, L. V. Trilisenko, J. Novotná, V. Běhal, I. S. Kulaev, and Z. Hošťálek, Subcellular localization of enzymes in *Streptomyces aureofaciens* and its alteration by benzyl thiocyanate. II. Anhydrotetracycline oxygenase and glucose-6-phosphate dehydrogenase, *Folia Microbiol.* 32:411 (1987).
5. J. Novotná, J. Neužil, V. Běhal, I. Vančurová, and Z. Hošťálek, A simultaneous assay for anhydrotetracycline oxygenase and tetracycline dehydrogenase using diode array spectrophotometry (to be published).
6. P. A. Miller, Cell-free studies on the biosynthesis of the tetracyclines, *Dev. Industr. Microbiol.* 8:96 (1967).
7. I. Vančurová, Anhydrotetracycline oxygenase and enzymes assimilating ammonium ions in *Streptomyces aureofaciens*, Ph.D. Thesis (in Czech), Institute of Microbiology, Prague (1988).
8. L. G. Howell, T. Spector, V. Massey, Purification and properties of p-hydroxybenzoate hydroxylase from *Pseudomonas fluorescens*, *J. Biol. Chem.* 247:4340 (1972).
9. M. Fukuyama, and Y. Miyake, Purification and some properties of cholesterol oxidase from *Schizophyllum commune* with covalently bound flavin. *J. Biochem.* 85:1183 (1979).
10. C. Binnie, M. Warren, and M. J. Butler, Cloning and heterologous expression in *Streptomyces lividans* of *Streptomyces rimosus* genes involved in oxytetracycline biosynthesis, *J. Bacteriol.* 171:887 (1989).
11. J. Hofsteenge, J. M. Vereijken, W. J. Weijer, J. J. Beintema, R. K. Wierenga, and J. Drenth, Primary and tertiary structure studies of p-hydroxybenzoate hydroxylase from *Pseudomonas fluorescens*. Isolation and alignment of the CNBr peptides; interactions of the protein with flavin adenine dinucleotide, *Eur. J. Biochem.* 113:141 (1980).

12. T. Ishizaki, N. Hirayama, H. Shinkawa, O. Nimi, and Y. Murooka, Nucleotide sequence of the gene for cholesterol oxidase from a *Streptomyces* sp., *J. Bacteriol.* 171:596 (1989).
13. V. Erban, V. Běhal, L. V. Trilisenko, J. Neužil, and Z. Hošťálek, Tetracycline dehydrogenase: spectrophotometric assay, properties, and localization in strains of *Streptomyces aureofaciens, J. Appl. Biochem.* 7:341 (1985).
14. J. Neužil, J. Novotná, I. Vančurová, V. Běhal, and Z. Hošťálek, A direct-injection reversed-phase liquid chromatographic micromethod for studying the kinetics of terminal reactions of tetracycline biosynthesis, *Anal. Biochem.* 181:125 (1989).
15. J. Novotná, J. Neužil, and Z. Hošťálek, Spectrophotometric identification of 8-hydroxy-5-deazaflavin:NADPH oxidoreductase activity in streptomycetes producing tetracyclines, *FEMS Microbiol. Lett.* 59:241 (1989).
16. A. P. M. Eker, J. K. C. Hessels, and R. Meerwaldt, Characterization of an 8-hydroxy-5-deazaflavin:NADPH oxidoreductase from *Streptomyces griseus, Biochim. Biophys. Acta* 990:80 (1989).
17. P. A. Miller, N. O. Sjolander, S. Nalesnyk, N. Arnold, S. Johnson, A. P. Doerschuk, and J. R. D. McCormick, Cosynthetic factor I, a factor involved in hydrogen-transfer in *Streptomyces aureofaciens, J. Am. Chem. Soc.* 82:5002 (1960).
18. J. R. D. McCormick, and G. O. Morton, Identity of cosynthetic factor 1 of *Streptomyces aureofaciens* and fragment F0 from coenzyme F420 of *Methanobacterium* species, *J. Am. Chem. Soc.* 104:4014 (1982).
19. A. P. M. Eker, A. Pol, P. van der Meyden, and G. D. Vogels, Purification and properties of 8-hydroxy-5-deazaflavin derivatives from *Streptomyces griseus, FEMS Microbiol. Lett.* 8:161 (1980).
20. L. Daniels, N. Bakhiet, and K. Harmon, Widespread distribution of a 5-deazaflavin cofactor in actinomyces and related bacteria, *System. Appl. Microbiol.* 6:12 (1985).
21. M. S. T. Kuo, D. A. Yurek, J. H. Coats, G, P. Li, Isolation and Identification of 7,8-didemethyl-8-hydroxy-5-deazariboflavin, an unusual cosynthetic factor in streptomycetes, from *Streptomyces lincolnensis, J. Antibiotics* 42:475 (1989).
22. J. H. Coats, G. P. Li, M. S. T. Kuo, D. A. Yurek, Discovery, production, and biological assay of an unusual flavenoid cofactor involved in lincomycin biosynthesis, *J. Antibiotics* 42:472 (1989).
23. L. G. M. Gorris, C. van der Drift, G. D. Vogels, Separation and quantification of cofactors from methanogenic bacteria by high-performance liquid chromatography: optimum and routine analyses, *J. Microbiol. Meth.* 8:175 (1988).
24. M. W. Peck, and D. B. Archer, Methods for the quantification of methanogenic bacteria, *Int. Industr. Biotechnol.* 9:5 (1989).

MOLECULAR GENETIC INVESTIGATION OF <u>STREPTOMYCES BAMBERGIENSIS</u> AND

<u>STREPTOMYCES FRADIAE</u> STRAINS, THE PRODUCERS OF FLAVOMYCIN AND TYLOSIN

A.V. Orekhov, S.B. Zotchev, G.Y. Florova, A.R. Foors,
L.B. Gul'ko, S.M. Rosenfeld, V.G. Zhdanov

Institute of Genetics and Selection of
Industrial Microorganisms,
Moscow, 113545, USSR

INTRODUCTION

   <u>Streptomyces bambergiensis</u> strains produce a phosphoglycolipid antibiotic complex flavomycin (moenomycin) [review Huber, 1978]. It is used in veterinary practice and is more active against Gram-positive bacteria. The antibiotics' action mechanism is based on inhibition of the cell wall peptidoglycan synthesis. The antibiotics interact with enzymes catalyzing the transglycosylation reaction, in which a disaccharide unit is transferred from a phospholipid precursor to the growning peptidoglycan chain. The flavomycin complex antibiotics have relatively high molecular weight – about 2 kD – and a complicated structure: they contain oligosaccharide, phosphoglyceric acid, a C_{25}-lipid alcohol moenocynol and a UV-chromophore. The genetics of <u>S.bambergiensis</u> and moenomycin biosynthesis is not studied well.
   On the contrary, the genetics of <u>S.fradiae</u> and the macrolide antibiotic tylosin produced by it, has been studied in detail, and a cluster of tylosin biosynthesis genes has been isolated [reveiw Baltz and Seno, 1988]. It is interesting that the producing strain contains several genes of tylosin resistance [review Baltz and Seno, 1988; Zalacain and Cundliffe, 1990] which are expressed differently in other Streptomyces strains. Earlier, from the <u>S.fradiae</u> B-45 strain we cloned three genes, <u>tlr1</u> – <u>tlr3</u>, determining increased inducible tylosin resistance in the <u>S.lividans</u> TK64 strain [Foors and Orekhov, 1989].

RESULTS AND DISCUSSION

<u>Instability of antibiotic production in S.bambergiensis strains</u>

   Many Streptomyces strains have the feature of genetic instability of several phenotypes, including antibiotic biosynthesis [review Huetter et al., 1988]. The frequency of spontaneous unstable mutations usually varies in the range of 0.1 – 1% and in several cases increases to more than 10% after mutagenesis. The genetic instability is sometimes associated with chromosome rearrangements such as deletions and amplifications which may be detected by the restriction fingerprint method.
   We have studied the stability of antibiotic biosynthesis in two <u>S. bambergiensis</u> strains and the influence of exogenous plasmids on their stability [Zotchev et al., 1989]. The S800 and S712 strains studied differ in stability (Table 1). The S800 strain is less stable in colony

Table 1. Stability of the S.bambergiensis S800 and S712 strains

Mutagenic treatment	Survival		Frequency of morfological mutations		Frequency of Ant⁻ mutations	
	S800	S712	S800	S712	S800	S712
EtBr, 5 /ug/ml	$5.2 \cdot 10^{-1}$	$8.1 \cdot 10^{-1}$	$1.2 \cdot 10^{-2}$	$3.1 \cdot 10^{-4}$	$4.9 \cdot 10^{-2}$	$<1.0 \cdot 10^{-3}$
UV-light	$2.2 \cdot 10^{-2}$	$3.0 \cdot 10^{-2}$	$1.3 \cdot 10^{-1}$	$8.0 \cdot 10^{-2}$	$2.5 \cdot 10^{-1}$	$<1.0 \cdot 10^{-3}$
Regeneration of protoplasts	1	1	$4.3 \cdot 10^{-2}$	$1.0 \cdot 10^{-4}$	$7.3 \cdot 10^{-2}$	$3.0 \cdot 10^{-3}$
Spontaneous	1	1	$5.0 \cdot 10^{-3}$	$1.0 \cdot 10^{-4}$	$1.0 \cdot 10^{-2}$	$<1.0 \cdot 10^{-3}$

morphology and antibiotic production ability both spontaneously and after being treated by mutagenic factors – EtBr, UV-light and protoplast regeneration.

One could propose that inactive mutants might contain DNA rearrangements. But we could not detect any changes in the restriction endonuclease fingerprints. The restriction fingerprints of S800 and S712 strains´ total DNAs are very similar; the differencies are detected only in several missing or extra bands.

Plasmid DNAs transform S800 and S712 strains with high efficiency [Zotchev et al., 1988; 1989]. Under nonselective conditions the S800 and S712 strain transformants by plasmid pIJ350 are very unstable in colony morfology; i.e. in colony form, size and colour, the sporulation. About 10% of transformants´ progeny are inactive Ant⁻ mutants. Under similar conditions, the transformants by the multicopy plasmid pVG101 [Birukova et al., 1985] and the low copy number plasmid pIJ943 [Lydiate et al., 1985] produce Ant⁻ mutants with frequencies typical for the wild type strains´ spotaneous production.

The pIJ350 plasmid inheritance stability in primary S800 and S712 strain transformants is lower (about 80%) as compared with stability of pVG101 and pIJ943 plasmids (more than 95%). But there is segregation of both Ant and inheritance properties in the progeny of the S800[pIJ350] transformed strain. One part of clones lost pIJ350 plasmid very efficiently (stability 0.2%), another part did not contain the plasmid at all, and the rest inherited the plasmid with normal stability of more than 95%. Several inactive Ant⁻ extransformants had suffered DNA deletion (the sum of the deleted bands on restriction fingerprints is more than 50 kb) (fig. 1).

Thus the S800 and S712 strains differ in stability of the antibiotic production Ant, and the S800 strain demonstrates spontaneous genetic instability. The S712 strain is stable in antibiotic production, but pIJ350 plasmid introduction induces genetic instability. The induction may be caused by an interaction of pIJ350 plasmid functions with a host Inc determinant linked genetically to the antibiotic production genes. In this case, the inactivation of the Inc determinant caused by long deletions is sometimes detectable by restriction fingerprinting.

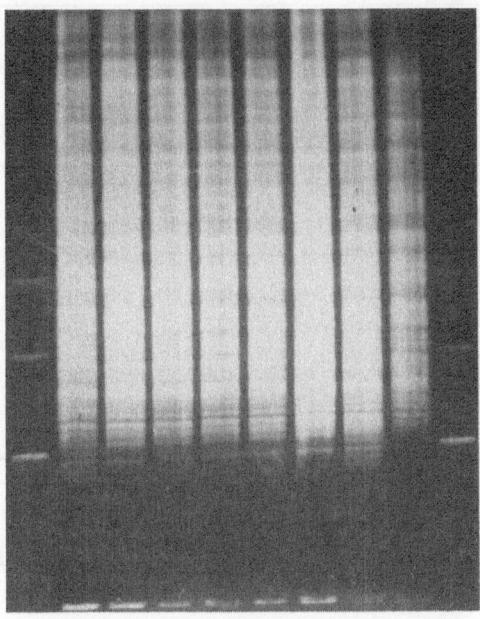

Fig. 1. Agarose gel (0.8% w/v) electrophoresis of BamHI digested total
DNAs isolated from S.bambergiensis strains. Tracks 1, 10 -
lambda DNA + HindIII; 2 - FT24 strain; 3 - S800; 4-8 - FD1-FD5
strains; 9 - S712.

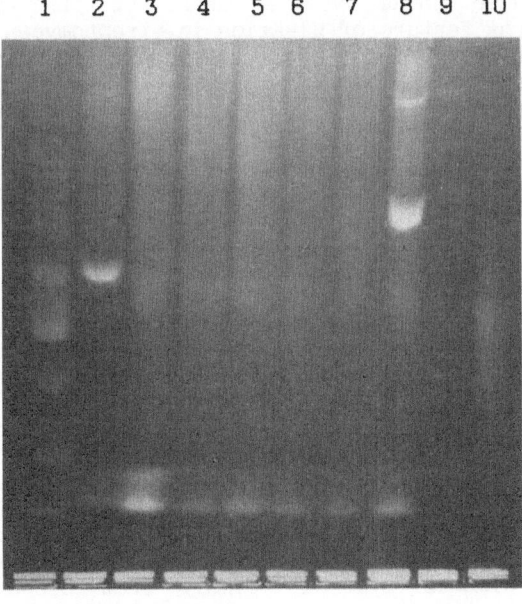

Fig. 2. Pulse-field agarose (1% w/v) electrophoresis of total DNAs of
the S.bambergiensis strains. Track 1 - lambda DNA oligomers;
2 - FT24 strain; 3 - S800; 4-8 - FD1-FD5 strains; 9 - S712;
10 - yeast DNA.

Table 2. Ltz phenotypes of S.bambergiensis strains

Strain	S712	FD1	FD2	FD3	FD4	FD5	S800	FT24
Lawn								
S712	–	–	–	–	–	–	–	–
FD1	–	–	–	–	–	–	–	–
FD2	+	+	–	–	–	–	+	+
FD3	+	+	–	–	–	–	+	+
FD4	+	+	–	–	–	–	+	+
FD5	+	+	–	–	–	–	+	+
S800	–	–	–	–	–	–	–	–
FT24	–	–	–	–	–	–	–	–

Detection of giant linear plasmids in S.bambergiensis strains and intraspecific mating

After S712 strain mutagenesis by diepoxyoctane we selected 4 inactive Ant⁻ mutants FD2 - FD5. All mutants contain the DNA deletions similar to those in the strains FT24 and FD1 obtained after transformation of the S800 and S712 strains by the pIJ350 plasmid, respectively (fig. 1). There is an absence of 3 characteristic bands on the BamHI-fingerprints.

Under conditions of joint growth of the S800 and S712 strains and their inactive Ant⁻ mutants we detected lethal zygosis reactions (table 2). The S800 and S712 strains and their extransformed mutants, FT24 and FD1 strains, produce pocks on the lawns of the FD2 - FD5 mutants. Phenotype Ltz⁺ is the feature of plasmids in Streptomyces strains [Hopwood et al., 1986]. But we could not detect any cccDNA in many S.bambergiensis strains tested.

Many Streptomyces strains contain giant linear plasmids which may be detected by pulse-field electrophoresis [Kinashi, 1988]. The results of pulse-field electrophoresis of the S800, S712, FT24, FD1, FD2 - FD5 strains´ total DNAs are presented on fig. 2. The S800, S712 and FT24 strains contain GLPs of different sizes. The S712 strain contains a GLP of 640 kb in size, named PSB1. The S800 strain contains 2 GLPs of 370 kb and 50 kb, named PSB2 and PSB3. FD1 and FD2 - FD5 strains contain no GLPs. The FT24 strain contains a 50 kb GLP. The preliminary results of DNA-DNA blot hybridization show that PSB3 has strong homology to PSB1 and to the 50 kb GLP of FT24 stain, and the FD1 strain with an Ltz⁺ phenotype on the lawns of the FD2 - FD5 strains and containing no GLP has homology to PSB3 in total DNA.

Thus the S.bambergiensis S800 and S712 strains contain GLPs which seem to determine an Ltz⁺ phenotype on the plasmidless strains´ lawns. Spontaneously unstable in antibiotic production, the S800 strain contains 2 GLPs - PSB2 (370 kb) and PSB3 (50 kb). The more stable S712 strain contains one GLP - PSB1 (640 kb). Plasmid PSB3 has a strong homology with plasmid PSB1. We suppose that PSB1 is a product of recombination between the PSB2 and PSB3 plasmids. It is likely that diepoxyoctane has an eliminating property against S.bambergiensis GLPs, as well as plasmid pIJ350. There is a correlation between changes in GLPs and biosynthesis of flavomycin.

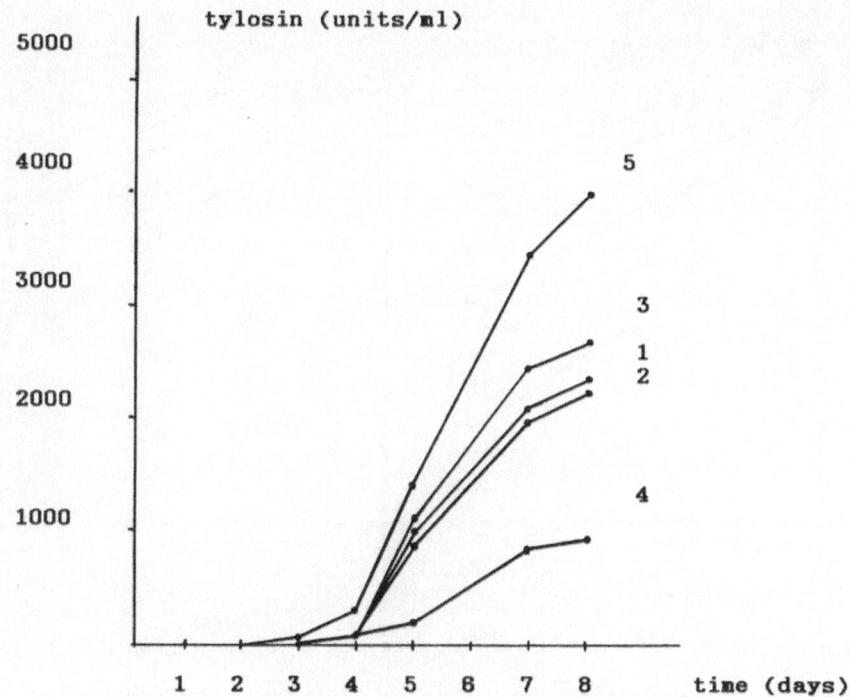

Fig. 3. Efficiency of plating of the S.fradiae 25A strains containing
the plasmids pVG251 (2), pVG252 (3), pVG253 (4), pIJ702 (5), no
plasmid (1) on different concentrations of tylosin in the absence
of exogenous induction by tylosin (A) and after induction by
1 /ug/ml of tylosin (B).

Antibiotic biosynthesis control by GLPs is confirmed by the results of
crosses between S712 strain and inactive mutants, the FP8-1 Ant⁻Str^r and
FT24-1 Ant⁻Rif^{r}Del strains [Zotchev et al., 1988b]. Strain FP8 was
obtained after strain S712 protoplast regeneration, and strain FT24 as the
S800 strain pIJ350 plasmid extransformant. The restoration of Ant⁺
phenotype in selected Str^r or Rif^r colonies occurs with high frequencies –
$6.2 \cdot 10^{-2}$ and $4.8 \cdot 10^{-2}$, respectively. The Ant⁺Rif^r recombinants are
simultaneously restored in deletion bands, and 640 kb GLP appears, instead
of 50 kb one.

Interspecific mating between S712 and S.lividans TK64 strains

The GLPs seem to be capable of very efficient transfer. This is
supported by the results of interspecific cross between the S712 and S.
lividans TK64 strains. In SBC8, Ltz⁺Str^r transconjugant, after a
segregation analysis we have identified two Ltz factors. As they were
isolated by alkaline lysis procedure for plasmid DNA, we conclude that
these factors (named pBL1 and pBL2) are not located in the chromosome.
One of them, pBL1, has strong homology to GLPs PSB1 and PSB3.

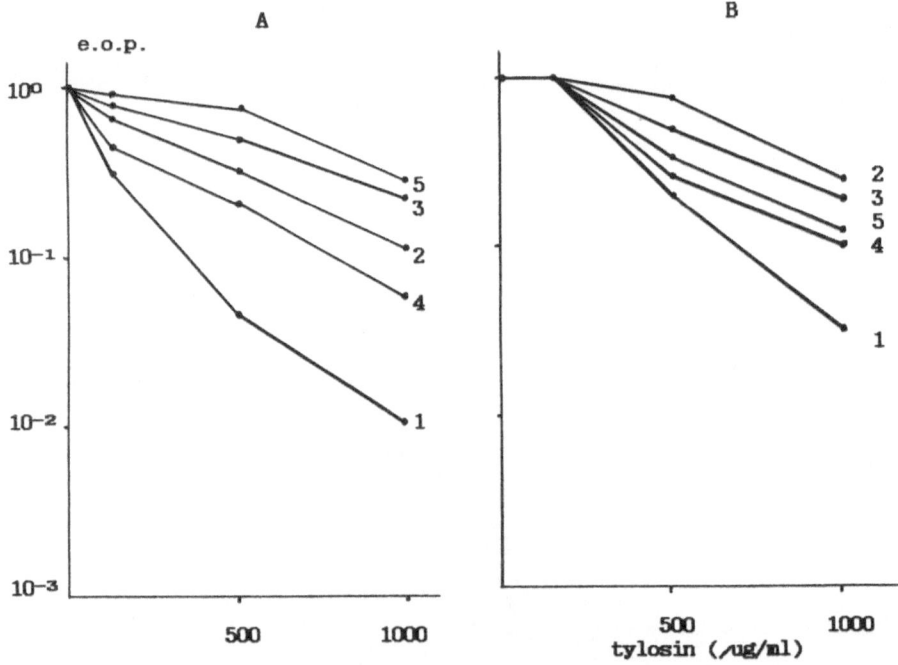

Fig. 4. Tylosin production by S.fradiae 25A strain (5) and S.fradiae
25A strains, containing plasmids pVG251 (1), pVG252 (2),
pVG253 (3), pIJ702 (4).

Resistance of S.lividans and S.fradiae B-45 strains to tylosin and related antibiotics

It is known that tylosin producer, S.fradiae, has inducible resistance to macrolides, linkosamides, streptogramin B (MLS) including its own antibiotic product, tylosin [Fujisawa and Weisblum, 1981]. And this is true for S.fradiae B-45 strain studied [Florova et al., 1988]. Using the antibiotics diffusion method we confirmed high inducible ML resistance of B-45 strain. The inducers are ML antibiotics and related to tylosin compounds, but not chloramphenicol.

The S.lividans TK64 strain has inducible ML resistance too, but its level is not very high. Chloramphenicol seems to be also involved in this resistance as it is the inducer of ML resistance. The chloramphenicol sensitive mutant of S.lividans 66 strain is at the same time more sensitive to ML antibiotics, but it still has inducibility of ML resistance. One could suppose that chloramphenicol sensitivity of S. lividans depends on chloramphenicol and ML antibiotics transport.

The tlr1 - tlr3 genes

The tlr1 - tlr3 genes on multicopy plasmids, derivatives of plasmid pIJ702, determine an increased inducible resistance to tylosin [Foors and Orekhov, 1989]. The role of tlr1 - tlr3 genes in producing strain is not clear. All tested inactive strains, including tylosin sensitive ones, contain no deletions of these genes: the tlr1 and tlr3 genes are unique, and the tlr2 gene has 7 homologs in total DNAs. The introduction of the tlr1 - tlr3 recombinant plasmids into S. fradiae tylosin-sensitive mutants increases their tylosin resistance, but introduction of the pIJ702 vector plasmid itself increases resistance.

Attempts to use tlr1 - tlr3 plasmids to increase antibiotic production of the high producing strain was unsuccessful. The multicopy plasmids used slightly increased tylosin resistance (fig. 3). But they influenced tylosin production negatively (fig. 4). There was a delay in the start of antibiotic production and 2-fold less productivity on the 8th day of fermentation. Thus, multicopy plasmids seem to be a burden for the high producing strain.

There is still an interestig fact to be investigated further. The inheritance stabilities of the tlr1 - tlr3 plasmids in S.fradiae strains are 100% as compared to 3-10% of the vector plasmid pIJ702. We suppose that the stabilizations are caused by the integration of a portion of the plasmids DNAs into chromosome.

REFERENCES

Baltz R.H. and Seno E.T., 1988, Genetics of Streptomyces fradiae and tylosin biosynthesis, Ann. Rev. Biochem.,42: 547-574.

Birukova I.V., Bolotin A.P. and Lomovskaya N.D., 1985, Designing of plasmid vector molecule for Streptomyces system, Biotechnologia (in Russian), 1, No 2, 54-61.

Florova G.Y., Chinenova T.A., Lomovskaya N.D., Zotchev S.B., Yustratova L.S. and Zhdanov V.G., 1988, Identification of genes responsible for tylosin biosynthesis and resistance to tylosin in Streptomyces fradiae B-45 strain, Biotechnologia (in Russian), 4, No 1, 37-45.

Foors A.R. and Orekhov A.V., 1989, Genes defining increased inducible resistance to tylosin in Streptomyces lividans TK64 and their molecular cloning from tylosin-producing organism Streptomyces fradiae B-45, Antibiotiki i Khimiotherapia (in Russian), 34, 332-337.

Fujisawa Y. and Weisblum B., 1981, A family of r-determinants in Streptomyces spp. that specifies inducible resistance to macrolide, lincosamide, and streptogramin type B antibiotics, 1981, J. Bacteriol., 146, 621-631.

Hopwood D.A., Kieser T., Lydiate D.J. and Bibb M.J., 1986, Streptomyces plasmids: their biology and use as cloning vectors, in: "The bacteria: a treatese on stucture and function, vol. 9. Antibiotic producing Streptomyces", Queener S.W., Day L.E., ed., pp. 159-220, Academic Press, Inc., Orlando, Fla.

Huber G., 1979, Moenomycin and related phosphorus-containing antibiotics, in:" Antibiotics. Vol. V/1", Hahn F.E., ed., pp. 135-153, Springer, Berlin.

Huetter R., Birch A., Hausler A., Voegtli M., Madon J. and Krek W., 1988, Genome fluidity in streptomycetes, in:" Biology of actinomycetes '88", pp. 111-116, Japan Scientific Societies Press, Tokyo.

Kinashi H., 1988, Giant linear plasmids in Streptomyces and antibiotic production, in: "Biology of actinomycetes '88", pp. 117-122, Japan Scientific Societies Press, Tokyo.

Lydiate D.J., Malpartida F. and Hopwood D.A., 1985, The Streptomyces plasmid SCP2*: its functional analysis and development into useful cloning vectors, Gene, 35, 223-235.

Zalacain M. and Cundliffe E., 1990, Cloning of a fourth resistance gene from the tylosin producer, in: Abstracts 19th Ann. Meetings UCLA Symp. on Mol. & Cell. Biol., J. Cell. Biochem., Suppl. 14A, 123.

Zotchev S.B., Rosenfeld S.M. and Zhdanov V.G., 1988, Transformation of Streptomyces bambergiensis S712 protoplasts with plasmid DNA, Antibiotiki i Khimiotherapia (in Russian), 33, 200-203.

Zotchev S.B., Orekhov A.V., Rosenfeld S.M. and Zhdanov V.G., 1988b, High frequency recombination in crosses between Streptomyces bambergiensis strains, <u>in</u>: Abstracts 2nd Internat. Symp. on Overproduction of Microbial Products, p. 152, Ceske Budejovice.

Zotchev S.B., Orekhov A.V., Rosenfeld S.M. and Zhdanov V.G., 1989, Influence of pIJ350 plasmid on genetic instability of antibiotic production property in <u>Streptomyces bambergiensis</u>. <u>Antibiotiki i Khimiotherapia</u> (in Russian), 34, 180-186.

ANTIBIOTIC RESISTANCE: PRESENT STATE AND PROSPECTS

Wolfgang Piepersberg

Chemische Mikrobiologie
Bergische Universität GH Wuppertal
Gauss-Str. 20, D-5600 Wuppertal 1

INTRODUCTION

The current state of antibiotic research is characterized by contradictory movements, such as decreasing interest of industrial companies in R&D efforts on classical antibiotic screening and further derivatisation of known substances versus increasing interest in new applications and new means of production (e.g. hybrid antibiotics or biotransformations), or booming academical interest in basic research on genetics and regulation of antibiotic biosynthesis and resistance versus slow-down of so-called "defense-research" of companies marketing major antibiotic groups. In this context it is interesting to note that recent progress on our understanding of the mechanisms, evolution and distributing genetic forces of antibiotic resistance is enormous, as is manifested in several reviews and books (Davies and Smith, 1978; Foster, 1983; Piepersberg et al., 1988; Cundliffe, 1989; Wiedemann et al., 1986; Levy and Novick, 1986) and that, in contrast, our knowledge on the interplay between antibiotic action, resistance development and regulation of antibiotic production is very poor. Our current view on the scenery might change considerably, whenever we know more about the origin and function of antibiotics (also "secondary metabolites" or "natural products") in nature (Zähner et al., 1982; Hütter, 1986; Williams et al., 1989; Davies, 1990; Piepersberg et al., 1988 and this volume).

RESISTANCE DETERMINANTS AND MECHANISMS

Generally there are two possibilities how antibiotic resistance can be aquired by a given type of sensitive cell: (i) by mutating resident genes, and (ii) by taking up additional, preformed resistance genes. The second path is the one, which is more relevant in spreading of clinically relevant resistance determinants, however the first possibility also occurs in clinical situations. In producing organisms a similar situation must have existed long times ago, and also seems to have been solved on both ways (cf. Cundliffe, 1989; Tab. 1), which than led to generally spreadable or individualized solutions to the resistance problem. Today the hypothesis that most re-

Table 1.　Antibiotic resistance mechanisms and distribution.

mechanism	example(s) (antibiotics)	occurence in	
		pathogens	producers
degradation	betalactamases (betalactams)	+	- (?)
trapping	binding proteins (bleomycin)	+	+
export	TetA-like membrane proteins (tetracyclines, chloramphenicol, methylenomycin, etc.)	+	+
modification (antibiotic)	phosphotransferases (aminoglycosides, viomycin, macrolides)	+	+
	acetyltransferases (aminoglycosides, chloramphenicol, peptide antibiotics)	+	+
	adenylyltransferases (aminoglycosides)	+	- (?)
modification (target site)	methyltransferases (MLS, thiostrepton)	+	+
target site alteration	nucleotide exchange (macrolides)	-	+
"bypass" (insensitive 2nd target site)	gyrase (novobiocin)	-	+
	glutamine synthetase (phosphinothricin)	-	+
	nucleotide biosynthesis (sulfonamides, trimethoprim)	+	(synthetic)

sistance determinants could have originated from the producing organisms (Walker and Walker, 1970; Benveniste and Davies, 1973) can be regarded as proven for part of the wide-spread mechanisms. However, exeptions to this rule might also occur, as for instance in the case of chloramphenicol acetylation (see below). In the following we will regard the best studied resistance mechanisms for some of the well known antibiotic groups individually (see also Tab. 1).

　1. Betalactams. Betalactamases and impermeability of outer layers of bacterial cell walls (e.g. the outer membranes of gram-negative bacteria) are the main resistance mechanisms. Betalactamases can be regarded as decendents of or parallel developments to the original target molecules of betalactams, the eubacterial peptidoglycan metabolizing peptidases ("penicillin-binding proteins"), which all seem to be related to a very ancient enzyme family, including many esterases and the

serine proteases (Brenner, 1988). In accord with this is the large variety and diversity of betalactamases, many of which seem to be encoded by chromosomal genes and seem to occure in all groups of eubacteria. Scince the betalactams (betalactones) themselves seem to be a very old group of non-ribosomal peptide molecules (Miller and Ingolia, 1989; Cohen et al., 1990) it is tempting to speculate that they could be modulators of peptidoglycan metabolism. Therefore, the excreted betalactamases could be the balance holders for concentrations of these extracellularly active molecules.

Polyketides. The occurence of this large and diverse group of natural products seems to be based on biosynthetic pathways with a common basic pattern, namely, the polymerisation of short, CoA-activated acyl residues via enzymes related to fatty acid synthases. However further processing and modification of the originally linear chains leads to very different end products. Similarly diverse is the spektrum of resistance mechanisms observed with polyketide antibiotics, including transport phenomena, target site modification or exchange, and modification of the antibiotic, as illustrated by the following examples.

1. Tetracyclines. The most wide-spread tetracycline resistance mechanism is the active export of the antibiotic by TetA-like membrane proteins (Foster, 1983). TetM-type resistance proteins act on the ribosomal level, probably constituting a changed target site on sensitive ribosomes by replacing the elongation factor Tu. Both resistance mechanisms seem to occur also in the tetracycline producers (Cundliffe, 1989).

2. Macrolides. The clinically relevant bacteria resistant against macrocyclic lactones of the erythromycin type are generally modified in their large (23S) ribosomal RNAs, always in the same position and either by di- or monomethylation of an adenine residue (Foster, 1983; Cundliffe, 1989). The same mechanism also causes cross-resistance to lincosamides and streptogramins, so-called MLS resistance. Again, in producers this same type of resistance determinants occur besides an unknown mechanism of other type, which is only known from cloned resistance genes (Cundliffe, 1989). A third group of resistance determinants, specific for a given group qf macrolides, was detected in gram-negative bacteria, which encoded macrolide sugar phosphotransferases (O'Hara et al., 1989). Also, the aquirement of insensitive ribosomes by replacement of an otherwise conserved nucleotide in 23S rRNA (A to G in pos. 2058) in spiramycin-producing S.ambofaciens constitutes another MLS resistance mechanism. In this species one out of four rRNA gene clusters contains this exchange and thereby allows to switch to resistant ribosomes under inducing production conditions (Pernodet et al., 1988).

3. Others. Other members of the polyketide family of substances such as acthinorhodin-like pigments, anthracyclines, polyethers and ansamycins didnot similarly cause dissemination of resistance determinants in the human environment since they were mostly not used as antibiotics, or not to a large extent. However, resistance mechanisms of various function seem to exist in the producers in some studied cases.

Chloramphenicol. This old and widely used antibiotic is

mostly inactivated by O-acetyltransferases (CAT) in most clinically isolated resistant bacteria (Foster, 1983). Interestingly, this resistance mechanism seems not to be used in the
producing streptomycetes, but in other streptomycetes (Murray
et al., 1989). Also, a resistance mechanism similar to the
TetA-type export mechanism was found for chloramphenicol in
corynebacteria, recently (A. Pühler, pers. comm.).

Aminoglycosides. Also these antibiotics are normally inactivated by enzymatic modification in resistant pathogens of
the nosocomial type (Davies and Smith, 1978; Foster, 1983).
However, a large variety of resistance enzymes either N-acety-
lating (AAC), or O-phosphorylating (APH), or O-adenylylating
(ANT, AAD) the aminocyclitol aminoglycosides in varying positions have been described (see also Piepersberg et al., this
volume). In producers another type of resistance, namely ribosomal (16S) RNA methylation has been described besides the
above mentioned (Cundliffe, 1989).

Others. For a larger group of self-toxic antibiotic-like
substances resistance mechanisms have been identified in both
producers and non-producers. Most of the detected mechanisms
fall into one of the functional groups alredy mentioned, e.g.
the acetylation of peptide-like antibiotics such as nourseothricin or phosphinothricin (cf. Krügel et al. and Wohlleben
et al., this volume). Interesting further examples are the
"bypass" mechanisms like the novobiocin resistance mediated by
a second, insensitive form of gyrase, expressed only when the
antibiotic is produced (Cundliffe, 1989), or the phosphinothricin resistance via an insensitive glutamic acid synthase
(W. Wohlleben et al., this volume). Also, the occurence of
antibiotic-"trapping", specifically binding proteins for the
bleomycin/phleomycin group of secondary metabolites as well in
clinical isolates as in producers besides modifying enzymes
(acetyltransferases) is worth to be added here (G. Tiraby and
J. Davies, pers. comm.).

CLINICAL SITUATION

The state of the art in research on clinically relevant resistance is characterized firstly by extensive long-term studies surveying the spread of resistance determinants both geo-
graphically and dynamically (for instance dependent on changes
in antibiotic use) and secondly by the introduction of molecular biologic tools, mainly gene-specific hybridization probes
(cf. Levy et al., 1986; Wiedemann et al. 1986). This improvement in diagnostic means has forced us into the situation of
mere data collectors. More and more phenotypic and genotypic
variants of resistance determinats come to our knowledge where
obviously the now known set of resistance determinants is only
the visible "tip of an iceberg".

For instance, a study carried out recently at Schering Corp.
(G. Miller and R. Hare, pers. comm.) including more than 5,000
clinical isolates of gram-negative bacteria resistant to 2-de-
oxystreptamine aminoglycosides from 93 hospitals in three con-
tinents (North and South America, Europe) and collected over
the five recent years 14 different DNA probes from relevant
aminoglycoside resistance genes between 20 and 60 percent of
the strains didnot hybridize to any known gene for a given resistance phenotype. This obviously means, that we have to ex-

pect an even much larger set of not closely related resistance determinants, which are running around in natural populations of microorganisms, than those already known. Even the number of basically different resistance mechanisms may not yet be fully known for the well studied groups of antibiotics. Also, it is still a matter of speculation which are the most efficient versus the more rare transfer and dissemination mechanisms in nature for the resistance genes, transposition, conjugation, mobilization, or transduction. Probably combinations thereof are realized in clinical situations.

MOLECULAR BIOLOGY AND EVOLUTION

The presently accumulating data on resistance mechanisms also allow a first hypothezising about the currently existing functions of toxic secondary metabolites and the interplay between antibiotic-target site interaction and resistance mechanism(s), or about the evolutionary sources of both antibiotics and resistance mechanisms (see also the introductory remarks). It is most obvious that the most recent impulses came from studies in antibiotic producers. But even here the question what functions these systems have for the producers or their environment is not clarified in a single case. Nor is the question yet to be answered where these complex and highly conserved biosynthetic, regulatory and resistance-determining apparatuses, only occuring scattered in minorities of strains of various species, could have originated.

Besides the rule that most resistance derterminants should originally come from the antibiotic producers several other possibilities could be envisaged: e.g. the transfer of bypass mechanisms from eucaryotic sources into bacteria, for which the unnatural situation in use of some synthetic antimetabolites, such as sulfonamides and trimethoprim-like substances, could represent an example (Foster, 1983). Similar paths could have been followed in much earlier evolutionary times by introduction from eucaryotes of genes similar to the eucaryote-like gene for glutamine synthetase II in streptomycetes, which seems to confer additional resistance to phosphinothricin in its producers (see above and Wohlleben et al., this volume).

The findings that all antibiotic phosphotransferases and all eucaryotic protein kinases constitute a distinct family of proteins, and similarly that some groups of antibiotic acetyl-tranferases are related to bacterial ribosomal protein acetylating enzymes (Piepersberg et al., 1988; this volume), and that MLS resistance conferring 23S rRNA methyltransferases are related to essential 16S rRNA methyltransferases (van Buul and van Knippenberg, 1985) might give clues towards a more general understanding of the above mentioned questions. Especially the facts that the targets of these enzymes are very similar in each case where studied (e.g. the protein modifying enzymes recognize small conserved peptides as substrates resembling some peptide or aminoglycoside antibiotics; cf. Edelman et al., 1987) and that their evolutionary age is obviously very high, could mean that resistance and biosynthesis of secondary metabolites was evolved in parallel to metabolism of proteins (or RNA). The main blooming period in which antibiotic-like natural products might have had general and essential functions could have occured in pre-cellular times (Davies, 1990).

ACKNOWLEDGEMENT

The communication of work prior to publication by many collegues is gratefully acknowledged. Especially, I have to thank J.E. Davies, J. Distler and G. Miller for many helpful discussions. The work on antibiotic resistance in the authors group was granted by the Deutsche Forschungsgemeinschaft.

REFERENCES

Benveniste, R, and Davies, J., 1973, Aminoglycoside antibiotic -inactivating enzymes in actinomycetes similar to those present in clinical isolates of antibiotic-resistant bacteria, Proc. Natl. Acad. Sci. USA, 70:2276.

Brenner, S., 1988, The molecular evolution of genes and proteins: a tale of two serines, Nature, 334:528.

Van Buul, C.P.J.J., and van Knippenberg, P.H.,1985, Nucleotide sequence of the ksgA gene of Escherichia coli: comparison of methyltransferases effecting dimethylation of adenosine in ribosomal RNA, Gene, 38:65.

Cohen, G., Shiffman, D., Mevarech, M., and Aharonowitz, Y., 1990, Microbial isopenicillin N synthase genes: structure, function, diversity and evolution, Trends Biotechnol., 8:105.

Cundliffe, E., 1989, How antibiotic-producing organisms avoid suicide, Ann. Rev. Microbiol., 43:207.

Davies, J., 1990, What are antibiotics? Ancient functions for modern activities, Mol. Microbiol., submitted.

Davies,J., and Smith,D.I., 1978, Plasmid-determined resistance to antimicrobial agents, Ann. Rev. Microbiol., 32:469.

Edelman,A.M., Blumenthal, D.K., and Krebs, E.G., 1987, Protein serine/threonine kinases, Ann. Rev. Biochem., 56:567.

Foster,T.J., 1983, Plasmid-determined resistance to antimicrobial drugs and toxic metal ions in bacteria, Microbiol. Rev., 47:361.

O'Hara,K., Kanda,T., Ohmiya,K., Ebisu, T., and Kono, M., 1989, Purification and characterization of macrolide 2'O-phosphotransferase from a strain of Escherichia coli that is highly resistant to erythromycin, Antimicrob. Agents Chemother., 33:1354.

Hütter, H., 1986, Overproduction of microbial metabolites, in: "Biotechnology", Vol. 4., H.J. Rehm, and G. Reed, ed., VCH Verlagsges., Weinheim.

Levy,S.B., and Novick,R.P.,1986, "Antibiotic Resistance Genes: Ecology, Transfer, and Expression," Cold Spring Harbor Laboratory, Cold Spring Harbor, New York.

Mansouri,K., Pissowotzki, K., Distler, J., Mayer, G., Heinzel, P., Braun, C., Ebert, A., and Piepersberg, W., 1989, Genetics of streptomycin production, in: "Genetics and Molecular Biology of Industrial Microorganisms," C.L. Hershberger, S.W. Queener, and G. Hegeman,ed., American Society for Microbiology, Washington DC.

Miller,J.R., and Ingolia,T.D., 1989, Cloning beta-lactam genes from Streptomyces spp. and fungi, in: "Genetics and Molecular Biology of Industrial Microorganisms," C.L. Hershberger, S.W. Queener, and G. Hegeman,ed., American Society for Microbiology, Washington DC.

Murray, I.A., Gil, J.A., Hopwood, D.A., and Shaw, W.V., 1989, Nucleotide sequence of the chloramphenicol acetyltransferase gene of Streptomyces acrimycini, Gene, 85:283.

Pernodet,J.-L., Boccard, F., Alegre, M.-T., Blondelet-Rouault, M.-H., and Guerineau,M.,1988, Resistance to macrolides, lincosamides and streptogramin type antibiotics due to a mutation in an rRNA operon of <u>Streptomyces ambofaci</u> - <u>ens,</u> EMBO J., 7:277.

Piepersberg, W., Distler, J., Heinzel, P., and Perez-Gonzalez, J.A., 1988, Antibiotic resistance by modification: Many resistance genes could be derived from cellular control genes in actinomycetes. - A hypothesis, Actinomycetol., 2:83.

Walker,M.S., and Walker, J.B., 1970, Streptomycin biosynthesis and metabolism. Enzymatic phosphorylation of dihydro-streptobiosamine moieties of dihydrostreptomycin-(streptidino)phosphate and dihydrostreptomycin, J.Biol. Chem., 245:6683.

Wiedemann, B., Bennett, P.M., Linton, A.H., Sköld, O., and Speller,D.C.E., 1986, "Evolution, Ecology and Epidemiology of antibiotic resistance," Academic Press, London.

Williams,D.H., Stone,M.J., Hauk, P.R., and Rahman, S.K., 1989, Why are secondary metabolites (natural products) biosynthesized?, J. Nat. Prod., 52:1189.

Zähner, H., Drautz, H., and Weber, W., 1982, Novel approaches to metabolite screening, <u>in:</u> "Bioactive microbial products: Search and discovery," J.D. Bu'Lock, L.J. Nisbet, and D.J. Winstanley, ed., Academic Press, London.

EVOLUTION OF ANTIBIOTIC RESISTANCE AND

PRODUCTION GENES IN STREPTOMYCETES

Wolfgang Piepersberg, Peter Heinzel, Kambiz
Mansouri, Ulrike Mönnighoff, and Klaus Pissowotzki

Chemische Mikrobiologie
Bergische Universität GH Wuppertal
Gauss-Str. 20, D-5600 Wuppertal 1

INTRODUCTION

Increasing amounts of DNA and protein sequence data became
available recently from genetic studies on antibiotic production and resistance in both producing and resistant bacteria.
This sequence information mirrors the current state of a
long-term evolution which obviously very early have lead to
complete pathways, which in later stages have diversified or
degenerated, or became individualized especially in the actinomycete group of microorganisms. Examples are the pathways
for betalactams polyketides, and aminoglycosides (Hershberger
et al., 1989; Cundliffe, 1989; Martin and Liras, 1989). Also,
convergently evolved genetic traits have to be postulated. The
resistance genes coding for antibiotic or target site modifying enzymes (phospho-, acetyl-, adenylyl-, and methyltransferases) seem to have a central position in the overal development which created the secondary metabolic pathways for the
respective - mostly ribosomal targeted - antibiotics and the
concomitant gathering of genes to larger clusters (Piepersberg
et al., 1988). Also, they could be derived from other control
genes such as for regulatory protein kinases or for ribosomal
processing.

The clusters of genes for antibiotic production are frequently in instable genomic segments, are compactly organized
similar to phage or plasmid genomes and are strictly regulated
via complex and hierarchicaly organized regulatory networks
responding to environmental stimuli (cf. articles of A.L. Demain and D.A. Hopwood in Hershberger et al., 1989; Martin and
Liras, 1989). However, we do not know yet wether secondary metabolism is a late invention of nature, meaning that it occured in individual situations and for individual solutions after
protein-dominated metabolism and cellular live had been established. Alternatively, so-called secondary metabolites could
be of similar age as the essential "primary" metabolites, but
could have lost their crucial functions because these where
replaced later by other, mostly macromolecular structures (a
hypothesis of J.E. Davies).

Genetics and Product Formation in Streptomyces
Edited by S. Baumberg *et al.*, Plenum Press, New York, 1991

Scheme 1. Conserved differences in APH(3") (streptomycin) and
           APH(3') (neomycin-kanamycin) resistance enzymes.

A. Sequences around the SstI site conserved in S.griseus and
   S.fradiae genes.

```
(1)    A  R  V  E  R  E  L  P  V  R  L  D  Q  E  R  T  D   177
(2)    A  R  L  T  G  E  L  A  R  R  R  D  Q  E  A  A  D   184
       GCCCGGCTCACCGGGGAGCTCGCCCGTCGGCGCGATCAGGAGGCCGCCGAC
                        ------
                         SstI
                        ------
       GCG-----------GAGCTCGACCGCACCCGT---CCCGAGAAGGAGGAC
(3)    A  -  -  -  -  E  L  D  R  T  R  -  P  E  K  E  D   178
(4)    A  -  -  -  -  E  L  D  L  T  R  -  P  E  K  E  D   172
(5)    G  -  -  -  -  L  L  L  E  S  K  -  P  V  T  E  D   176
(6)    N  -  -  -  -  E  L  T  E  T  R  -  V  E  -  E  R   176
(7)    D  -  -  -  -  D  F  L  K  T  E  -  P  E  -  E  E   175
(8)    Q  -  -  -  -  W  L  C  E  N  Q  -  P  Q  -  E  E   171
(9)    A  -  -  -  -  R  L  K  A  R  M  -  P  D  G  E  D   180
(10)   K  -  -  -  -  E  M  H  K  L  L  -  P  F  S  P  D   191
(11)   E  -  -  -  -  A  M  H  R  L  L  -  P  L  A  P  D   191
```

B. Amino-terminal segment.

```
(1)    L  D  P  L  T  W  G     267    APH(3")  RSF1010
(2)    L  D  P  L  T  W  G     272    APH(3")  S.griseus
(3)    L  D  E  F  F           268    APH(3')  S.fradiae
(4)    L  D  E  F  F           263    APH(3')  S.ribosidificus
(5)    L  D  E  F  F           262    APH(3')  B.circulans
(6)    L  D  E  L  N           259    APH(3')  A.baumannii
(7)    L  D  E  L  F           263    APH(3')  S.aureus
(8)    L  D  E  M  F           250    APH(3')  C.jejuni
(9)    L  D  E  F  F           264    APH(3')  Tn5
(10)   L  D  E  F  F           271    APH(3')  Tn903
(11)   L  D  E  L  F           271    APH(3')  RP4
```

Data are compiled from Piepersberg et al. (1988; op. cit.) for
sequences (2), (3), (5), (7), (9), (10), and (11); other data
are from: (1) Scholz et al. (1989), (4) Hoshiko et al. (1988),
(6) Martin et al. (1988), (8) Tenover et al. (1989). Note that
the main differences in the two subfamilies of enzymes consist of additional amino acid residues in the APH(3") proteins.

RESULTS AND DISCUSSION

Antibiotic phosphotransferases

    In the course of the investigation of two different strep-
tomycin phosphotransferases from S.griseus we have shown earlier that antibiotic and protein phosphotransferases are related to each other and unrelated to other phosphotransferases
of intermediary metabolism (Distler et al., 1987; Heinzel et
al., 1988; Piepersberg et al., 1988). They seem to form an ancient protein family. The similarities were noticed mainly in
three sequence motifs called G/K (N-terminal), I (central protein chain), and II/III (C-terminal half). The G/K motif is
assumed to be involved in the binding of ATP. The most highly
conserved motif II/III was postulated to comprise the main
part of the catalytic and substrate binding sites. It was very

Table 1. Resistance phenotypes of aminoglycoside phosphotransferases (APH) in S.lividans 66.

strain/plasmid[a]	MIC[b]		APH-type
	neomycin	streptomycin	
TK23/-	≤ 0.1	≤ 1.0	none
TK23/pIJ61	16	≤ 1.0	APH(3')
TK23/pPHD405	≤ 0.1	100	APH(3")
TK23/pF1	4	≤ 1.0	APH(3')?

a: the replicons in pIJ61 and pF1 are the same (SLP1.2), in
   pPHD405 the replicon is derived from pIJ101
b: MIC = minimal inhibitory concentration; tests were carried
   out in MM medium of Hopwood et al. (1985).

interesting to note that only two significantly different C-
terminal sequence motifs were oberservable in the respective
sequences of the highly related subfamilies consisting of the
APH(3') (substrates: neomycin-kanamycin group aminoglycosides)
and the APH(3") (substrates: streptomycin group aminoglycosides) aminoglycoside-modifying resistance enzymes (Scheme 1).
This became especially evident when the sequence of a second
APH(3")-encoding gene was published which was from a gram-negative bacterial source (Scholz et al., 1989).

   The Sst I site conserved in both the aphA gene from S.fra-
diae and the aphE gene from S.griseus (Heinzel et al., 1988;
see Scheme 1) could therefore be used to test for this hypothesis by forming hybrid genes. We have tested one of these
gene fusions, in which the N-terminus of the aphE gene was
joined to the C-terminal segment (containing the II/III motif)
of the aphA gene, for phosphorylation of neomycin and streptomycin, respectively. The result was that only neomycin was a
substrate of the hybrid enzyme both in vivo (Tab. 1; plasmid
pF1) and in vitro (not shown). This allows the preliminary
conclusion, that the substrate specificity is really in the
C-terminal segment.

   Both, the bacterial antibiotic phosphotransferases and the
eucaryotic protein kinases recognize similar substrates (cf.
Piepersberg et al., 1988): preferably cationic substances with
free hydroxylgroups. Also the protein kinases recognize short
primary structure elements of proteins as well as the derived
short peptides of same sequence with mostly positively charged
side groups and with either serine/threonine or tyrosine as
hydroxyl donors. Therefore, the questions rize wether both enzyme classes were evolved from each other or in parallel from
a common precursor (probably an enzyme having low-molecular
weight substrates), and wether their modifying activity had
somthing to do with the primordial function of this (these)
hypothetical substrate(s). The conclusions which can be drawn
by now and from the above results, are the following: The
mentioned kinase family and their antibiotic substrates are
very old. Maybe, even the antibiotic-like substrates could be
the more ancient ones, and, therefore, precursors of (sub-)
structures of proteins.

Antibiotic acetyltransferases

A similar situation becomes evident when the antibiotic acetyltransferases are compared to each other and to other acetyltransferases for low- and high-molcuar weight substances (Piepersberg et al., 1988). The classes difineable this way comprise again members of related antibiotic- and protein-acetylating enzymes, however this time bacterial protein-modifying acetyltransferases (of N-terminal amino groups of ribosomal proteins in E.coli). We became, therefore, interested in the question wether or not antibiotic or protein acetyltransferases even more related to each other could be found in streptomycetes. For this approach we used an oligonucleotide probe (30-mer) for the detection of a conserved stretch in class I acetyltransferase genes and mimicking the hypothetical DNA sequence of a rimI -like gene in streptomycetes (Scheme 2). In the genomic DNA of several streptomycete species one or two significant signals could be detected by hybridization (not shown).

One of two genomic sequences hybridizing in the DNA of a paromomycin producing strain of S.rimosus was isolated from a DNA bank in phage lambda EMBL3 and sequenced (not shown). As a result a second puromycin acetyltransferase (PuAT2) gene was detected, as evidenced by the sequence similarity of the two genes and the enzymatic activity (U. Mönnighoff, R. Rebollo, A. Jimenez, and W. Piepersberg, unpublished observations). In the sequences, both DNA and protein, corresponding to the oligonucleotide probe used the PuAT2 gene of S.rimosus an intermediary position between the RimI and PuAT1 sequences can be observed, which for the protein is even more similar to the RimI sequence (cf. Scheme 2). This could either mean that this

Scheme 2. Identification in S.rimosus of a puromycin acetyltransferase (PuAT2)-encoding gene via an oligonucleotide probe combining characteritics of streptomycete genes and the rimI gene of E.coli

```
-----------------------------------------------------------------
StAT   112        V   A   P   G   H   R   G   K   G   I   G
                 GTC GCC CCC GGT CAC CGC GGC AAG GGC ATC GGC
PhAT   93         V   S   P   R   H   Q   R   T   G   L   G
                 GTC TCC CCC CGC CAC CAG GGC ACG GGA CTG GGC
PuAT1  130        V   S   P   D   H   Q   G   K   G   L   G
                 GTC TCG CCC GAC CAC CAG GGC AAG GGT CTG GGC
RimI   71         V   D   P   D   Y   Q   R   Q   G   L   G
                 GTC GAT CCT GAC TAT CAG CGT CAG GGA TTG GGA
-----------------------------------------------------------------
probe,       3' CAG AGG GGG CTG GTG GTC GCC TTC CCG GAC CCG 5'
mixed            C           A           G   T
-----------------------------------------------------------------
found:
PuAT2  130        V   D   P   G   R   Q   G   Q   G   L   G
                 GTC GAC CCC GGC CGC CAG GGG CAG GGA CTG GGC
-----------------------------------------------------------------
mismatches to:
probe  5          **          *   *       *
PuAT1  8          ***         *   *           * *       *
RimI   10         *   *   *  ***      * *           *       *
-----------------------------------------------------------------
```

Data from Piepersberg et al. (1988) or unpublished. St = strep-tothricin; Ph = phosphinothricin; other abbrev. see text.

is merely a convergent trait or that some of the putative ribosomal RimI-like acetyltransferases do not exist in streptomycetes and are replaced in function by the acetylated secondary metabolite, which also could bind to the ribosome. At
least in earlier evolution the latter speculative relationship
could have been the true function of the antibiotic modification of this kind, which in the case of puromycin and phosphinothricin is still at the same time a step in biosynthesis,
similar to the 6-phosphorylation in streptomycin biogenesis.
To prove some of these hypotheses we will have to test further
for the presence of essential ribosomal protein acetyltransferases in streptomycetes, for instance for the RimL analogue
acetylating the N-terminus of the highly conserved ribosomal
protein L12, which seems to occur in all eubacteria.

Also, by comparing the available sequences of acetyltrans-

Scheme 3. Conserved sequence motif in various N- or O-ace-
           tyltransferases transfering acetyl groups from
           acetyl-coenzyme A to antibiotics and other sub-
           strates.
----
                        *   * *   * *
(1)   148 PWDDPHGPDSPLARLVAMGGRVLLLGAPLEALTLLHHAEALAD-APGKRF
(2)   148 PWDDPHGPDSPLARLAGAGGRVLLLGAPLDTLTLLHHAEARAE-APGKRF
(3)   148 PWDHPHGPDTPLARLIAHSGRVLLLGAPLDTMTLLHHAEALAD-VRSKRF
(4)   135 ELGHALGEGSPVERFVRLGGKALLLGAPLNSVTALHYAEAVAD-IPNKRW
(5)   120 PLPP-HSPASPVARVHELDGQVLLLGVGHDANTTLHLAELMAK-VP---Y
(6)   134 IGRHAWIGGGAIILPGVTIGDHAVIGAG-SVVTRDVPAGSTAMGNPA-RV
(7)   134 IGNNVWIGSHVVINPGVTIGDNSVIGAG-SIVTKDIPPNVVAAGVPC-RV
(8)   172 IREGVMIGAGAKILGNIEVGRGAKIGAG-SVVLQPVPPHTTAAGVPA-RI
(9)   357 LREDVQAYVKEAIKRAEAAPAATGGGIP-GMLPWPKVD-FSKFGEIEEVE
(10)   76 PWKARNAYDWTAESTVYVSPRHQRTGLG-STLYTHLLKSLEAQGFKSVVA
(11)   113 LAPHRPKDPAWFLATVGVSPDHQGKGLG-SAVVLPGVEAAERAGVPA-FL
(12)   113 MGPHRPTEPVWFLGSVGVDPGRQGQGLG-GAVIRPGLEAAEQAGVPA-FL
(13)   54 FAITQVVLDEATLFNIAVDPDYQRQGLG-RALLEHLIDELEKRGV-ATLW
(14)   96 LPRFEQPRSEIYIYDLAVSGEHRRQGI-ATALI-NLLKH-EANALGAYVI
(15)   95 AVSYSAWNQRLTIEDIEVAPGHRGKGIG-RVLMRHAADFARERGAG-HLW
(16)   67 IGLRPMYKETWELHPLVVRPDYQNLGIG KILLKELENRAREQGIIGIAL
(17)   86 YKYPKTDEIVYGMDQFIGEPNYWSKGIG-TRYIK-LIFEFLKKERERNAN
(18)   104 WWEEETDPGVRGIDQLLADASQLGKGLG-TKLVR-ALVELLFNDPEVTKI
(19)   87 WWEDETDPGVRGIDQSLADPTQLGKGLG-TRLVR-ALVELLFSDPTVTKI
(20)   95 SNVVRGSFHACYLGYSIGQ-KWQGKGL----MFE-AL-TAAIR-YMQRTQ
(21)   84 FNRIEPLNKTAEIGYWLDE-SHQGQGI----ISQ-AL-QALIHHYAQSGE
----
* = conserved glycine or proline residues.
According to the classification of Piepersberg et al. (1988)
the acetyltransferases were grouped in classes. Class III:
(1) AAC(3)-VII Sri, (2) AAC(3)-VIII Sfr, (3) AAC(3)-IX Mch,
(4) AAC(3)-III Eco, (5) AAC(3)-IV Eco; non-classified: (6)
NodL Rle, (7) LacA Eco, (8) CysE Eco, (9) DHLipAT Eco; class
I: (10) PhAT Shy, (11) PuAT1 Sal, (12) PuAT2 Sri, (13) RimI
Eco, (14) AAC(3)-I Eco, (15) StAT Sla, (16) AAC(6')-I Cdi;
class II: (17) AAC(6') Sfa, (18) AAC(6')-IV Eco, (19) AAC(6')
-II Pae, (20) RimJ Eco, (21) RimL Eco. Data are from the following sources: (2) Perez-Gonzalez and Piepersberg, unpublished; (3) Salauze and Davies, pers. commun.; (6) Surin and
Downie (1988); (7) Hediger et al. (1985); (9) Stephens et al.
(1983); (12) Mönnighoff and Piepersberg, unpublished; (19)
Shaw et al. (1989); (21) Tanaka et al. (1989); all other data
are compiled from Piepersberg et al. (1988).

ferase proteins we found a distant relationship in most of them around a few glycine and/or proline residues highly conserved in the more related classes (Scheme 3). This motif was most distantly similar in the class II acetyltransferases and absent only from the chloramphenicol acetyltransferases which obviously form a subclass of the class II enzymes (Piepersberg et al., 1988). The presence of this motif also in the primary metabolic acetyltransferases ("non-classified" in Scheme 3), which are much more related among each other than to any of the antibiotic and protein acetyltransferases (not shown), could mean either that the proteins are all derived from the same very ancient precursor. Alternatively a common acetyl-coenzyme A utilising domain could have been used which also could have evolved separately and convergently. As for most other functionally related protein groups this will have to be reinvestigated whenever much more sequence data are available.

Streptomycin production genes

We have cloned and partially analyzed a cluster of streptomycin production genes from S.griseus (Mansouri et al. 1989). The current state in sequence analysis has revealed the presence of at least 14 genes including a resistance gene, aphD encoding APH(6), and a regulatory gene, strR encoding a positive regulator which possibly acts via antitermination of transcription (Distler et al., 1987 and 1989). The order of the known genes is strK-I-H-G-F-B1-A-R-D-E-L-M-B2-N-O where all genes but the strDEL operon and the putative strO gene (only partially sequenced) are oriented from right to left.

When the organization of these genes is compared to the gene cluster for 3'-hydroxy streptomycin production from S.glaucescens a high degree of conservation is observed (Mayer et al., 1988). This has prompted us to investigate the organization and structure of some of the genes in more detail, because several evolutionary aspects can be investigated on such gene clusters responsible for the production of very closely related substances (G. Mayer, K. Mansouri, and W. Piepersberg, unpublished). Some further results are:
1. In S.glaucescens a rearrangement relative to the S.griseus gene cluster occures in that the segment comprising the strK gene and some possible further genes downstream thereof (about 4.5 kb) are inserted in opposite orientation between the aphD and strR gens in S.glaucescens (not shown). Since this set of genes has a partially coordinated transcriptional mode in S.griseus (Distler et al., 1987) it will be interesting to see, wether and if so how the regulation of transcription is coordinated for the S.glaucescens genes.
2. The conservation in DNA sequences is restricted only to the coding regions where the order and approximate distance of genes is retained. Inside the reading frames strong conservation seems to be achieved by strong selection for the functional requirements of the mature protein products. For example, in the StrK gene product, which most likely is an extracellular streptomycin (6- or 3"-)phosphate-specific phosphatase, only the mature portion is highly conserved, whereas the signal peptides are already much more distant (Scheme 4).
3. Of special interest for evolutionary aspects are such genes which obviously had been duplicated within the cluster and then divergently evolved for giving slight differences in substrate specificities of their respective enzyme products, such

166

Scheme 4. Comparison of the N-terminal segments of the StrK
        proteins of <u>S.griseus</u> (1) and <u>S.glaucescens</u> (2).

A. Signalpeptides (/ = putative processing sites).

```
(1)              MRFAYGRLPWRRGAVLGSALLVLVTAPAA/STA/        32
(2)              MTAVIPGTGRAARRSARRWGGVAAALVTALTVTSA/APA/  38
identity                  *   **    **    *   *   *
```

B. N-terminus of mature StrK proteins.

```
(1)              ----TERSGPAPARSVILLIGDGMGDAEITAARNYSVGA   67
(2)              PRIGVTGTEPARPRNVILLIGDGMGDSEITAARNYTVGA   77
identity            **  * ********** ******* ***
```

----

as the two amidinotransferases StrB1 and StrB2. Sequence comparison of the two enzymes from <u>S.griseus</u> showed 69 percent
sequence identity and very similar chain lengths of 347 and
349 amino acid residues, respectively. Preliminary sequence
comparisons with the StrB1 protein from <u>S.glaucescens</u> indicate
that this kind of data can give answeres to both types of questions: What are the distances in age and what are the functional differences among sets of related proteins, which
clearly originated from a common ancestor?
4. In our context another interesting notion is the one that
in the putative gene product StrN a sequence motif occures,
also similarly placed, which is very similar to the above mentioned II/III-motif of antibiotic and protein phosphotransferases. A distant similarity can also be interpreted in the rest
of the StrN polypeptide chain relative to the primary structures of the APH(6) and APH(3") enzymes of <u>S.griseus</u> (not
shown). However, the conserved lysine residue in the putative
ATP-binding site of all the other phosphotransferases in this
family is absent. This could mean that the StrN gene product
has either lost the phosphotransferase function, e.g. in exchange against a phosphatase activity, or that the above mentioned substrate binding domain for aminoglycosides (and other
cationic substrates) was used for another enzymatic function
by modular domain exchange.

CONCLUSIONS

    Many questions concerning the evolutionary sources and for-
ces in secondary metabolism are still not to be answered. The
main reason for this is that the bulk of the data, predominan-
tly sequence data for biosynthetic, regulatory and resistance
genes, allowing to test for the relevance of different hypotheses is still lacking. However, a first overview allows to
state the following:
1. Secondary metabolites and their productive pathways are basically very old, rather than products of a recent activity in
a "playground of evolution" (Zähner et al., 1982; Piepersberg
et al., 1988).
2. The gathering of production genes to firm clusters, similarly as observed in plasmids or in viruses, seems to have
been a mean for conservation under strong selective pressures
and for easy horizontal gene transfer of functional but individualized units (whole pathways). We do not know yet, however,
of what nature those selective forces are, which have stabili-
zed the existence of secondary metabolic traits.

3. Genes for the production of secondary metabolites have evolved continuously and allowed to produce variants of end products by the modular behaviour of single genes or groups thereof. This means that the obvious occurence of such events as the aquirement of additional production genes, the exchange of such genes horizontally, and the loss of single production genes could have been of major importance. Examples are the closely related groups of aminoglycosides and macrolides such as the streptomycin (streptomycins and bluensomycin) or neomycin (neomycins, paromomycins and lividomycins) groups or the uniform group of tylosin, carbomycin, and spiramycin, respectively.

4. The branching-off and the regulatory connections between so-called secondary metabolism and the intermediary metabolism in producing cells seems to be much stronger interlaced than was believed before.

5. Most enzyme families involved in secondary metabolism, especially evident now for the resistance proteins, were derived from few ancient protein groups which were evolved both divergently and sometimes convergently in their substrate specificity to form various catalytic functions. This is no principal difference towards evolution in primary metabolism, but the variance seems to be much larger (e.g. the variance in polyketide synthases versus the related and rather uniform fatty acid synthases).

6. The antibiotic- or target site-modifying enzymes and their counterparts with other substrates will probably be also in future one of the best sources for evolutionary studies. For instance, the target-site methyltransferases causing MLS-resistance (cf. Piepersberg, this volume) are related to essential and most likely universal rRNA methyltransferases, for which the KsgA enzyme of E.coli is an example (van Buul and van Knippenberg, 1985). Also, not in every case such modifying enzymes necessarily should originate from the producing organisms, as was evidenced by analysis of a streptomycete chloramphenicol acetyltransferase, for which no counterpart was detectable in the chloramphenicol producer (Murray et al., 1989). An interesting point to investigate in future will also be the question wether the antibiotic adenylyltransferases observed in cliniclal isolates do also occure in some producers and wether these are also related to protei-adenylylating enzymes known to regulate the activity of certain enzymes.

7. There are many evidences for micromolecular mimicry of macromolecular substructures in the structures of secondary metabolites which also could give indications on a function for many of these substances (cf. Piepersberg et al., 1988). In evolutionary terms, however, this mimicing process in effect could have happened in the opposite direction in the precellular periods. Thus secondary metabolites could be molecular fossils similar to coenzymes, effectors or other putatively low-molecular weight components of early catalytic complexes (J. Davies, pers. commun.).

ACKNOWLEDGEMENT: We have to thank many collegues for helpful discussions, especially J. Davies and C. Thompson. The main support for the work reported herein came via grants from the Deutsche Forschungsgemeinschaft.

References

Van Buul, C.P.J.J., and van Knippenberg, P.H., 1985, Nucleotide sequence of the ksgA gene of Escherichia coli : comparison of methyltransferases effecting dimethylation of adenosine in ribosomal RNA, Gene, 38:65.

Cundliffe, E., 1989, How antibiotic-producing organisms avoid suicide, Ann. Rev. Microbiol., 43:207.

Distler, J., Ebert, A., Mansouri, K., Pissowotzki, K., Stockmann, M., and Piepersberg, W., 1987, Gene cluster for streptomycin biosynthesis in Streptomyces griseus : nucleotide sequence of three genes and analysis of transcriptional activity, Nucleic Acids Res., 15:8041.

Distler, J., Mansouri, K., Pissowotzki, K., Piepersberg, W., 1989, Genetics and regulation of streptomycin production in streptomycetes. in: DECHEMA Biotechnology Conferences 3, p.307, VCH Verlagsgesellschaft, Weinheim.

Hediger, M.A., Johnson, D.F., Nierlich, D.P., and Zabin, F., 1985, DNA sequence of the lactose operon: the lacA gene and the transcription termination region, Proc. Natl. Acad. Sci. USA, 82:6414.

Heinzel, P., Werbitzky, O., Distler, J., and Piepersberg, W., 1988, A second streptomycin resistance gene from Streptomyces griseus codes for streptomycin-3"-phosphotransferase. Relationships between antibiotic and protein kinases, Arch. Microbiol., 150:184.

Hershberger, C.L., Queener, S.W., and Hegeman, G.,1989, "Genetics and Molecular Biology of Industrial Microorganisms," American Society for Microbiology, Washington DC.

Hopwood, D.A., Bibb, M.J., Chater, K.F., Kieser, T., Bruton, C.J., Kieser, H.M., Lydiate, D.J., Smith, C.P., Ward, J.M., and Schrempf, H., 1985, "Genetic Manipulation of Streptomyces. A Laboratory Manual," The John Innes Foundation, Norwich.

Hoshiko, S., Nojiri, C., Matsunaga, K., Katsumata, K., Satoh, E., and Nagaoka, K., 1988, Nucleotide sequence of the ribostamycin phosphotransferase gene and its control region in Streptomyces ribosidificus , Gene, 68:285.

Mansouri, K., Pissowotzki, K., Distler, J., Mayer,G., Heinzel, P., Braun, C., Ebert, A., and Piepersberg, W., 1989, Genetics of streptomycin production, in: "Genetics and Molecular Biology of Industrial Microorganisms," C.L. Hershberger, S.W. Queener, and G. Hegeman, ed., American Society for Microbiology, Washington DC.

Martin, P., Julien, E., and Courvalin, P., 1988, Nucleotide sequence of Acinetobacter baumannii aphA-6 gene: evolutionary and functional implications of sequence homologies with nucleotide binding proteins, kinases and other aminoglycoside modifying enzymes, Mol. Microbiol., 2:615.

Martin, J.F., and Liras, P., 1989, Organization and expression of genes involved in the biosynthesis of antibiotics and other secondary metabolites, Ann. Rev. Microbiol., 43:173.

Mayer, G., Vögtli, M., Pissowotzki, K., Hütter, R., and Piepersberg, W., 1988, Colinearity of streptomycin production genes in two species of Streptomyces. Evidence for occurrence of a second amidinotransferase gene, Mol. Genet. (Life Sci. Adv.), 7:83.

Murray, I.A., Gil, J.A., Hopwood, D.A., and Shaw, W.V., 1989, Nucleotide sequence of the chloramphenicol acetyltransferase gene of Streptomyces acrimycini, Gene, 85:283.

Piepersberg, W., Distler, J., Heinzel, P., and Perez-Gonzalez, J.A., 1988, Antibiotic resistance by modification: Many resistance genes could be derived from cellular control genes in actinomycetes. - A hypothesis, Actinomycetol., 2:83.

Scholz, P., Haring, V., Wittmann-Liebold, B., Ashman, K., Bagdasarian, M. and Scherzinger, E., 1989, Complete nucleotide sequence and gene organization of the broad-host-range plasmid RSF1010, Gene, 75:271.

Shaw,, K.J., Cramer, C.A., Rizzo, M., Mierzwa, R., Gewain, K., Miller, G.H., and Hare, R.S., 1989,Isolation, characterization, and DNA sequence analysis of an AAC(6')-II gene from Pseudomonas aeruginosa , Antimicrob. Agents Chemother., 33:2052.

Stephens, P.E., Darlison, M.G., Lewis, H.M., and Guest, J.R., 1983, The pyruvate dehydrogenase complex of Escherichia coli K12. Nucleotide sequence encoding the dihydrolipoamide acetyltransferase component, Eur. J. Biochem., 133:481.

Surin, B.P., and Downie, J.A., 1988, Characterization of the Rhizobium leguminosarum genes nodLMN involved in efficient host-specific nodulation, Mol. Microbiol., 2:173.

Tanaka, S., Matsushita, Y., Yoshikawa, A., and Isono, K., 1989, Cloning and molecular characterization of the gene rimL which encodes an enzyme acetylating ribosomal protein L12 of Escherichia coli , Mol. Gen. Genet., 217:289.

Tenover, F.C., Gilbert, T., and O'Hara, P., 1989, Nucleotide sequence of a novel kanamycin resistance gene, aphA-7 , from Campylobacter jejuni and comparison to other kanamycin phosphotransferase genes. Plasmid 22: 52-58, 1989.

Zähner, H., Drautz, H., and Weber, W., 1982, Novel approaches to metabolite screening, in: "Bioactive Microbial Products: Search and discovery," J.D. Bu'Lock, L.J. Nisbet, and D.J. Winstanley, ed., Academic Press, London.

GENETIC ANALYSIS OF DIFFERENT RESISTANCE MECHANISMS AGAINST

THE HERBICIDAL ANTIBIOTIC PHOSPHINOTHRICYL-ALANYL-ALANINE

Wolfgang Wohlleben, Walter Arnold, Iris Behrmann, Inge Broer,
Doris Hillemann, Alfred Pühler, Eckhard Strauch

Universität Bielefeld, Fakultät Biologie, Lehrstuhl Genetik
Postfach 8640
D-4800 Bielefeld 1, F.R.G.

SUMMARY

Streptomyces viridochromogenes Tü494, the producer of phosphinothricyl-alanyl-alanine (Ptt), is sensitive to its own antibiotic. Two
phenotypically discernible Ptt-resistant *S. viridochromogenes* mutants, ES1
and ES2, were isolated and employed to clone resistance genes. Thus, two
different DNA fragments both conferring Ptt resistance could be detected.
The DNA regions including the resistance genes were sequenced and the gene
products were investigated. The first gene (*pat*) encodes a phosphinothricin
N-acetyltransferase which inactivates the antibiotically effective component
phosphinothricin. Following modification of the 5' region, the *pat* gene was
transferred into plants and phosphinothricin-resistant transgenic plants
were obtained. The second gene, a glutamine synthetase (GS) gene mediated
Ptt resistance in multi-copy state only. The gene product is heat-labile,
and the deduced amino acid sequence was shown to be highly homologous to
eucaryotic and to *Rhizobiaceae* GSII-type enzymes. Therefore the gene was
named *glnII*. Southern hybridizations with different *Streptomyces* strains
suggest that they all carry two types of GS genes, *glnA* and *glnII*.

INTRODUCTION

In 1972 Bayer et al. reported the isolation of the antibiotic
phosphinothricyl-alanyl-alanine (phosphinothricin-tripeptide, Ptt) from the
culture broth of *S. viridochromogenes* Tü494 (Fig. 1). This tripeptide is
identical to bialaphos which was identified as a secondary metabolite from
S. hygroscopicus (Kondo et al., 1973). Ptt is transported into bacterial
cells via the oligopeptide transport system (Diddens et al., 1976) and
hydrolyzed intracellularly by peptidases. The released phosphinothricin
(Pt) is a structural analogue of glutamic acid (Fig. 1) and inhibits com-
petitively bacterial glutamine synthetases (Bayer et al., 1972).

Pt which can directly be transported into plant cells inhibits the glutamine synthetase of plants too. Since Ptt is also cleaved in plant cells,
both, Pt and Ptt can be used as non-selective herbicides (Lea et al., 1984).

In order to isolate genes involved in Ptt biosynthesis, we were
interested in the identification of possible Ptt resistance genes, because

$$CH_3$$
$$O=P-OH$$
$$CH_2$$
$$CH_2$$
$$CHNH_2$$
$$CO-ala-ala$$

Phosphinothricyl-alanyl-alanine
(Phosphinothricin - tripeptide)

$$CH_3$$
$$O=P-OH$$
$$CH_2$$
$$CH_2$$
$$CHNH_2$$
$$COOH$$

Phosphinothricin

$$O=C-OH$$
$$CH_2$$
$$CH_2$$
$$CHNH_2$$
$$COOH$$

glutamic acid

Fig. 1. Chemical structure of phosphinothricyl-alanyl-alanine (Ptt), phosphinothricin (Pt) and glutamic acid.

antibiotic resistance genes are known to be physically linked to biosynthetic genes (Martin and Liras, 1989). Such resistance genes may also be used for the construction of herbicide-resistant plants or as marker genes in plant transformation experiments.

In this manuscript we describe the isolation and characterization of two genes from *S. viridochromogenes* which confer Ptt resistance.

RESULTS AND DISCUSSION

Identification of Ptt-Resistant *S. viridochromogenes* Mutants

S. viridochromogenes Tü494 wild-type bacteria were shown to be sensitive to their own antibiotic. Low Ptt concentration (5 μg/ml) prevents both, germination of spores and mycelial growth from producing cultures on minimal agar (E. Strauch et al., 1988). But with a high frequency (10^{-4}), colonies arose on minimal agar containing up to 100 μg Ptt/ml, after two to three days. Following further incubation also Ptt-sensitive bacteria can grow, probably taking advantage of metabolites excreted by Ptt-resistant bacteria (Fig. 2). Two phenotypes of Ptt-resistant colonies could be distinguished after an incubation of one week. Type A colonies showed an inhibition zone in the surrounding lawn (Fig. 2A), and type B colonies were overgrown by a lawn of sensitive bacteria (Fig. 2B).

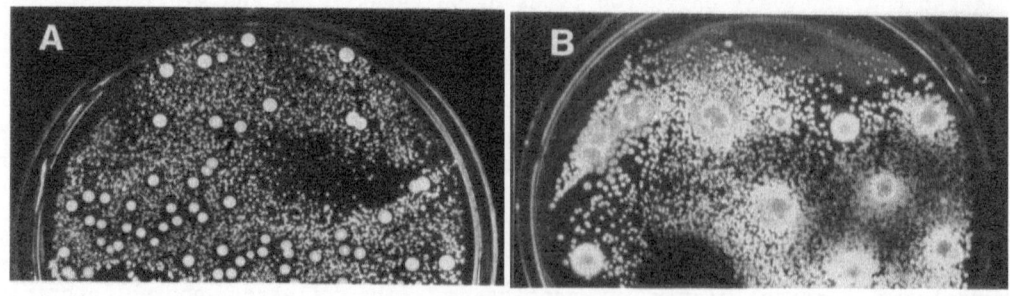

Fig. 2. Phenotype of Ptt-resistant *S. viridochromogenes* mutants ES1 and ES2. A few spores of spontaneously Ptt-resistant mutants ES1 or ES2, respectively, were plated together with sensitive wild-type spores on minimal agar containing 100 μg Ptt/ml. Following an incubation of one week at 28°C the two types of Ptt-resistant mutants could be distinguished. (A) Type A colonies inhibited growth of surrounding bacteria. (B) Type B colonies were overgrown by a lawn of sensitive cells.

Bacteria of both types were further investigated and it could be demonstrated that they were resistant against Ptt throughout the whole life cycle. We therefore assumed that mutations have occurred leading to constitutive expression of Ptt resistance (E. Strauch et al., 1988). The Ptt-resistant mutants were named *S. viridochromogenes* ES1 (type A) and ES2 (type B), respectively, and used for cloning of genes conferring Ptt resistance. Since *S. lividans* is sensitive to 5 μg Ptt/ml, it was chosen as host for shot-gun cloning experiments.

Identification of a 0.8-kb *BglII* Fragment from *S. viridochromogenes* ES1 Conferring Ptt Resistance

For a cloning experiment total DNA of *S. viridochromogenes* ES1 was digested with *Bam*HI and ligated with the *Bam*HI-cleaved *Streptomyces* high-copy vector pEB2 (Wohlleben et al., 1986). So, plasmid pES1 (Fig.3) could be identified which carried a 4.0-kb *Bam*HI fragment conferring Ptt resistance (E. Strauch et al., 1988). By subcloning, the coding region was localized on a 0.8-kb *Bgl*II fragment (Fig. 3). Following insertion of this 0.8-kb fragment into the promoter probe plasmid pIJ425 (Ward et al., 1986) it mediated Ptt resistance independent of the orientation of the fragment. For this reason, we suggested that the resistance gene is transcribed from its own promoter in *Streptomyces* (E. Strauch et al., 1988).

When the fragment was fused with the *lacZ* promoter of the *E. coli* plasmid pSVB20 (Arnold and Pühler, 1988) and transformed into *E. coli* which is sensitive to 2 μg Ptt/ml, expression was observed in one orientation only. Therefore, it could be concluded that the *Streptomyces* promoter cannot be recognized in *E. coli*, and transcription in *E. coli* occurred from the *lacZ* promoter. The gene is transcribed in the direction from the *Sma*I site to the *Nco*I site (Fig. 3). Since the resistance level of transformants in

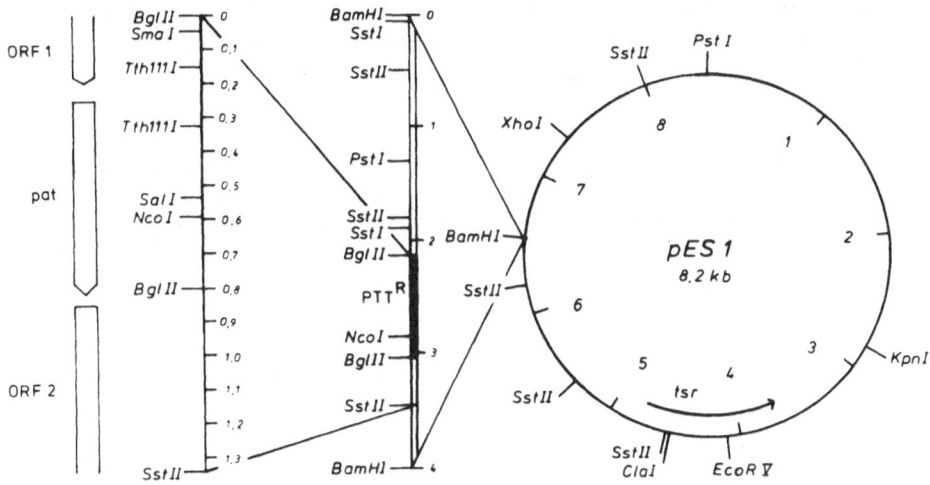

Fig. 3. Restriction map of plasmid pES1. Plasmid pES1 consists of the *Streptomyces* vector pEB2 and a 4.0-kb *Bam*HI fragment of *S. viridochromogenes* ES1 DNA. The black bar indicates the 0.8-kb *Bgl*II fragment carrying the *pat* gene. The 1.3-kb *Bgl*II-*Sst*II fragment which has been sequenced is further magnified and the position of the *pat* gene and two incomplete ORFs (ORF1, ORF2) is given. *Sst*I and *Nco*I sites were not determined for the vector part, *Tth*111I and *Sal*I sites were only mapped on the 1.3-kb fragment. The vector contains the thiostrepton resistance gene (tsr) of *S. azureus* (Thompson et al., 1980) as marker.

E. coli was determined to be 100 μg Ptt/ml, translation of the *Streptomyces* gene is obviously not disturbed in *E. coli*.

The Ptt Resistance Gene Located on the 0.8-kb *Bgl*II Fragment Encodes a Pt *N*-Acetyltransferase

Bioassays indicated that Ptt is not inactivated by crude extracts prepared from *S. lividans* harbouring fragments conferring Ptt resistance. However the antibiotically active component Pt can be modified by these extracts. The formation of a new substance, comigrating with *N*-acetyl-Pt on thin layer chromatography plates was observed. Therefore, we concluded that the Ptt resistance gene, located on the 0.8-kb *Bgl*II fragment, codes for a Pt *N*-acetyltransferase (*pat*)(E. Strauch et al., 1988).

Sequence Analysis of a 1.3-kb *Bgl*II-*Sst*II DNA Fragment Containing the *pat* Gene

After sequencing the 1.3-kb *Bgl*II-*Sst*II fragment including the 0.8-kb *Bgl*II fragment (Fig. 3) one complete ORF which represents the *pat* gene could be identified (Wohlleben et al., 1988). The sequence data showed two possible translational start codons. Therefore we constructed a 44-bp insertion between these codons and studied Ptt resistance in *E. coli* and *Streptomyces*. Since in bacteria the expression of the *pat* gene is almost totally prevented by the insertion, it is likely that translation is initiated at the first GTG codon (Wohlleben et al., 1988). The deduced amino acid sequence revealed a protein of 183 amino acids of 20.6 kDal.

The *bar* gene of *S. hygroscopicus* which also codes for a Pt *N*-acetyltransferase has been sequenced by Thompson et al. (1987). Comparison of the two deduced amino acid sequences showed an identity of 85% (Fig. 4), the sequence homology on nucleotide level was calculated to be 87%. This indicated a narrow evolutionary relationship of these two genes. Further comparisons between the Pt *N*-acetyltransferase and other acetyltransferases revealed a significant degree of homology between various enzymes of this type (Wohlleben et al., 1989).

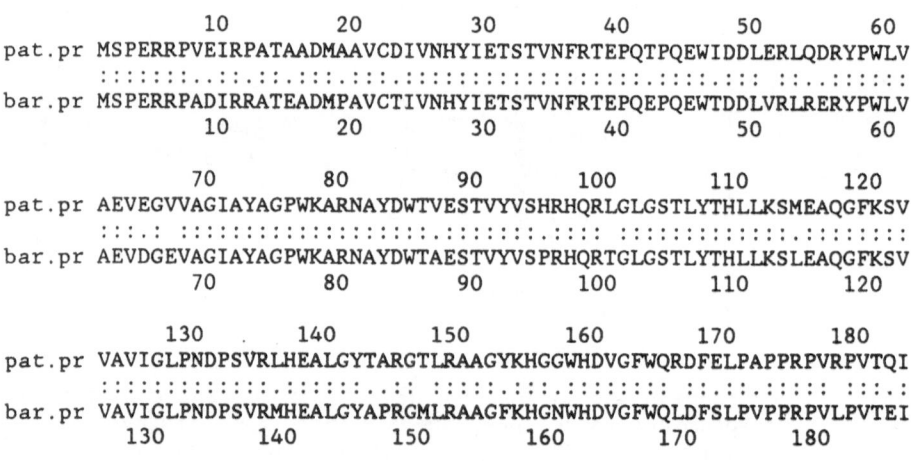

Fig. 4. Comparison of the amino acid sequences specified by the *pat* gene of *S. viridochromogenes* (pat.pr) and the *bar* gene of *S. hygroscopicus* (bar.pr). The amino acid sequences given in the one letter code were compared using the LFASTA-program (Lipman and Pearson, 1985). Identical amino acids are marked by two points, conservative changes by one point.

Expression of a Modified *pat* Gene in Plants

To test whether the *pat* gene can be expressed in plants, the *pat* gene was modified and cloned into the binary vector pROK1 (Baulcombe et al., 1986). First, the 5' part of the *S. viridochromogenes pat* gene was replaced by a synthetic DNA fragment which delivers an ATG start codon and four nucleotides for optimal translation in plants. This modified *pat* gene was then inserted between the 35S CaMV promoter and the 3' end of the *nos* gene (Fig. 5). The resulting plasmid pIB16.1 was introduced into *Nicotiana tabacum*, *Dauca carota* and *Medicago sativa* by Agrobacterium-mediated gene transfer. Transformed plants were resistant to at least 60 μg Pt/ml, whereas untransformed plants were sensitive to 1-2 μg Pt/ml. The integration of the *pat* gene into the plant genome was demonstrated by Southern analysis.

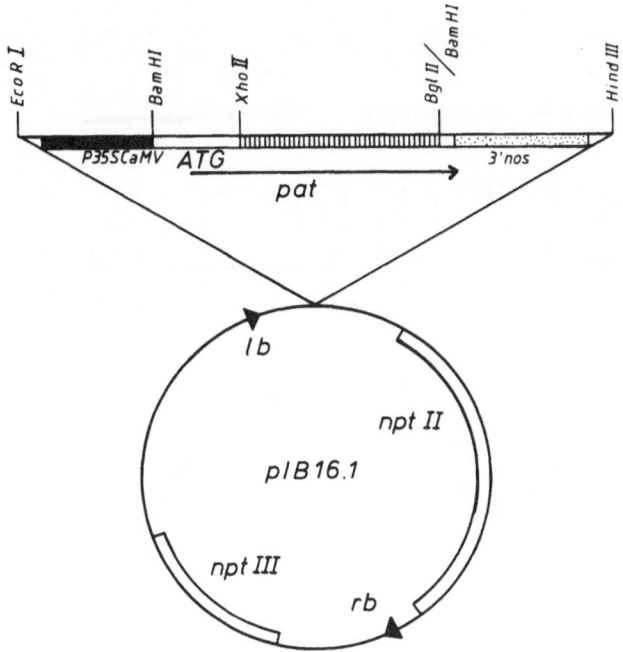

Fig. 5. Restriction map of the binary vector pIB16.1. A synthetic *Bam*HI-*Xho*II fragment (open box) was fused to the *Xho*II-*Bgl*II fragment of the *pat* gene (hatched box), thus providing the *pat* gene with plant translational signals. This fusion carrying the modified *pat* gene was inserted into the single *Bam*HI site of pROK1 between the 35S CaMV promoter (blackened box) and the 3' end of the *nos* gene (stippled box). Plasmid pIB16.1 contains also the *npt*II gene consisting of the Tn*5 aph*II coding region flanked by the *nos* promoter and terminator between the right (rb) and left (lb) borders of the T-DNA. The *npt*III gene from *Streptococcus faecalis* can be used as selection marker in bacteria.

Furthermore the radiochemical *pat*-assay which has been developed for bacteria (E. Strauch et al., 1988) could be applied to test the expression of the resistance gene in plants. Selection was possible in different stages, e.g. for leaf-discs, calli, shoots or regenerated plants. Thus, the *pat* gene proved to be a suitable marker gene for plant transformation experiments. It allows a fast, clean, and efficient selection. At least for *N. tabacum* the *pat* gene delivers more definite results than the usually applied *npt*II gene (Broer et al., 1989).

Use of the *pat* Gene for the Isolation of Biosynthetic Genes

Two different proceedings led to the identification of Ptt-biosynthetic genes in *S. viridochromogenes*. Firstly, a cosmid library of *S. viridochromogenes* has been screened using the *pat* gene as a probe. One hybridizing cosmid, pPTcos1, was shown to carry Ptt-biosynthetic genes: By subcloning cosmid-DNA into *Streptomyces* vectors and transformation of Ptt non-producing *S. viridochromogenes* mutants, complementation has been proven (Alijah et al., 1990).

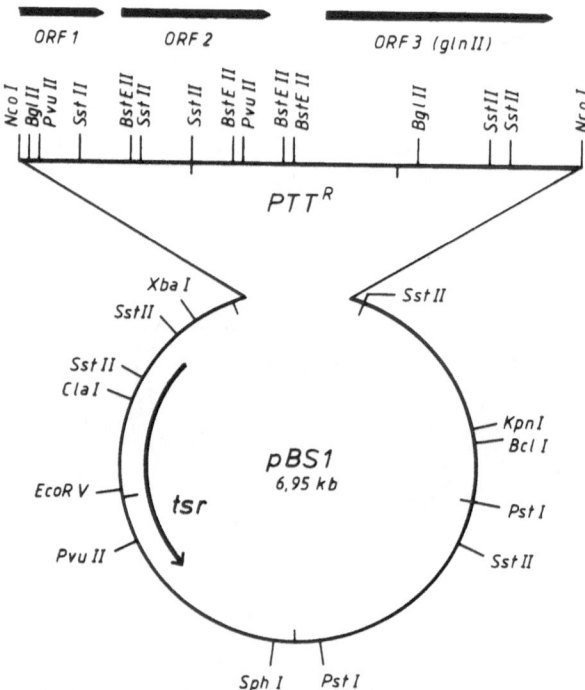

Fig. 6. Restriction map of plasmid pBS1. Plasmid pBS1 consists of the temperature-sensitive *Streptomyces* vector pGM4 and a 2.75-kb *NcoI* fragment of *S. viridochromogenes* ES2 DNA which replaced a 0.5-kb *NcoI* fragment of the vector and conferred Ptt resistance (Ptt^R). The position of three open reading frames (ORF1, ORF2 and ORF3/*glnII*) identified by sequence analysis is given. *Bst*EII sites are not mapped in the vector part.

Secondly, internal fragments of the *pat* gene have successfully been applied for mutational cloning and gene rescuing experiments using the temperature-sensitive *Streptomyces* vector pGM160 (Muth et al., 1989). Thus fragments adjacent to the *pat* gene in the *S. viridochromogenes* chromosome have been isolated. They could also be used for the complementation of non-producing mutants (Muth et al., 1989).

Identification of a 2.75-kb *Nco*I Fragment from *S. viridochromogenes* ES2 Conferring Ptt Resistance in Multi-Copy State

In order to enable the isolation of a Ptt resistance gene, different from the *pat* gene, the mutant *S. viridochromogenes* ES2 was chosen. Furthermore, the restriction enzyme *Nco*I which has a recognition site in the *pat* gene (Wohlleben et al., 1988) was used for the construction of a gene library in *Streptomyces* vector pGM4 (Wohlleben et al., 1986), which possesses a copy number of about 55 copies per chromosome (Labes et al., 1990). Plasmid pBS1 (Fig. 6) habouring a 2.75-kb insert was obtained. It conferred Ptt resistance to *S. lividans* and *S. viridochromogenes* wild-type cells (Behrmann et al., 1990). In contrast to the *pat* gene which mediated high-level Ptt resistance in single-copy state (E. Strauch et al., 1988) the *Nco*I fragment conferred resistance in multi-copy state only. In order to prove this, the temperature sensitivity (Muth et al., 1989) of the basic vector pGM4 was utilized to integrate plasmid pBS1 into the chromosome (Fig. 7). The resulting bacteria carried only two copies of the *Nco*I fragment and were Ptt-sensitive (Behrmann et al., 1990).

Sequence Analysis of the 2.75-kb *Nco*I Fragment

Crude extracts of *S. lividans* carrying pBS1 could neither inactivate Ptt nor Pt. Therefore, we determined the nucleotide sequence of the *Nco*I fragment to elucidate the resistance mechanism. Applying the "codon usage" program of Staden and McLachlan (1982), 3 open reading frames (ORFs) could be identified (Behrmann et al., 1990). Subcloning experiments revealed that ORF3 (Fig. 6.) mediates Ptt resistance. The deduced OFR3 amino acid sequence was subject of a homology search in the NBFR protein sequence data library. A surprisingly high homology (> 50%) was found to glutamine synthetase (GS) sequences from eucaryotic organisms and to GSII sequences from *Rhizobiaceae* (Behrmann et al., 1990). However, only little overall homology could be detected to GSI-type sequences from procaryotes as *E. coli* or *S. coelicolor*. But the 5 regions proposed to be active sites of the GS (Rawlings et al., 1987) are ubiquitously conserved between all GS sequences and the ORF3 polypeptide (Fig. 8). We therefore assumed that ORF3 codes for a GS.

Ptt Resistance Mediated by the 2.75-kb *Nco*I Fragment is Caused by an Increased Glutamine Synthetase Activity

Two possible resistance mechanisms mediated by the *Nco*I fragment can be considered: Either the cloned GS is resistant against Pt or overexpression of the cloned GS gene compensates the antibiotic effect of Pt.

Since the resistance depended on the multi-copy state of the 2.75-kb *Nco*I fragment (Fig. 7) it was more likely that overexpression of a GS gene is the reason for Ptt resistance than the expression of a Pt-resistant GS. Evidence for an overexpression has been obtained by measuring the GS activities by the γ-glutamyl-transferase assay (Bender et al., 1977). GS activity was significantly higher in crude extracts of *S. lividans*, containing the ORF3 region on a multi-copy plasmid, than in control extracts. Furthermore, it could be shown that this activity can be inhibited by addition of Pt. Moreover the GS activity expressed by the cloned fragment was shown to be heat-labile (Behrmann et al., 1990) as it is known for eucaryotic GS and

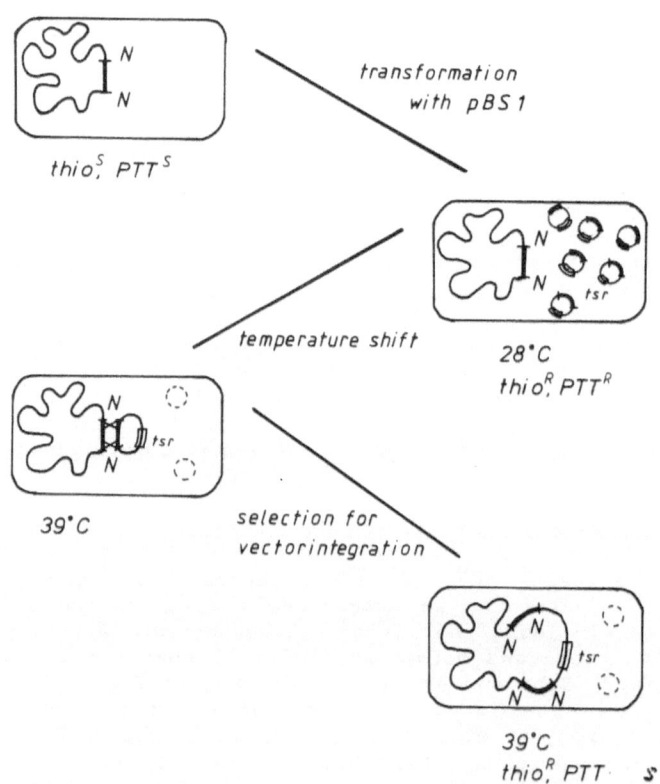

Fig. 7. Scheme of the integration of plasmid pBS1 into the
S. viridochromogenes genome. Plasmid pBS1 was
transformed into Ptt-sensitive (Ptt^S) and
thiostrepton-sensitive (thio^S) S. viridochromogenes wild-type bacteria. The selected transformants carried about 55 plasmid copies per chromosome. They were Ptt-resistant (Ptt^R) and
thiostrepton-resistant (thio^R) at 28°C. After
rising the incubation temperature to 39°C pBS1
plasmids were lost because of the temperature-
sensitivity of their pSG5 replicon (Muth et al.,
1989). Only plasmids integrated into the chromosome via homologous recombination could survive.
The 2.75-kb NcoI fragment (black bar flanked by
"N") served as region of homology. Thiostrepton-
resistant colonies growing at 39°C contained an
integrated plasmid pBS1 (proven by hybridization).
Therefore, they carry the thiostrepton resistance
gene (tsr) and two copies of the 2.75-kb NcoI
fragment. Such bacteria are Ptt-sensitive.

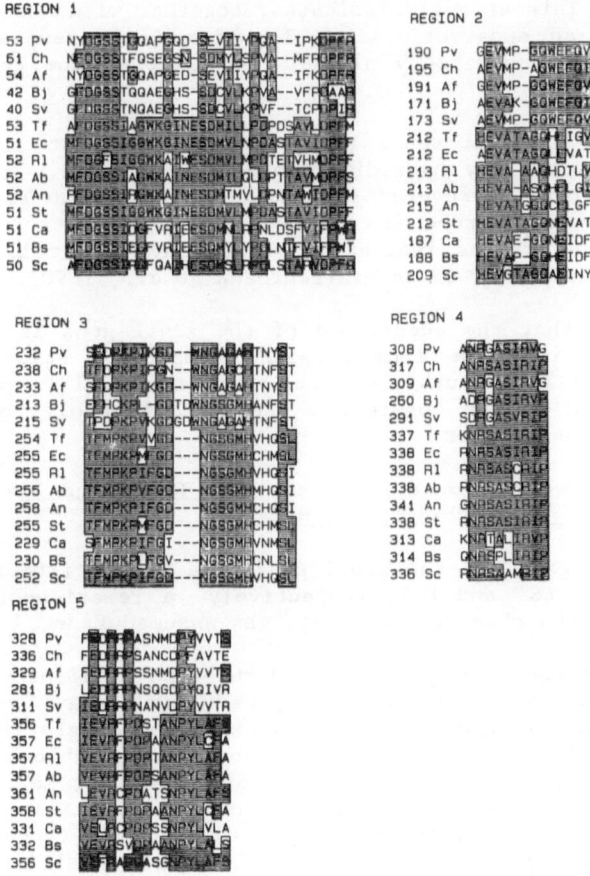

Fig. 8. Comparison of *S. viridochromogenes* GSII amino acid sequence with other GSII sequences. The five protein regions proposed to be active sites of the enzyme (Almassy et al., 1986, Rawlings et al., 1986) were compared. The *S. viridochromogenes* sequence (Sv) was aligned with the GS sequences of *Phaseolus vulgaris* (Pv) (Gebhardt et al., 1986), *Creticulus longicaudatus* (Ch) (Hayward et al., 1986), *Medicago sativa* (Af) (Tischer et al., 1986), *Bradyrhizobium japonicum* (Bj) (Carlson and Chelm, 1986), *Thiobacillus ferrooxidans* (Tf) (Rawlings et al., 1986), *Escherichia coli* (Ec) (Colombo and Villafranca, 1986), *Rhizobium leguminosarum* (Rl) (Colonna-Romano et al., 1987), *Azospirillum brasilense* (Ab) (Bozouklian and Elmerich, 1986), *Anabaena* sp. strain 7120 (An) (Tumer et al., 1983), *Salmonella typhimurium* (St) (Janson et al., 1986), *Clostridium acetobutylicum* (Ca) (Janssen et al., 1988), *Bacillus subtilis* (Bs) (M.A. Strauch et al., 1988) and *Streptomyces coelicolor* (Sc) (Wray and Fisher, 1988). The positions of the amino acids written in the one letter code are given. Identical amino acids and isoleucine/valin changes are boxed and hatched. A "-" markes a gap introduced into the sequence by the LFASTA program.

Rhizobiaceae GSII. This strongly indicates, together with the sequence data, that ORF3 encodes an eucaryotic-like GS, which is overexpressed in cells containing ORF3 on a multi-copy plasmid. Therefore, the gene encoded by ORF3 was termed *glnII*.

Interestingly the occurrence of *glnII* genes is not limited to Ptt-producing *Streptomyces*. By hybridization studies it could be demonstrated that all tested strains harboured sequences homologous to *glnII*. Furthermore, signals to a *glnA*-probe from *S. coelicolor* (Wray and Fisher, 1988) were found in these strains indicating, that many if not all *Streptomyces* species carry both types of GS genes (Behrmann et al., 1990).

Further hint, that the occurrence of the *glnII* gene is not correlated with Ptt biosynthesis, came from detailed comparison between the cosmid pPTcos1, carrying Ptt-biosynthetic genes, and the *glnII* gene. The *glnII* gene is not located on the cosmid and therefore probably not part of the Ptt-biosynthetic gene cluster.

Relationship Between the Cloned Ptt Resistance Genes and the Phenotypes of the Ptt-Resistant *S. viridochromogenes* Mutants

Since the *pat* gene and the *glnII* gene were cloned from the *S. viridochromogenes* mutants ES1 and ES2, respectively, a relationship between the isolated genes and the observed phenotype was assumed first.

S. viridochromogenes ES2 which formed the type B colonies (Fig. 2B) has been used for cloning the *glnII* gene. One may speculate that (over)expression of this gene leads to the excretion of glutamine, which can antagonize the antibiotic effect of Ptt. So it may enable sensitive wild-type colonies to grow in the neighbourhood of the mutant ES2. To test this hypothesis, heat-sensitive and heat-resistant GS activities were determined to compare the expression of the *glnII* gene in the wild-type *S. viridochromogenes* and the mutants ES1 and ES2. In all cases, similar values were obtained suggesting that the *glnII* gene is not overexpressed in *S. viridochromogenes* ES2. Thus, the observed phenotype is not caused by the *glnII* gene. Therefore, further investigations are necessary to elucidate, how growth of sensitive bacteria near ES2 mutants can occur.

S. viridochromogenes ES1 prevents growth of sensitive bacteria in its neighbourhood (Fig. 2A). This may be explained by constitutive expression or overexpression of Ptt-biosynthetic genes. An increase in antibiotic production may lead to the observed inhibition zones. In order to verify this speculation, Ptt production was measured. However, neither *S. viridochromogenes* ES1 nor ES2 were Ptt overproducer. Furthermore, the *pat* gene of the wild-type was isolated using the cloned gene as a probe. Then the 5' non-coding region was sequenced. This region is postulated to be involved in the regulation of the highly homologous *bar* gene in *S. hygroscopicus* (Thompson et al., 1987). But no differences in the sequences of the *pat* genes of the wild-type and the mutant ES1 could be detected. This suggests that the phenotype of type A colonies is not directly connected with the cloned *pat* gene itself.

To examine the role of the *pat* gene in *S. viridochromogenes* the 0.8-kb *Bgl*II fragment was replaced by the 1.05-kb *Bcl*I fragment containing the thiostrepton resistance gene (tsr) of *S. azureus*. The resulting mutants were not able to form spontaneously Ptt-resistant colonies of any phenotype. Therefore, it is obvious that the *pat* gene is at least necessary for the appearance of the different colony types. But the mechanisms generating type A and type B phenotypes have still to be discovered. In future, mutant ES1 will be used for further investigations of the *pat* gene and its regulation, whereas *S. viridochromogenes* ES2 may be a candidate for cloning a further Ptt resistance gene perhaps involved in Ptt-degradation.

ACKNOWLEDGEMENT

We thank K. Aufderheide and A. Pielsticker for preparing the illustrations and W. Jäger for typing the manuscript. The work was supported in part by a grant from Hoechst AG, Frankfurt am Main (F.R.G.), and by the BMFT (318787A, 19374A). I. Behrmann and D. Hillemann acknowledge the receipt of scholarships from the Studienstiftung des deutschen Volkes.

REFERENCES

Alijah, R., Dorendorf, J., Müller, S., Pühler, A., and Wohlleben, W., 1990, Analysis of phosphinothricyl-alanyl-alanine biosynthesis in *Streptomyces viridochromogenes* Tü494: Characterization of non-producing mutants and isolation of a biosynthetic gene, manuscript in preparation.

Almassy, R. J., Janson, C. A., Hamlin, R., Xuong, N. H., and Eisenberg, C., 1986, Novel subunit-subunit interaction in the structure of glutamine synthetase, *Nature*, 323:304-309.

Arnold, W., and A. Pühler, 1988, A family of high-copy-number plasmid vectors with single end-label sites for rapid nucleotide sequencing, *Gene*, 70:171-179.

Baulcombe, D. C., Saunders, G. R., Bevan, M. W., Maya, A. M., and Harrison, B. D., 1986, Expression of biologically active viral satellite DNA from the nuclear genome of transformed plants, *Nature*, 321:446-449.

Bayer, E., Gugel, K. H., Hagele, K., Hagenmeier, H., Jessipow, S., König, W. A., and Zähner, H., 1972, Stoffwechselprodukte von Mikroorganismen: Phosphinothricin und Phosphinothricyl-Alanyl-Alanin, *Helv. Chim. Acta*, 55:224-239.

Behrmann, I., Hillemann, D., Pühler, A., Strauch, E., and Wohlleben, W., 1990, Overexpression of a *Streptomyces viridochromogenes* gene (*glnII*) encoding an eucaryotic-like glutamine synthetase confers resistance against the antibiotic phosphinothricyl-alanyl-alanine, *J. Bacteriol.*, manuscript submitted

Bender, R. A., Janssen, K. A., Resnick, A. D., Blumenberg, M., Foor, F., and Magasanik, B., 1977, Biochemical parameters of glutamine synthetase from *Klebsiella aerogenes*, *J. Bacteriol.*, 129:1001-1009.

Bozouklian, H., and Elmerich, C., 1986, Nucleotide sequence of the *Azospirillum brasilense* Sp7 glutamine synthetase structural gene, *Biochimie*, 68:1181-1187.

Broer, I., Arnold, W., Wohlleben, W., and Pühler, A., 1989, The Phosphinothricin *N*-Acetyltransferase Gene as a Selection Marker for Plant Genetic Engineering, *in*: "Proceedings Appl. Plant Mol. Biol.," G. Galling, ed., pp. 240-246, Zentralstelle für Weiterbildung, Braunschweig.

Carlson, T.A., and Chelm, B. K., 1986, Apparent eukaryotic origins of glutamine synthetase II from the bacterium *Bradyrhizobium japonicum*, *Nature*, 322:568-570.

Colombo, G., and Villafranca, J. J., 1986, Aminoacid sequence of *Escherichia coli* glutamine synthetase deduced from the DNA nucleotide sequence, *J. Biol. Chem.*, 261:10587-10591.

Colonna-Romano, S., Riccio, A., Guida, M., Defez, R., Lamberti, A., Iaccarino, M., Arnold, W., Priefer, U., and Pühler, A., 1987, Tight linkage of *glnA* and a putative regulatory gene in *Rhizobium leguminosarum*, *Nucleic Acids Res.*, 15:1951-1964.

Diddens, H., Zähner, H., Kraas, E., Göring, W., and Jung, G., 1976, On the transport of tripeptide antibiotics in bacteria, *Eur. J. Biochem.*, 66:11-23.

Gebhardt, C., Oliver, J. E., Forde, B. G., Saarelainen, R., and Miflin, B., 1986, Primary structure and differential expression of glutamine synthetase genes in nodules, roots and leaves of *Phaseolus vulgaris*, *EMBO J.*, 5:1429-1435.

Hayward, B. E., Hussain, A., Wilson, R. H., Lyons, A., Woodcock, V., McIntosh, B., and Harris, T. J. R., 1986, The cloning and nucleotide sequence of cDNA for an amplified glutamine synthetase gene from the Chinese hamster, <u>Nucleic Acids Res.</u>, 14:999-1008.

Janson, C. A., Kayne, P. S., Almassy, R. J., Grunstein, M., and Eisenberg D., 1986, Sequence of glutamine synthetase from *Salmonella typhimurium* and implications for the protein structure, <u>Gene</u>, 46:297-300.

Janssen, P. J., Jones, W. A., Jones, D. T., and Woods, D. R., 1988, Molecular analysis and regulation of the *glnA* gene of the gram-positive anaerobe *Clostridium acetobutylicum*, <u>J. Bacteriol.</u>, 170:400-408.

Kondo, Y., Shomura, T., Ogawa, Y., Tsuruoka, T., Watanabe, H., Totukawa, K., Suzuki, T., Moriyama, C., Yoshida, J., Inouye, S., and Niida, T., 1973, Studies on a new antibiotic SF-1293, I. Isolation and physicochemical and biological characterization of SF-1293 substances, <u>Sci. Rep. Meiji Seika</u>, 13:34-41.

Labes, G., Simon, R., and Wohlleben, W., 1990, A rapid method for the analysis of plasmid content and copy number in various *Streptomyces* grown on agar plates, <u>Nucleic Acids Res.</u>, 18(8), in press.

Lea, P. J., Joy, K. W., Ramos, J. L. and Guerrero, M. G., 1984, The action of 2-amino-4-(methylphosphinyl)-butatonic acid (phosphinothricin) and its 2-oxo-derivative on the metabolism of cyanobacteria and higher plants, <u>Phytochem.</u>, 23:1-6.

Lipman, D. J., and Pearson, W. R., 1985, Rapid and sensitive protein similarity searches, <u>Science</u> 227:1435-1444.

Martin, J. F., and Liras, P., 1989, Organisation and expression of genes involved in the biosynthesis of antibiotics and other secondary metabolites, <u>Annu. Rev. Microbiol.</u>, 43:173-206.

Muth, G., Nußbaumer, B., Wohlleben, W., and Pühler, A., 1989, A vector system with temperature-sensitive replication for gene disruption and mutational cloning in Streptomycetes, <u>Mol. Gen. Genet.</u>, 219:341-348.

Rawlings, D. E., Jones, W. A., O'Neill, E. G., and Woods, D. R., 1987, Nucleotide sequence of the glutamine synthetase gene and its controlling region from the acidophilic autotroph *Thiobacillus ferrooxidans*, <u>Gene</u>, 53:211-217.

Staden, R., and McLachlan, A. D., 1982, Codon preference and its use in identifying protein coding regions in long DNA sequences, <u>Nucleic Acid Res.</u>, 10:141-156.

Strauch, E., Wohlleben, W., and Pühler, A., 1988, Cloning of a phosphinothricin *N*-acetyltransferase gene from *Streptomyces viridochromogenes* Tü494 and its expression in *Streptomyces lividans* and *Escherichia coli*, <u>Gene</u>, 63:65-74.

Strauch, M. A., Aronson, A. I., Brown, S. W., Schreier, H. J., and Sonenshein, A. L., 1988, Sequence of the *Bacillus subtilis* glutamine synthetase gene region, <u>Gene</u>, 71:257-265.

Thompson, C. J., Ward, J. M., and Hopwood, D. A., 1980, DNA cloning in *Streptomyces*: Resistance genes from antibiotic-producing species, <u>Nature</u>, 286:525-527.

Thompson, C. J., Movva, N. R., Tizard, R., Crameri, R., Davies, J., Lauwereys, M., and Botterman, J., 1987, Characterization of the herbicide-resistance gene *bar* from *Streptomyces hygroscopicus*, <u>EMBO</u>, 6:2519-2523.

Tischer, E., DasSarma, S., and Goodman, D. M., 1986, Nucleotide sequence of an alfalfa glutamine synthetase gene, <u>Mol. Gen. Genet.</u>, 203:221-229.

Tumer, N. E., Robinson, S. J., and Haselkorn, R., 1983, Different promoters for the *Anabaena* glutamine synthetase gene during growth using molecular or fixed nitrogen, <u>Nature</u>, 306:337-342.

Wohlleben, W., Muth, G., Birr, E., and Pühler, A., 1986, A vector system for cloning in *Streptomyces* and *E. coli*, <u>in</u>: "Sixth Int. Sym. on Actinomycetes Biology,"G. Szabó, S. Biró, and M. Goodfellow, eds., pp. 99-101, Akadiémiai Kiadó, Budapest.

Wohlleben, W., Arnold. W., Broer, I., Hillemann, D., Strauch, E., and Pühler, A., 1988, Nucleotide sequence of the phosphinothricin *N*-acetyltransferase gene from *Streptomyces viridochromogenes* Tü494 and its expression in *Nicotiana tabacum*, Gene, 70:25-37.

Wohlleben, W., Arnold, W., Bissonnette, L., Pelletier, A., Tanguay, A., Roy, P. H., Gamboa, G. C., Barry, G. F., Aubert, E., Davies, J., and Kagan, S. A., 1989, On the evolution of Tn*21*-like multiresistance transposons: Sequence analysis of the gene (*aacC1*) for gentamicin acetyltransferase-3-I (AAC(3)-I), another member of the Tn*21*-based expression cassette, Mol. Gen. Genet., 217:202-208.

Ward, J. M., Janssen, G. R., Kieser, T., Bibb, M. J., Buttner, M. J., and Bibb, M. J., 1986, Construction and characterization of a series of multi-copy promoter probe plasmid vectors for *Streptomyces* using the aminoglycoside phosphotransferase gene from Tn*5* as indicators, Mol. Gen. Genet., 203:468-478.

Wray, L. V. jr., and Fisher, S. H., 1988, Cloning and nucleotide sequence of the *Streptomyces coelicolor* gene encoding glutamine synthetase, Gene, 71:247-256.

INVESTIGATIONS OF EXPRESSION AND PROTEIN SEQUENCE OF THE AMINOACETYLTRANSFERASE GENE NAT1 FROM STREPTOMYCES NOURSEI

Hans Krugel, Gisela Fiedler, Frank Hanel,
Colin Smith[1], and S. Baumberg[2]

Dept. Microbiol Genetics. Central Institute
Microbiol. and Exp. Therapy, Acad. Sci. GDR, Jena
[1]Dept. Biochem., UMIST, Manchester, U.K.
[2]Dept. Genetics, University, Leeds, U.K.

INTRODUCTION

Nourseothricin (Nc) belongs to the group of streptothricin (St) antibiotics that includes grisin and racemomycin, which exhibit antibacterial, antifungal and antiviral activity (Bocker and Bergter, 1986). Although they are not used therapeutically they give remarkable effects as fodder additives in animal husbandry (Bocker and Bergter, 1986). Resistance against the streptothricins is mediated via acetylation (Keeratipibul, 1983, Tschaepe et al., 1984, Haupt et al., 1986). From NMR studies Kobayashi et al. (1987) found that in Streptomyces lavendulae this involves a monoacetylation of the beta-amino-group of the beta-lysine moiety of the streptothricin. The gene (stat) for this enzymatic activity was cloned and sequenced and its transcription start shown to be near the translational start (Horinouchi et al., 1987). In addition, a transposon-specified St-acetyltransferase (SAT) was identified in enterobacteria isolated from animals treated with Nc (Tschaepe et al., 1984). The nucleotide sequence of this gene was determined by Hein et al. (1989). From Streptomyces noursei a gene (nat1) conferring resistance to Nc via acetylation was cloned as plasmid pNAT1 and analyzed (Krügel et al., 1988). Here we present the nucleotide sequence of nat1, its transcriptional start, a putative promoter and regulatory sequence, the deduced amino acid sequence and its comparative analysis via a protein data base. Transcriptional activity is detected by the expression of a promoterless

BglII/BamHI   **Sau3A**   **Sal GI**

1 AGATCCTTGAGGCACGTTGGTCCGCCAGATCGACGAACAGTCGTCGAAATGGGCCCCATGTCGTTCGTCGACG

-35   — · · · · **sIR** · · — **-10**   **mIR**

74 AGGTGTACCGGAAGGTCCGGCCGGACACACTGACCGGTGCCCATAGCCTCGGCGGCTCCGCGGGGCGGGACAG

                                      lIR

147 GAGTGCGGTCGGCCGAAACATGCCGTCAGGCGGTCAGGGTGGCCGTAGCGTCGCGCC ATG ACC ACT CTT
                                                              M   P   S   G   G   Q   G   G   R   S   V   A   P   M- T   T   L

                    KpnI
216 GAC GAC ACG GCT TAC CGG TAC CGC ACC AGT GTC CCG GGG GAC GCC GAG GCC ATC
     D   D   T   A   Y   R   Y   R   T   S   V   P   G   D   A   E   A   I

270 GAG GCA CTG GAT GGG TCC TTC ACC ACC GAC ACC GTC TTC CGC GTC ACC GCC ACC
     E   A   L   D   G   S   F   T   T   D   T   V   F   R   V   T   A   T

324 GGG GAC GGC TTC ACC CTG CGG GAG GTG CCG GTG GAC CCG CCC CTG ACC AAG GTG
     G   D   G   F   T   L   R   E   V   P   V   D   P   P   L   T   K   V

378 TTC CCC GAC GAC GAA TCG GAC GAC GAA TCG GAC GAC GGG GAG GAC GGC GAC CCG
     F   P   D   D   E   S   D   D   E   S   D   D   G   E   D   G   D   P

432 GAC TCC CGG ACG TTC GTC GCG TAC GGG GAC GAC GGC GAC CTG GCG GGC TTC GTG
     D   S   R   T   F   V   A   Y   G   D   D   G   D   L   A   G   F   V

486 GTC ATC TCG TAC TCG GCG TGG AAC CGC CGG CTG ACC GTC GAG GAC ATC GAG GTC
     V   I   S   Y   S   A   W   N   R   R   L   T   V   E   D   I   E   V

540 GCC CCG GAG CAC CGG GGG CAC GGG GTC GGG CGC GCG TTG ATG GGG CTC GCG ACG
     A   P   E   H   R   G   H   G   V   G   R   A   L   M   G   L   A   T

594 GAG TTC GCC GGC GAG CGG GGC GCC GGG CAC CTC TGG CTG GAG GTC ACC AAC GTC
     E   F   A   G   E   R   G   A   G   H   L   W   L   E   V   T   N   V

                    Sau3A
648 AAC GCA CCG GCG ATC CAC CAC GGC TAC CGG CGG ATG GGG TTC ACC CTC TGC GGC CTG
     N   A   P   A   I   H   H   G   Y   R   R   M   G   F   T   L   C   G   L

702 GAC ACC GCC CTG TAC GAC GGC ACC GCC TCG GAC GGC GAG CGG CAG GCG CTC TAC
     D   T   A   L   Y   D   G   T   A   S   D   G   E   R   Q   A   L   Y

     SphI
756 ATG AGC ATG CCC TGC CCC TAG
     M   S   M   P   C   P   --

Fig. 1. Nucleotide sequence of natl. In the precoding region
the open arrow shows the start of transcription;
possible -10 and -35 regions are underlined, and
small (sIR), medium (mIR) and large (lIR) inverted
repeat sequences are indicated. A hypothetical ORF
in the precoding region overlaps the start codon
of natl (Krügel et al., in preparation).

APH II gene (neo) under the direction of promoters of nat1 and a further unidentified ORF.

RESULTS AND DISCUSSION

The nt sequence of the proximal part of the original BamHI-fragment from S. noursei encoding a Nc-acetyltransferase is shown in Fig. 1. The ORF for the enzyme was identified from subcloning experiments where insertions in the KpnI and SphI sites led to inactivation of the resistance phenotype, from expression in E. coli (Krügel et al., 1988) and from comparison with the nt sequence of the stat gene of S. lavendulae. This potential ORF codes for a protein of 190 amino acid residues, slightly smaller than the Mr deduced from Coomassie blue-stained PA-SDS gel electropherograms (Krügel et al., 1988). Function of the putative promoter region was confirmed as follows. A Sau3A/KpnI fragment containing it was obtained after digestion with Asp 718, end-labelled with polynucleotide kinase, recut with PstI in the polylinker near the Sau3A site, isolated and used for S1 nuclease mapping of the transcription start site. The results were different for RNA obtained from S. lividans (pNAT1) or S. lividans (pIJ487.N1), where pIJ487.N1 contains the original BamHI fragment from pNAT1 deleted for the 5' BamHI/Sau3A fragment (Krügel et al., 1988). Both RNAs protect one identical fragment, indicating the presence of an identical promoter and start site utilized in both plasmids. For pIJ487(N1) an additional, start far upstream, was observed. The potential promoter overlaps the polylinker and is probably an artificial creation during insertion of fragment N1 into the BamHI-site of the polylinker sequence in the promoter probe plasmid. Fig. 1 presents the detailed analysis of the hypothetical regulatory region. Around the - 18 position of this sequence there is a perfect inverted repeat structure, which might serve as an additional recognition signal for the polymerase or the binding site of a hypothetical regulatory protein. The first 23 nucleotides of the mRNA are again of imperfect symmetry; their function might also be a recognition one like an operator. In front of the translational start codon no ribosome binding site can be found, but there is a possible one 55 nucleotides upstream. The sequence between is able to form a loop structure with a free energy change of about G = -36 kcal/mol. When the RNA

forms this structure, the RBS and ATG are in a suitable position. This kind of posttranscriptional regulation is well
documented for erythromycin-induced drug resistance (Gryczan
et al., 1980, Horinouchi and Weisblum, 1981). A similar
structure can be found in the precoding region of stat - a potential RBS is separated from the ATG by an IR sequence. This
is in contradiction to the determination of the transcriptional start near the ATG of stat by Horinouchi et al. (1987).
Within the nontranslated sequence one may postulate an ORF of
42 bp, coding for a hypothetical oligopeptide with 13 amino
acid residues of the following sequence:

M P S G G Q G G R S V A P.

The high flexibility obtained from the glycines and the hydrogen bonds and salt bridges from the serines, glutamine and
arginine residues respectively, could enable this peptide to
participate in specific interactions. Further experimental
work will be requiered to develop a model integrating the observed speculiarities of the sequence.

To prove the hypotheses based on sequence analysis of
the precoding region, we conducted the following experiment.
Clones of S. lividans containing one of the promoter probe
plasmids pIJ486 or pIJ487 together with one of fragments N1
or S11 (see Fig. 2) in the correct orientation were grown in
5 ml M79 (Krügel et al., 1988) with thiostrepton for 48 h and
then shifted into 75 ml M79 with or without 50 µg/ml nourseothricin for another 24 h. N1 contains three transcriptional
starts, where pnat is the promoter of the nat1 gene, px the
artificial promoter created during insertion into the polylinker, and pURF the promoter of the divergent unidentified
reading frame (Krügel et al., 1988, Krügel et al., in preparation). In S11, px and pURF are deleted. After 24 h cells
were collected, sonicated 3 times for 30 seconds at 100 Watts
in ice, and centrifuged for 20 min at 15 krpm at 4 °C. Supernatants and their appropriate dilutions were used in aminoglycoside phosphotransferase II (AphII) assays (Seno et al.,
1984, Ward et al., 1986). The high AphII activity found with
N1 emphasizes the high promoter activity of this fragment. The
lack of direct correlation between the m.i.c. values (Krügel

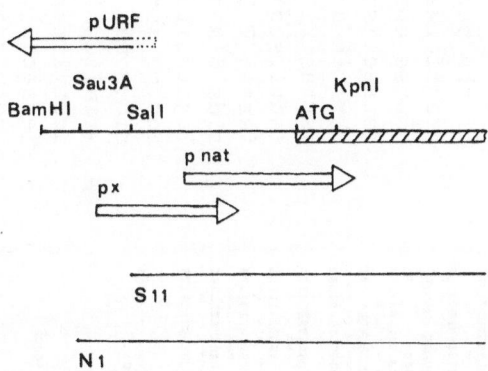

Fig. 2. Complication of the precoding region of <u>nat</u>l and the
directions and starts of transcription. Subfragment
.Sll (from <u>Sal</u> GI site) is deleted for px and pURF
activities.

Table 1. Aminoglycoside phosphotransferase assays

Strain	vector	fragment	Nc	Km^R μg/ml	APH specific activity (μmol NADH oxidized $\min^{-1}$ mg protein^{-1})
TK24	pIJ486	N1	+	500	130
	"	N1	−		130
	"	S11	+	150	4
	"	S11	−		9
	pIJ487	N1	+	1500	110
	"	N1	−		120
	"	S11	+	50	0
	"	S11	−		0

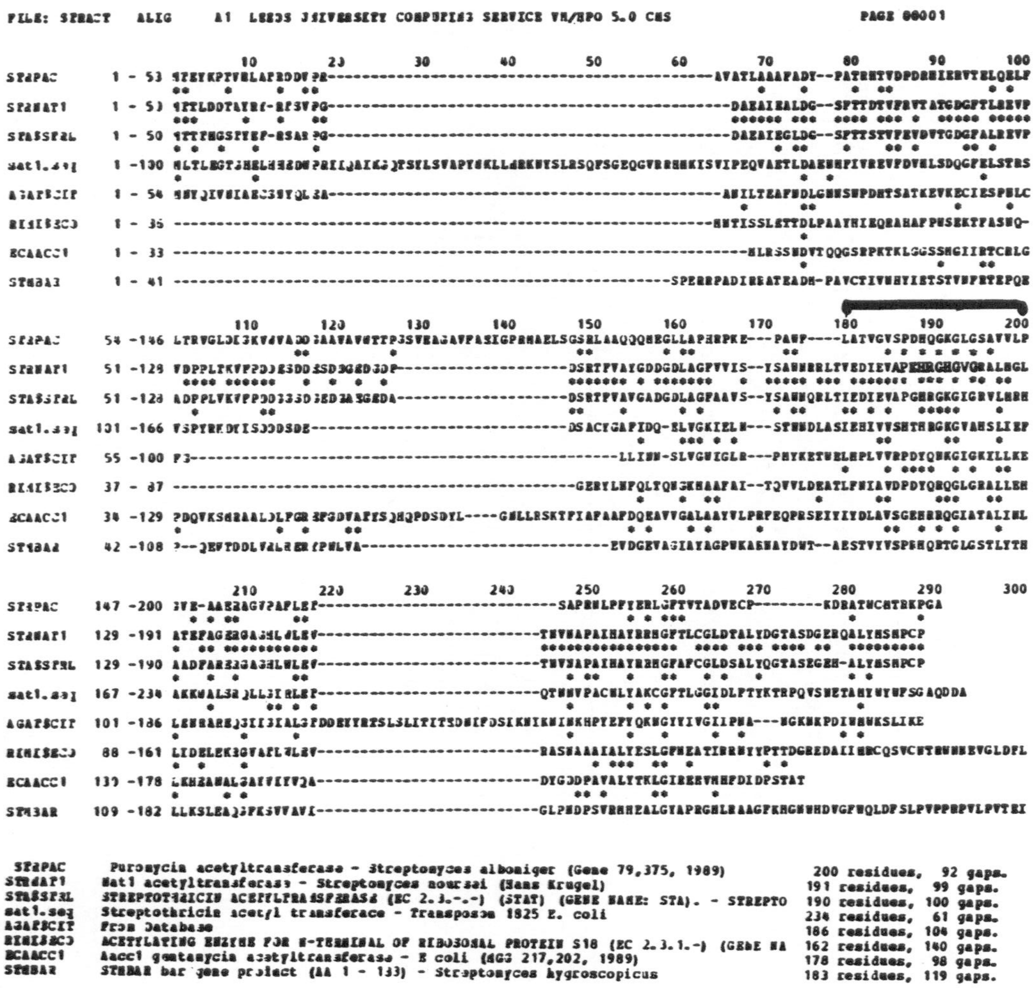

Fig. 3. Protein matrix alignment of the puromycin acetyltransferase (strpac), nourseothricin acetyltransferase (strnat), streptothricin acetyltransferase (strastrl), streptothricin (sat1.seq) acetyltransferase Tn1825, aminoglycoside 6-N-acetyltransferase (agatscitdi), acetylating enzyme for N-terminal of ribosomal protein S18 (rimiseco), gentamycin acetyltransferase Aacc1 (ecaaccl), and bialaphos acetyltransferase (stmbar). The thick horizontal line delimits the commonmotif in aminoacyltransferases.

et al., 1988) for Km and the specific activities may be due to assay conditions. In no case did we observe a correlation between precultivation with or without nourseothricin for 24 h. Therefore we reduced the induction period with nourseothricin from 24 h to 2 h (data not shown). Under these conditions we still could not detect any increase in aph phosphotransferase activity. This means, that pURF and (pnat, px) located on fragment N1 and also pnat, the sole promoter located on fragment S11, (Fig. 2) are expressed constitutively in Streptomyces lividans. Further in situ investigations in S. noursei, the original strain from which the nat1 was derived, should answer the question whether the sequences around the start of transcription are involved in regulation of expression of this gene.

By use of a protein sequence database (Akrigg et al., 1988) with pairwise and multiple alignments, eight aa sequences were compared (Fig. 3). In extension of previously published alignments (Piepersberg et al., 1989, Wohlleben et al., 1989) we added the sequences of the deduced nat1 and sat (Hein et al., 1989) products, which show a high degree of similarity to Stat and, to a less extent, the other acetyltransferases. From our alignment we postulate a motif (Fig. 3) for amino acetyltransferases; this has been used in a matrix identification search (Krügel et al., in preparation).

The proposed origin of resistance determinants from antibiotic-producing and target organisms from "housekeeping enzymes" (Piepersberg et al., 1989) seems to be supported by our data since a common motif of 14/20 amino acid-residues from amino-acetyltransferases occurs in a phytochrome from rice, two proteins from Agrobacterium and a fatty acid synthetase (acyl carrier function?).

REFERENCES

Akrigg, D., Bleasby, A. J., Dix, N. I. M., Findlay, J. B. C., North, A. C. T., Parry-Smith, D., Wootton, J. C., Blundell, T. L., Gardner, S. G., Haves, F., Islam, S., Sternberg, M. J. E., Thornton, J. M., Tickle, I. J., 1988, A protein sequence /structure database, Nature, 335: 745-746.

Bocker, H., Bergter, F., 1986, Nourseothricin : properties, biosynthesis, preparation, Arch. Exper. Vet. Med., Leipzig, 40: 646-657.

Gryczan, T. J., Grandi, G., Hahn, J., Grandi, R., and Dubnau, D., 1980, Conformational alteration of mRNA structure and the posttranscriptional regulation of erythromycin-induced drug resistance, Nucl. Acids Res., 8: 6081-6097.

Haupt, I., Thrum, H., and Noack, D., 1986, Self-resistance of the nourseothricin-producing strain Streptomyces noursei, J. Basic Microbiol., 26: 323-328.

Hein, U., Tietze, E., Weschke, W., Tschäpe, H., and Wobus, U., 1989, Nucleotide sequence of a plasmid-borne streptothricin-acetyltransferase gene (sat-1), Nucl. Acids Res., 17: 7103.

Hoffmann, H., Haertl, A, Bocker, H., Kahnel, H.-J., Hesse, G., Flemming, J., 1986, Pharmacokinetics of nourseothricin in laboratory animals, Arch. Exper. Vet. Med., Leipzig, 40: 699-709.

Hopwood, D. A., Bibb, M. J., Chater, K. F., Kieser, T., Bruton, C. J., Kieser, H. M., Lydiate, D. J., Smith, C. P., Ward, J. M., Schrempf, H., 1985, Genetic manipulation. A laboratory Manual. John Innes Foundation, Norwich.

Horinouchi, S., and Weisblum, B., 1981, The control region for erythromycin resistance : free energy changes related to induction and mutation to constitutive expression, Mol. Gen. Genet., 182: 341-348.

Horinouchi, S., Furuya, K., Nishiyama, M., Suzuki, H., and Beppu, T., 1987, Nucleotide sequence of the streptothricin acetyltransferase gene from Streptomyces lavendulae and its expression in heterologous hosts, Jour. Bacteriol., 169: 1929-1937.

Keeratipibul, S., Sugiyama, M., and Nomi, R., 1983, Mechanism of resistance to streptothricin of a producing microorganism, Biotechnol. Lett., 5: 441-446.

Kobayashi, T., Uozomi, T., and Beppu, T., 1986, Cloning and characterization of the streptothricin resistance gene which encodes streptothricin acetyltransferase from Streptomyces lavendulae, J. Antibiot., 39: 688-693.

Krügel, H., Fiedler, G., Gase, K., and Haupt, I., 1989, Streptothricin resistance, in: "Bioactive metabolites from Microorganisms", M. E. Bushell and U. Gräfe, eds., Elsevier Publ., Amsterdam

Krügel, H., Fiedler, G., Haupt, I., Sarfert, E., and Simon, H., 1988, Analysis of the nourseothricin-resistance gene (nat) of Streptomyces noursei, Gene, 62: 209-217.

Lacalle, R. A., Pulido, D., Vara, J., Zalacain, M., Jiminez, A., 1989, Molecular analysis of the pac gene encoding a puromycin N-acetyltransferase from Streptomyces alboniger, Gene, 79: 375-380.

Piepersberg, W., Distler, J., Heinzel, P., and Perez-Gonzales, J.-A., 1988, Antibiotic resistance by modification: many resistance genes could be derived from cellular control genes in Actinomycetes. - A Hypothesis, Actinomycetol., 2: 83-98.

Tschäpe, H., Tietze, E., Prager, R., Voigt, W., Wolter, E., and Seltmann, G., 1984, Plasmid-borne streptothricin resistance in Gram-negative bacteria, Plasmid, 12: 189-196.

Wohlleben, W., Arnold, W., Broer, I., Hillemann, D., Strauch, E., and Puehler, A., 1988, Nucleotide sequence of the phosphinothricin N-acetyltransferase gene from Streptomyces viridochromogenes Tü 494 and its expression in Nicotinia tabacum, Gene, 70: 25-37.

Wohlleben, W., Arnold, W., Bissonnette, L., Pelletier, A., Tanguay, A., Roy, P. H., Gamboa, G. C., Barry, G. F., Aubert, E., Davies, J., and Kagan, S. A., 1989, On the evolution of Tn-21 like multiresistance transposons: Sequence analysis of the gene (aacC1) for gentamycin acetyltransferase -3-I(AAC(3)-I), another member of the Tn21-based expression cassette, Mol. Gen. Genet., 217: 202-208.

BETA-LACTAMASE GENES FROM STREPTOMYCES SPECIES

Hiroshi OGAWARA

Department of Biochemistry, Meiji College of Pharmacy
Nozawa-1, Setagaya-ku, Tokyo 154, Japan

I. INTRODUCTION

Beta-lactamases are produced by a wide range of different prokaryotic cells with great variety in chemical, physical and enzymatic properties (1). Beta-lactamases are referred to as such on the basis of only one common property: they catalyze the hydrolysis of the beta-lactam ring of penicillins and cephalosporins to yield antibacterially inactive products, penicilloic acids and cephalosporoic acids. This causes the pathogenic bacteria to be resistant against beta-lactam compounds. In fact, beta-lactamase is the main mechanism of clinical resistance against many beta-lactam compounds, even at the present time. At the same time, however, these enzymes are also produced by nonpathogenic bacteria such as Streptomyces (2,3) and the cyanobacteria (4). In Streptomyces, beta-lactamases are produced constitutively and abundantly irrespective of their resistance to beta-lactam antibiotics (2,5). Resistance to beta-lactams in Streptomyces is due to the changes of the targets of beta-lactams, penicillin-binding proteins (3).

Furthermore, beta-lactamases are hydrolyzing enzymes and are not involved in the biosynthesis of beta-lactams in Streptomyces. This contrasts with the other enzymes implicated in the resistance such as aminoglycoside- and ribosome-modifying enzymes (6,7,8). At least some genes encoding these enzymes form a cluster with their respective biosynthetic genes and are expressed coordinately (9). The roles of beta-lactamases in Streptomyces therefore are not completely understood. However, at least a part of beta-lactamases in pathogenic bacteria may be derived from or may be closely related evolutionarily to those in Streptomyces, microorganisms producing beta-lactams (10). This is supported by the comparative study of nucleotide sequences of these enzymes: beta-lactamases in Streptomyces form members of a superfamily of evolutionarily related active-site serine beta-lactam-interacting proteins (11).

One group of beta-lactamase in Streptomyces has a peculiar property of binding blue dextran (5). This type of beta-lactamase is very interesting evolutionarily because such proteins, in general, have a dinucleotide fold and bind ATP, NAD^+ or $NADP^+$ (12). Beta-lactamase from Streptomyces cellulosae is one such enzyme which binds blue dextran and $NADP^+$ and can be purified by blue Sepharose (13). Similar beta-lactamases such as OXA-2-type beta-lactamases are also found clinically (14). The first part of this paper is concerned with the gene of this beta-lactamase. The gene hybridized not only to DNA of S. cellulosae, a

DNA donor, but also to DNAs of several Streptomyces species, irrespective of their production of beta-lactamase, suggesting that this gene could have some homology to other genes than beta-latamases.

Previously, we cloned the genes encoding beta-lactamase of Streptomyces cacaoi (15) and the nucleotide sequence was determined (16). This enzyme belongs to the class A beta-lactamase but can hydrolyze methicillin as well (17). The second part of this paper describes the analysis of the promoter region of the gene and drastic changes of beta-lactamase activities of various clones depending on the lengths of DNA sequences upstream of the structural gene, suggesting the function of a positive regulator in this region.

II. GENE FROM STREPTOMYCES CELLULOSAE KCCS127

Cloning of the gene

Using Streptomyces lividans 1326 (18) and pIJ385 (19) as a host-vector system, a gene from S. cellulosae encoding beta-lactamase, a blue dextran binding protein, was cloned. Chromosomal DNA from S. cellulosae was completely digested with SacI or PstI and the unfractionated fragments were used in a shotgun cloning experiment. In the case of SacI, three beta-lactamase positive clones were obtained among 1725 recombinants and with PstI, one positive clone was isolated among 1518 recombinants. However, only one clone from the former clones showed beta-lactamase activity in a liquid culture. A plasmid named pMCP180 which contained a 2.3kb DNA insert from S. cellulosae was isolated from the positive clone. When this plasmid was transformed into S. lividans 1326, the liquid culture of the transformant again revealed beta-lactamase activity. Thus, once one obtains the gene, it is rather stable. The restriction enzyme map of the 2.3kb insert in pMCP180 is shown in Fig. 1. It is clear that the right part of the insert in the figure had no SacI site, although the SacI enzyme was used for the cloning. As the size of the corresponding SacI fragment in the chromosome was 6.0kb, as will be described later, it is suggested that the SacI fragment in the chromosome gave rise to the deletion of the 3.7kb fragment during the course of cloning, and because of this unstable character, only one beta-lactamase positive clone might be isolated eventually among four positive clones at the start. The size of the colonies possessing the gene was, in general, smaller than that of the other colonies.

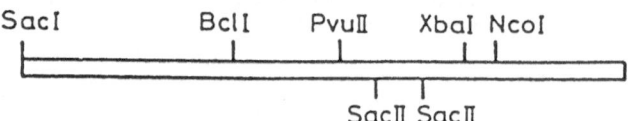

Fig. 1. Restriction enzyme map of 2.3kb
insert in pMCP180

The time course of beta-lactamase production in S. lividans 1326 containing pMCP180 (clone CMA180) was similar to that in S. cellulosae. Clone CMA180 produced beta-lactamase enzyme at 14.3 units/ml in E medium (2), which was about 6 times higher production than that of S. cellulosae KCCS127. On the other hand, no beta-lactamase activity was detected in S. lividans 1326 and S. lividans 1326 transformed with pIJ385.

In order to confirm that the 2.3kb DNA fragment was derived from S. cellulosae DNA, a Southern hybridization experiment (20) was performed by using 1.9kb SacI-XbaI fragment (Fig.1) as a probe. The probe hybridized at 1.9kb to pMCP180 or chromosomal DNA of S. cellulosae digested with SacI and XbaI. It also hybridized at 6.0kb to chromosomal DNA of S. cellulosae digested with SacI, although it hybridized at 7.6kb to pMCP180 digested with SacI. These results, together with the fact that the cloned DNA segment lacks one SacI end indicate that the 2.3kb DNA fragment in pMCP180 was surely derived from S. cellulosae and that, in addition, during the course of cloning, the 3.7kb fragment was lost to yield pMCP180. At the same time, a part of pIJ385 was also sacrificed.

When the beta-lactamase enzyme from clone CMA180 was precipitated with ammonium sulfate and applied to a column of Sephadex G75 equilibrated with 0.1M phosphate buffer, pH7.0, the peak of the enzyme activity came in fractions similar to that from S. cellulosae. On the other hand, when the same preparation was applied to a column of Sephadex G75 together with blue dextran, the peak of the enzyme activity appeared in the void volume position. The beta-lactamase from clone CMA180 therefore has a property of binding to blue dextran and being excluded from the gel matrix, which is a particular property of the beta-lactamase from S. cellulosae (3). The isoelectric point in the alkaline region of the beta-lactamase from clone CMA180 was also similar to that of S. cellulosae.

Properties of the gene

As described above, when 1.9kb SacI-XbaI DNA fragment (Fig. 1) was used as a probe, it hybridized to DNA fraction from S. cellulosae KCCS127 but not from S. lividans 1326. However, it also hybridized to DNAs from S. fradiae Y59, S. lavendulae KCCS055, S. lavendulae KCCS263, S. coelicolor KCCS006 and S. diastaticus KCCS128 but not to S. phaeochromogenus KCCS070, S. lavendulae KCCS057, S. lavendulae KCCS985 and S. cacaoi KCCS352. S. phaeochromogenus KCCS070, S. fradiae Y59, S. cacaoi KCCS352, S. coelicolor KCCS006 and S. diastaticus KCCS128 produce beta-lactamase constitutively (5), while S. lavendulae KCCS055, S. lavendulae KCCS057, S. lavendulae KCCS263 and S. lavendulae KCCS985 do not. In addition, beta-lactamases from S. phaeochromogenus KCCS070 and S. fradiae Y59 have a high affinity to blue dextran like that from S. cellulosae KCCS127, while others do not. In contrast, neither beta-lactamase gene from S. lavendulae ATCC 8664 (21) nor that from S. cacaoi

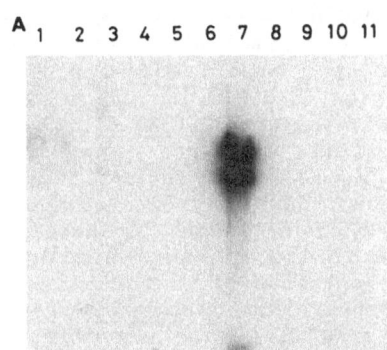 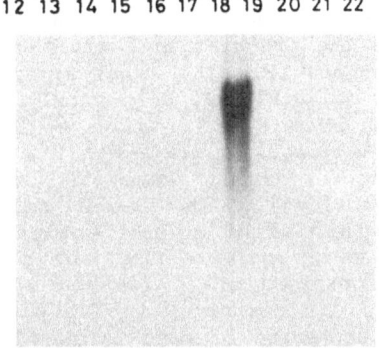

Fig. 2. Southern blot hybridization. A, Probe used was the beta-lactamase gene from S. lavendulae ATCC8664. B, Probe used was the beta-lactamase gene from S. cacaoi KCCs352. 1,12: S. diastaticus KCCS128; 2,13: S. coelicolor KCCS006; 3,18: S. cacaoi KCCS352; 4,14: S. lavendulae KCCS985; 5,15: S. lavendulae KCCS263; 6,16: S. lavendulae KCCS057; 7,17: S. lavendulae KCCS055; 8,19: S. cellulosae KCCS127; 9,20: S. fradiae Y59; 10,21: S. phaeochromogenus KCCS070; and 11,22: S. lividans 1326.

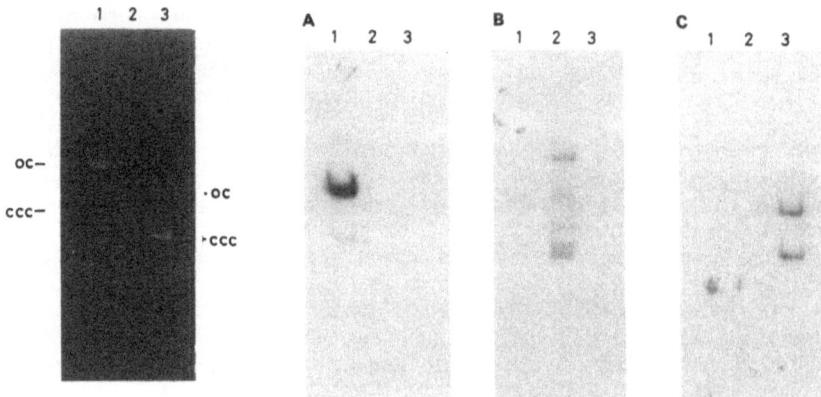

Fig. 3. Southern blot hybridization. A, S. lavendulae; B, S. cacaoi; and C, S. cellulosae. 1, pMCP182 containing beta-lactamase gene from S. lavendulae ATCC8664; 2, pMCP126 containing beta-lactamase gene from S. cacaoi KCCS352; and 3, pMCP180 containing beta-lactamase gene from S. cellulosae KCCS127.

KCCS352 (15) hybridized to DNAs of these strains except its own DNA under the same hybridization condition (Fig. 2). Moreover, these three beta-lactamase genes did not cross-hybridize with each other (Fig. 3). These results suggest strongly that the hybridization by 1.9kb SacI-XbaI DNA fragment took place without connection to the presence of beta-lactamase activity, and even if the probe bound to DNAs of the beta-lactamase-producing strains, it was not directly related to the property of the beta-lactamase of binding blue dextran. As described previously (13), it is possible that these beta-lactamases might be derived from NAD^+ or $NADP^+$-dependent proteins like dehydrogenases. It is expected that the evolutional relationship of these beta-lactamases and other proteins and their physiological role in the cells will be clarified by further study.

III. GENE FROM STREPTOMYCES CACAOI KCCS352

The transcription start site of the beta-lactamase gene from S. cacaoi was determined by S_1 mapping to be located about 90bp upstream from the translational start site (Fig. 4). When compared with other sequences (Fig. 5), that in the -10 region (GAGAGT) is very similar to that of gylP1 (GAGACT), gylP2 (TAGAGT) (22) and ermEP2 (GAGGAT) (23), but that in the -35 region (TTCGAC) is little different from that of gylP1(TTGACG), gylP2(TCGAAC) or ermEP2 (TTGACG) and is instead similar to that of cellulase gene (TTCACC) (24) (the underline indicates identical nucleotides). The leader sequence is 88bp in length. These facts, together with the nucleotide sequence in the ribosome binding site (AGGAGG), indicate that the regulatory mechanism of the beta-lactamase gene expression is rather legitimate prokaryotic or Streptomyces type (25). In contrast, in CMA39 containing pMCP39, the beta-lactamase gene was expressed by read-through from the vector promoter very rich in A+T (26). This may reflect the different time of the highest beta-lactamase activity, indicating the use of different sigma factors.

During the course of subcloning, the beta-lactamase activity was found to change drastically depending on the lengths of the upstream sequence. Even if one compares clone CMA142 (pMCP142) and clone CMA146 (pMCP146)(Fig. 6), beta-lactamase activity varied about 50 times, although pMCP146 has a 600bp nucleotide sequence upstream of the transcription start site. Therefore, we analyzed whether a factor(s), which

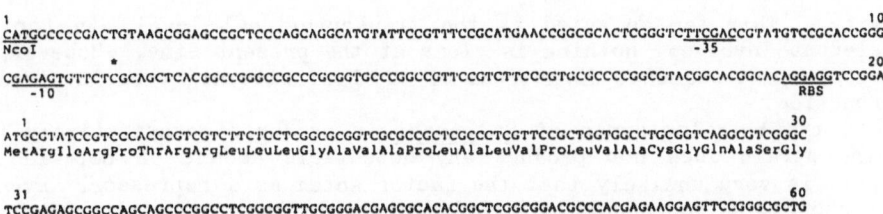

```
1                                                                                                    100
CATGGCCCCGACTGTAAGCGGAGCCGCTCCCAGCAGGCATGTATTCCGTTTCCGCATGAACCGGCGCACTCGGGTCTTCGACCGTATGTCCGCACCGGGA
NcoI                                                                           -35
                            *
                                                                                                     200
CGAGAGTGTTCTCGCAGCTCACGGCCGGGCCGCCCCGCGGTGCCCGGCCGTTCCGTCTTCCCGTGCGCCCCGGCGTACGGCACGGCACAGGAGGTCCGGAC
   -10                                                                                     RBS

1                                                                  30
ATGCGTATCCGTCCCACCCGTCGTCTTCTCCTCGGCGCGGTCGCGCCGCTCGCCCTCGTTCCGCTGGTGGCCTGCGGTCAGGCGTCGGGC
MetArgIleArgProThrArgArgLeuLeuLeuGlyAlaValAlaProLeuAlaLeuValProLeuValAlaCysGlyGlnAlaSerGly

31                                                                 60
TCCGAGAGCGGCCAGCAGCCCGGCCTCGGCGGTTGCGGGACGAGCGCACACGGCTCGGCGGACGCCCACGAGAAGGAGTTCCGGGCGCTG
SerGluSerGlyGlnGlnProGlyLeuGlyGlyCysGlyThrSerAlaHisGlySerAlaAspAlaHisGluLysGluPheArgAlaLeu
```

Fig. 4. A part of nucleotide and amino acid sequences of the beta-lactamase gene from S. cacaoi. The transcription start site as determined by S₁ mapping is indicated by an asterisk.

					Leader (bp)
Beta-lactamase		CCGGCGCACTCGGGTCTTCGACCGTATGTCCGCACCGGGACGAGAGTGTTCTCGCA			88
First Group	gylP1	TTGACG	-17bp-	GAGACT-8bp-U	149
	gylP2	TCGAAC	-19bp-	TAGAGT-6bp-A	99
	ermEP2	TTGACG	-18bp-	GAGGAT-6bp-C	?
	casA	TTCACC	-17bp-	TACCGT-4bp-T	74
Second Group	tsrP1	GTCAGGGCAGCCAT	-14bp-	TAGGGT-6bp-A	45
	tsrP2	CGCAGCCCAGAAAT	-14bp-	AATACT-6bp-U	173
Third Group	vphP1	GCAGCGCCGTGTGCGGCCTG	-7bp-	CCGGCGGGAGCGA-6bp-U	356
	endoH	ATTGACTTGATTGA-8bp-		CGGCGGGCAGGG-7bp-G	ca140

Fig. 5. Comparison of the nucleotide sequence of the promoter region of the beta-lactamase gene and the other genes. The underline under the nucleotide sequence of beta-lactamase indicates the inverted repeat centered at the dot. The underlines in other genes indicate the identical nucleotide with that of beta-lactamase gene. An asterisk points out the transcription start site.

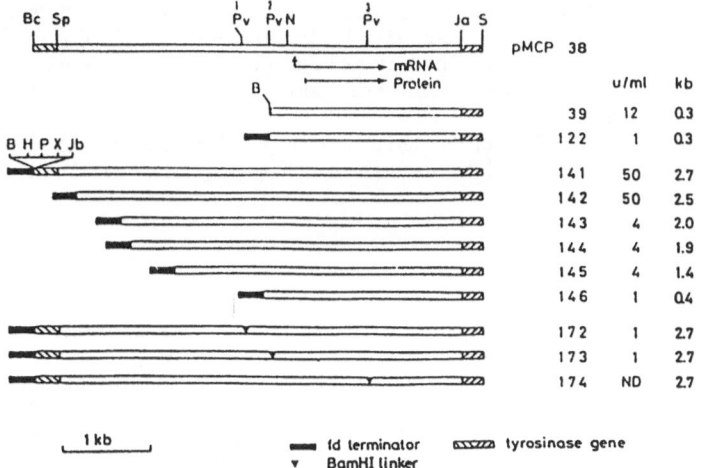

Fig. 6. Schematic representation of the DNA structures of various clones. The figures in kb are approximate lengths of DNA from the transcription start site to the upstream junction point with the vector plasmid, and those in units/ml are beta-lactamase activities in clones with various plasmids.

enhances the beta-lactamase activity, is present in this region by constructing various deletion and insertion plasmids. As a result, it was clarified that a trans-acting factor(s) was present in the upstream region and was involved in the two-step enhancement of the beta-lactamase activity. This factor acted at the transcriptional level. As for the cis-element involved, nothing is clear at the present time. However, an inverted repeat sequence downstream of the NcoI site might be involved in the function.

As the beta-lactamase of S. cacaoi is produced constitutively (17) and the strain does not produce any detectable amount of beta-lactams (27), it is very unlikely that the factor acted as a repressor. Even if the factor is a repressor, the inducer should be a constituent of the cell itself (self-inducible) or a metabolite but not a beta-lactam. In Bacillus licheniformis, the expression of the beta-lactamase gene (penP) is regulated by a repressor coupled with an antirepressor function (28,29,30), and these three genes locate very closely but penP is transcribed independently from the other two. The antirepressor is presumed to be a penicillin-binding protein as well as a DNA binding protein. A repressor is also implicated in the expression of beta-lactamase gene in Enterobacter cloacae (31) and Citrobacter freundii (32). Beta-lactamase in Escherichia coli, on the other hand, is regulated by a growth rate-dependent attenuator system (33). However, the case of S. cacaoi described here is assumed to be different from these systems.

Recently, enhancing elements or activator proteins have also been detected in bacteria (34,35). In Streptomyces, activator proteins are expected to be involved in the expression of genes implicated in antibiotic biosynthesis (36) and differentiation (37), but no quantitative result has been published yet. Although nothing is known about the structure of the factor concerned in this study, it may belong to these types of proteins.

ACKNOWLEDGMENTS

I am grateful to A. Seino, D. A. Hopwood, M. J. Bibb and B. Jaurin for their kind gifts. The second part of this paper was carried out as a joint work with Drs. M. V. Lenzini, J. Dusart and J. M. Ghuysen in Liege, Belgium and my laboratory. This work was supported in part by the Grant-in-Aids for Scientific Research from the Ministry of Education, Science and Culture of Japan.

I would like to dedicate this paper to the memory of one of my intimate friends, Dr. Bengtåke Jaurin in Sweden, who died on January 18, 1990.

REFERENCES

1) Hamilton-Miller, J. M. T. & J. T. Smith: Beta-lactamases. Academic Press, London, 1979.
2) Ogawara, H.: Production and property of beta-lactamases in Streptomyces. Antimicrob. Agents & Chemother. 8: 402-408, 1975.
3) Ogawara, H.: Antibiotic resistance in the pathogenic and the producing bacteria, with special reference to beta-lactam antibiotics. Microbiol. Rev. 45: 591-619, 1981.
4) Kushner, D. J. & C. Brenil: Penicillinase (beta-lactamase) formation by blue-green algae. Arch. Microbiol. 112: 219-223, 1977.
5) Ogawara, H.; S. Horikawa, S. Shimada-Miyoshi & K. Yasuzawa: Production and property of beta-lactamases in Streptomyces. Comparison of the strains isolated newly and thirty years ago. Antimicrob. Agents & Chemother. 13: 865-870, 1978.
6) Matsuhashi, Y.; T. Murakami, C. Nojiri, H. Toyama, H. Anzai & K. Nagaoka: Mechanism of aminoglycoside-resistance of Streptomyces harbaring resistant genes obtained from antibiotic producers. J. Antibiotics 38: 279-282, 1985.

7) Nakano, M. M.; H. Mashiko & H. Ogawara: Cloning of the kanamycin resistance gene from kanamycin-producing _Streptomyces_ species. J. Bacteriol. 157: 79–83, 1984.

8) Skeggs, P. A.; J. Thompson & E. Cundliffe: Methylation of 16S ribosomal RNA and resistance to aminoglycoside antibiotics in clones of _Streptomyces lividans_ carrying DNA from _Streptomyces tenjimariensis_. Mol. Gen. Genet. 200: 415–421, 1985.

9) Chater, K. F. & D. A. Hopwood: Antibiotic biosynthesis in _Streptomyces_, p. 129–150. In " D. A. Hopwood, and K. F. Chater (ed.), Genetics of bacterial diversity", Academic Press, London, 1989.

10) Ghuysen, J. M. & B. Joris: The bacterial active-site serine penicillin-interactive proteins family, p. 451–456. In "Y. Okami, T. Beppu, and H. Ogawara (ed.), Biology of actinomycetes '88", Japan Scientific Societies Press, Tokyo, 1988.

11) De Meester F; B. Joris, M. V. Lenzini, P. Dehottay, T. Erpicium, J. Dusart, D. Klein, J. M. Ghuysen, J. M. Frere & J. V. Beeumen: The active sites of the beta-lactamases of _Streptomyces cacaoi_ and _Streptomyces albus_ G. Biochem. J. 244: 427–432, 1987.

12) Thompson, S. T.; K. H. Cass & E. Stellwagen: Blue-dextran Sepharose: An affinity column for the dinucleotide fold in protein. Proc. Natl. Acad. Sci. USA 72: 669–672, 1975.

13) Ogawara, H. & S. Horikawa: Purification of beta-lactamase from _Streptomyces cellulosae_ by affinity chromatography on blue Sepharose. J. Antibiotics 32: 1328–1335, 1979.

14) Bush, K.: Characterization of beta-lactamases. Antimicrob. Agents & Chemother. 33: 259–263, 1989.

15) Lenzini, V. M.; S. Nojima, J. Dusart, H. Ogawara, P. Dohottay, J. M. Frere & J. M. Ghuysen: Cloning and amplified expression in _Streptomyces lividans_ of the gene encoding the extracellular beta-lactamase from _Streptomyces cacaoi_. J. Gen. Microbiol. 133: 2915–2920, 1987.

16) Lenzini, M. V.; H. Ishihara, J. Dusart, H. Ogawara, B. Joris, J. B. Beeumen, J. M. Frere & J. M. Ghuysen: Nucleotide sequence of the gene encoding the active site serine beta-lactamase from _Streptomyces cacaoi_. FEMS Microbiol. Lett. 49: 371–376, 1988.

17) Ogawara, H.; A. Mantoku & S. Shimada: Beta-lactamase in _Streptomyces cacaoi_. Purification and properties. J. Biol. Chem. 256: 2649–2655, 1981.

18) Lomovskaya, N. D.; N. M. Nkrtumian, N. L. Gostimskaya & V. N. Danilenko: Characterization of temperate actinophage c31 isolated from _Streptomyces coelicolor_ A3(2). J. Virol. 9: 258–262, 1972.

19) Hopwood, D. A.; M. J. Bibb, K. F. Chater, T. Kieser, C. J. Bruton, H. M. Kieser, D. J., Lydiate, C. P. Smith, J. M. Ward & H. Schrempf: Genetic manipulation of _Streptomyces_. A laboratory manual. The John Innes Foundation, Norwich, 1985.

20) Southern, E. M.: Detection of specific sequence among DNA fragments separated by gel electrophoresis. J. Mol. Biol. 98: 503–517, 1975.

21) Forsman, M.; B. Haggstrom, L. Lindgren & B. Jaurin: Molecular analysis of beta-lactamases from species of _Streptomyces_: comparison of amino acid sequences with those of other beta-lactamases. J. Gen. Microbiol. 136: in press.

22) Smith, C. P. & K. F. Chater: Structure and regulation of controlling sequences for the _Streptomyces coelicolor_ glycerol operon. J. Mol. Biol. 204: 569–580, 1988.

23) Bibb, M. J.; G. R. Janssen & J. M. Ward: Cloning and analysis of the promoter region of the erythromycin-resistance gene (_ermE_) of _Streptomyces erythreus_. Gene 41: E357–E368, 1986.

24) Nakai, R.; S. Horinouchi & T. Beppu: Cloning and nucleotide sequence of a cellulase gene, _casA_ from an alkalophilic _Streptomyces_ strain. Gene 65: 229–238, 1988.

25) Bibb, M. J.: Gene expression in _Streptomyces_ - Nucleotide sequences

involved in the initiation of transcription and translation, p. 25-34. In "G. Szabo, S. Biro, and M. Goodfellow (ed.), Biological, biochemical and biomedical aspects of actinomycetes", Akademiai Kiado, Budapest, 1985.

26) Bernan, V.; D. Filpula, W. Herber, M. Bibb & E. Katz: The nucleotide sequence of the tyrosinase gene from Streptomyces antibioticus and characterization of the gene product. Gene 37: 101-110, 1985.

27) Ogawara, H. & S. Horikawa: Penicillin-binding proteins of Streptomyces cacaoi, olivaceus and clavuligerus. Antimicrob. Agents Chemother. 17: 1-7, 1980.

28) Himeno, T.; T. Imanaka & S. Aiba: Nucleotide sequence of the penicillinase repressor gene penI of Bacillus licheniformis and regulation of penP and penI by the repressor. J. Bacteriol. 168: 1128-1132, 1986.

29) Imanaka, T.; T. Himeno & S. Aiba: Cloning and nucleotide sequence of the penicillinase antirepressor gene penJ of Bacillus licheniformis. J. Bacteriol. 169: 3867-3872, 1987.

30) Kobayashi, T.; Y. F. Zhu, N. J. Nicholls & J. O. Lampen: A second regulatory gene, blaR1, encoding a potential penicillin-binding protein required for induction of beta-lactamase in Bacillus licheniformis. J. Bacteriol. 169: 3873-3878, 1987.

31) Honore, N.; M. H. Nicolas & S. T. Cole: Inducible cephalosporinase production in clinical isolates of Enterobacter cloacae is controlled by a regulatory gene that has been deleted from Escherichia coli. EMBO J. 5: 3709-3714, 1986.

32) Lindberg, F.; L. Westman & S. Normark: Regulatory components in Citrobacter freundii ampC beta-lactamase induction. Proc. Natl. Acad. Sci. USA 82: 4620-4624, 1985.

33) Jaurin, B.; T. Brundstrom, T. Edlund & S. Normark: The E. coli beta-lactamase attenuator mediates growth rate-dependent regulation. Nature 290:221-225, 1981.

34) Henikoff, S.; G. W. Haughn, J. M. Calvo & J. C. Wallace: A large family of bacterial activator proteins. Proc. Natl. Acad. Sci. USA 85: 6602-6606, 1988.

35) Raibaud, O. & M. Schwartz: Positive control of transcription initiation in bacteria. Ann. Rev. Genet. 18: 173-206, 1984.

36) Distler, J.; A. Ebert, K. Mansouri, K. Pissowotzki, M. Stockman & W. Piepersberg: Gene cluster for streptomycin biosynthesis in Streptomyces griseus: nucleotide sequence of three genes and analysis of transcriptional activity. Nucl. Acids Res. 15: 8041-8056,1987.

37) Horinouchi, S.; H. Suzuki & T. Beppu: Nucleotide sequence of afsB, a pleiotropic gene involved in secondary metabolism in Streptomyces coelicolor A3(2) and "Streptomyces lividans". J. Bacteriol. 168: 257-267, 1986.

PROTEIN BIOSYNTHESIS AND SECRETION

Dieter Kluepfel

Institut Armand-Frappier
Université du Québec
Ville de Laval, Qué. Canada

Streptomyces are filamentous, Gram-positive microorganisms found generally in the soil and known mainly for their capacity to synthesize of numerous antibiotics and other secondary metabolites. They also produce a large number of extracellular enzymes which are essential to the degradation of biomass to assimilable carbon moieties. Generally these enzymes are hydrolases such as cellulases, xylanases, ligninases, amylases, lipases, nucleases etc. (Williams et al., 1983). Secretion is an essential and common feature to all these proteins.

In the last few years, secretion of both homologous as well as heterologous proteins and peptides from *Streptomyces* altered by genetic engineering techniques has been investigated in several laboratories. These studies were made possible largely by the fundamental advances in the molecular biology of the genus achieved by Dr. D.A. Hopwood and his colleagues at the John Innes Institute in Norwich (UK). Interest in this field is reflected in several reports that have been published on the cloning and expression of extracellular proteins. Thus, an agarase gene from *S. coelicolor* was cloned by Kendall and Cullum (1984) using the multicopy plasmid pIJ702 constructed by Katz et al. (1983) in *S. lividans* yielding a 500-fold overproduction of the enzyme. Similarly, Mondou et al. (1986) cloned a xylanase gene in the same species and Eckhardt et al. (1987) that of a β-galactosidase, while Henderson et al. (1987) expressed in the same system the genes of two proteases from *S. griseus*.

Recently, the possible use of the secretion mechanism in streptomycetes, particularly that of *S. lividans*, have been investigated for the expression of heterologous proteins or peptides (Chang and Chang, 1988). In Gram-positive bacteria such as *Streptomyces* proteins are secreted directly into the surrounding environment, whereas in Gram-negative organisms are generally retained in the periplasmic space of the cell walls. The advantages in producing proteins in their secreted form are significant. It allows easy recovery of the enzymes and separation from all intracellular proteins, often in the correctly folded structure and free of the N-terminal methionine, which is of particular importance for peptides used as therapeutic agents.

Studies on the protein secretion in bacteria show that the intracellular precursors of the exported proteins contain a N-terminal peptide chain called signal sequence which are essential in the translocation of the enzyme. The removal of such signal peptides is mediated by a peptidase which acts at the carboxyl end, frequently after a tripeptide sequence Ala-X-Ala. While these mechanisms of secretion have been well established in *Escherichia coli* (Ferenci and Silhavy, 1987), they are not as well documented in streptomycetes. Still, there are many similarities. Several signal sequences have been characterized. They were found of various length ranging from 23 to 56 amino acids and are also attached to the amino-terminal end of the precursor protein and presumed responsible for its translocation through the cell membrane where they are removed by specific peptidases similarly to those found in *E. coli*.

Using such secretion systems, several heterologous proteins have been expressed successfully in *S. lividans*. Thus, Bender et al. (1990) were able to express and secrete human interleukin-2 by fusing the DNA sequence coding for signal peptide of tendamistat, an α-amylase inhibitor, to the DNA encoding for the mature part of interleukin-2. Further progress on this expression system will be presented in this symposium by Koller and co-workers. Our late colleague Bengtåke Jaurin and co-workers (1990) reported just recently the use of the α-amylase promotor and signal sequence for the construction of a vector allowing the expression in *S. lividans* of such eukaryotic proteins as acetylcholinesterase and interferon. Finally, in the same symposium, Brawner et al. (1990) were able to express the human T cell receptor CD-4 in *S. lividans*. They demonstrated two particular advantages found in this streptomycetes: a) its ability to form the proper disulfide crosslinkages in the foreign peptide that are essential for its activity, and b) the synthesis of a proteinase inhibitor, called LEP-10 which occurs naturally in *S. lividans*. These findings should be very important factors for futur use of this microorganism for expressing of other heterologous peptides.

In the papers presented at the following Symposium several of these aspects will be developped further.

References

Bender, E., Koller, K.-P. and Engels, J. W., 1990, Secretory synthesis of human interleukin-2 by *Streptomyces lividans*. Gene 86: 227.

Brawner, M., Taylor, D. and Fornwald, J., 1990, Expression of the soluble CD-4 receptor in *Streptomyces*. J. Cellular Biochem. Suppl. 14A: 103, UCLA symp. on molecular & cellular biology, Abstr. CC036.

Chang, S.-Y. and Chang, S., 1988, Secretion of heterologous proteins in *Streptomyces lividans*. Biology of Actinomycetes'88, Proc. 7 th Int. Symp. p. 103. Y. Okami, T. Beppu, H. Ogawara, eds.

Eckhardt, T., Strickler, J., Gorniak, L., Burnett, W. V. and Fare, L. R., 1987, Characterization of the promotor, signal sequence, and amino terminus of a secreted ß-galactosidase from *Streptomyces lividans*. J. Bact. 169: 4249.

Ferenci, T. and Silhavy, T. J., 1987, Sequence information required for protein translocation from the cytoplasm. J. Bact. 169: 5339.

Henderson, G., Krygsman, P., Liu, C.J., Davey, C. C. and Malek, L. T., 1987, Characterization and structure of genes for proteases A and B from *Streptomyces griseus*. <u>J. Bact.</u> 169: 3778.

Jaurin, B., Granström, M. and Osterman, A., 1990, A recombinant system for extracellular production of polypeptides based on starch-inducible α-amylase promotor of *Streptomyces*. <u>J. Cellular Biochem.</u> Suppl. 14A: 126, UCLA Symp. on molecular & cellular biology, Abstr. CC408.

Katz, E., Thompson, C. J. and Hopwood, D. A., 1983, Cloning and expression of the tyrosinase gene from *Streptomyces antibioticus* in *Streptomyces lividans*. <u>J. Gen. Microbiol.</u> 129:2703.

Kendall, K. and Cullum, J., 1984, Cloning and expression of an extracellular gene from *Streptomyces coelicolor* A3(2) in *Streptomyces lividans* 66. <u>Gene</u>, 29:315.

Mondou, F., Shareck, F., Morosoli, R. and Kluepfel, D., 1986, Cloning of the xylanase gene of *Streptomyces lividans*. <u>Gene</u> 49: 323.

Williams, S.T., Goodfellow, M., Alderson, G., Wellington, E. M. H., Sneath, P. H. A. and Sackin, M. J., 1983, Numerical classification of *Streptomyces* and related genera. <u>J. Gen. Microbiol.</u> 129: 1743.

HOMOLOGOUS CLONING OF THE XYLANASE GENES AND THEIR EXPRESSION

IN *STREPTOMYCES LIVIDANS* 66

Dieter Kluepfel, Rolf Morosoli and François Shareck

Centre de microbiologie appliquée, Institut Armand-Frappier, Université du Québec, 531, boul. des Prairies, Ville de Laval, Qué., H7N 4Z3 Canada

INTRODUCTION

Over the last years, the microbial xylanases have been studied extensively due mainly to the various possibilities for industrial applications. Their potential in bioconversion of lignocellulosic biomass to fuels, feeds and chemicals was the principle interest for a number of years (Wong et al., 1988). More recently, the possible use of these enzyme systems as bleaching agent in the pulp and paper industry has gained a considerable attention (Paice et al., 1988). Replacing the presently used chemical bleaching sequences in paper manufacture totally or in part by enzyme treatments of the pulps would have a tremendous impact on the environmental pollution problems created by the discharge of waste. The concern caused recently by the detection of dioxin and furans in consumer goods such as milk cartons and baby diapers which has been attributed directly to the use of activated chlorine in the bleaching process.

Xylan, a ß-1,4-linked polymer of xylose, is hydrolyzed by the combined action of endo-xylanases as well as that of a ß-xylosidase and, less frequently, by an exo-xylanase (Reilly, 1981). Such enzyme systems are produced by many microorganisms, showing a wide range of substrate specificity (Lee et al., 1988). They are generally associated with the other lignocellulose-degrading enzymes such as cellulases and often ligninases as well as with a host of debranching enzymes capable of removing the various substituents of the xylan backbone (Dekker, 1985).

Over the last few years we have studied the production of xylanases by different *Streptomyces* (Ishaque and Kluepfel, 1981; Kluepfel et al., 1986). Since then we have isolated two of these enzymes, xylanase A and B, from *Streptomyces lividans* 66 (Morosoli et al., 1986; Kluepfel et al., 1990) and cloned their respective structural genes by functional complementation of a cellulase-negative mutant of *S. lividans* (Mondou et al., 1986; Vats-Mehta et al., 1990).

In the present paper we report the comparison of the two enzymes on the molecular as well as on their chemical and on their physiological levels.

MATERIALS AND METHODS

Microorganisms

Streptomyces lividans IAF18 and IAF42 are xylanase-producing clones carrying the xlnA and xlnB genes, respectively. They were obtained by a shotgun cloning of S. lividans 66 (strain 1326) by functional complementation in S. lividans 10-164 a phenotypic xylanase- and cellulase-negative mutant (Mondou et al., 1986) using the multicopy plasmid vector pIJ702 (Katz et al., 1983). S. lividans 3131 is the wild-type strain containing this plasmid vector.

The culture conditions for these strains have been described previously by Bertrand et al.(1989) and Kluepfel et al. (1990).

Enzyme recovery

The enzymes were recovered from the culture filtrates by ethanol precipitation. The precipitates were washed with acetone to obtain crude xylanase powder which can be stored over long periods at 4°C without loss of activity.

Enzyme purification

Both xylanases were purified by hplc on DEAE ion exchange and on gel columns as described by Morosoli et al. (1986) and Kluepfel et al. (1990).

Enzyme assay

Endo-ß-1,4-xylanase [EC 3.2.1.4] activities were determined by the dinitrosalicylic acid (DNS) method as described previously by Morosoli et al. (1986).

Rapid routine of xylanase activities in fermentation broths or during purification steps were carried out by a plate assay using the Remazol Brilliant Blue (RBB) covalently bound to xylan as described by Biely et al. (1985) and Kluepfel (1988).

Protein contents were determined by the wellknown Lowry method.

Antibodies and Western blotting

Rabbit antibodies were raised against the purified xylanases and Western blotting were carried out using anti-xylanase A and xylanase B antibodies coupled to ^{125}I-protein A as previously described (Mondou et al. (1986).

Enzymatic hydrolysis

Studies on the enzymatic hydrolysis of xylan were carried out in solutions of 1 international unit of xylanase in 1 ml of 0.1 M McIlvaine buffer, pH 6, containing 10 mg (1 %) of substrate. The mixtures were incubated at 60°C and samples withdrawn at intervals for analysis by hplc (Kluepfel et al., 1990).

Studies on sequential action of the xylanases was carried out in the same manner. After 20 h of incubation with one enzyme, the mixture was boiled to stop the reaction and the second xylanase added directly reincubating the whole for another 20 h.

208

Analysis of oligosaccharides

The hydrolysis products of the enzymatic reactions were analysed by hplc equipped with a refractometer as described by Kluepfel et al. (1990).

RESULTS AND DISCUSSION

Cloning of the xylanase genes

From a gene bank constructed in *S. lividans* mutant 10-164, with the chromosomal DNA digested with *Sau*3AI, a serie of putative xylanase positive clones were isolated by functional complementation. The individual clones were purified and their plasmid DNA was isolated. These hybrid plasmids when used to retransform the mutant 10-164, yielded 100% transformants expressing xylanase activity on RBB-xylan plates. These results showed that the enzyme activity was plasmid-linked. None of these xylanase producing clones showed any cellulase activity.

From preliminary examination of the enzyme activity on RBB plates three types of xylanases could be distinguished by their respective hydrolysis zones. So far two of these xylanases have been characterized in detail both on the molecular and on the biochemical levels. The plasmids coding for these enzymes pIAF18 and pIAF42 contained DNA fragments of 5.6 and 5.7 kb, respectively, inserted into the BglII site of pIJ702. In pIAF42 no hybridization was obtained with the *Bam*HI-*Sph*I fragment isolated from plasmid pIAF18 containing the entire *xln*A gene (Fig. 1). Sequencing data with pIAF42 showed that the gene coding for *xln*B was present in the 3.5 kb in the *Bgl*II fragment.

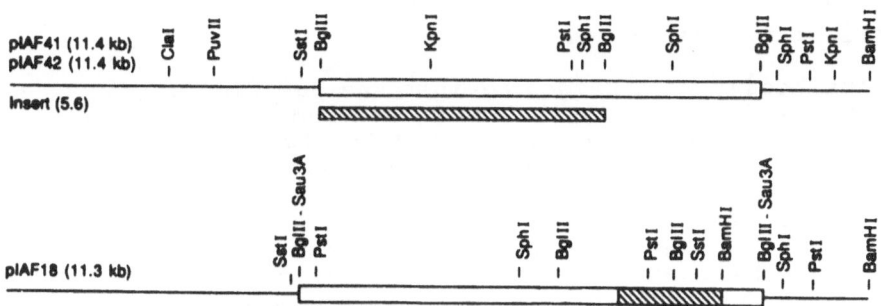

Fig.1. Restriction maps of the *xln*A (pIAF18) and *xln*B (pIAF42) clones. The single line presents the vector pIJ702. The box represents the chromosomal DNA inserts with the probable location of *xln* genes indicated by hatched regions.

Expression of the xylanase genes in *Streptomyces lividans* 10-164

The expression of the *xln*A and *xln*B genes cloned in *S. lividans* 10-164 was studied and compared to the wild-type strain *S. lividans* 3131, in submerged cultures using M13 medium

Table 1. Comparison of the xylanase activity detected in the supernatants obtained from submerged cultures of *S. lividans* 3131, IAF18, and IAF42 in TSB or M13 medium containing 1% xylan or xylose after 72 h of incubation at 34°C.

S. lividans strain	Xylanase activity IU		
	TSB	Xylose	Xylan
3131	0	0	7
IAF18	5	220	291
IAF42	25	389	273

(Bertrand et al., 1989) either 1% of xylose or oat spelts xylan as main carbon source (Table 1). None of the strains produced any detectable xylanase activity when grown on glucose indicating catabolic repression even so they had been cloned in a high-copy plasmid. The wild-type did not produced any xylanase on either TSB (Trypticase Soy Broth) or xylose containing medium, while on xylan the activity varied from 6-9 IU/ml of culture filtrate. In contrast, both the *xln* clones produced comparable high levels of enzyme activities both on xylose and on xylan. Thus enzyme synthesis is not repressed by xylose which can possibly be attributed to a gene dosage effect. Moreover, the higher activity found in xylose as compared to the xylan-grown cultures reflects the difference in growth on the two carbon sources. Xylose being readily assimilated by the cells promotes rapid growth accompanied by the synthesis of the xylanases. In comparison, in xylan medium, the cells grow slower depending on the assimilable sugars released by the hydrolytic action of the xylanases on the substrate. In addition, it was found that xylanase B binds to the only partially soluble oat spelts xylan and significant amounts of enzyme activity was found in the solids recovered after filtration.

Com·arison of the cloned xylanases

Antibodies were raised against the purified xylanases A and B. Western blot analysis of these two enzymes with their respective antibodies are presented in Fig. 2 and show a slight cross reaction between xylanase A and the anti-xylanase B antibodies.

The proteins secreted by the wild-type *S. lividans* 3131, the mutant 10-164 and the clones IAF18 and IAF42 grown in glucose, xylose and xylan medium were analyzed by SDS-PAGE followed by Western blotting (Fig.3). The mutant 10-164 did not produce any detectable xylanase A or B on either of the carbon sources. Glucose repressed xylanase synthesis in all strains. Strain 3131 produced as expected, certain amounts of both enzymes on xylanase medium only, while clones IAF18 and IAF42 each synthesized xylanase A and B respectively both on xylose and on xylan. Antibodies against xylanase A revealed the presence of a small amount of the 43 kDa protein in the supernatant of clone IAF42, while IAF18 did not produce any detectable xylanase B. This observation suggests the possibility of the presence of a positive activator on the insert of pIAF42 close to the *xlnB* gene, which could play a role in the expression of the chromosomal *xlnA* gene. Since streptomycetes are found in the soil in the presence of abundant quantities of

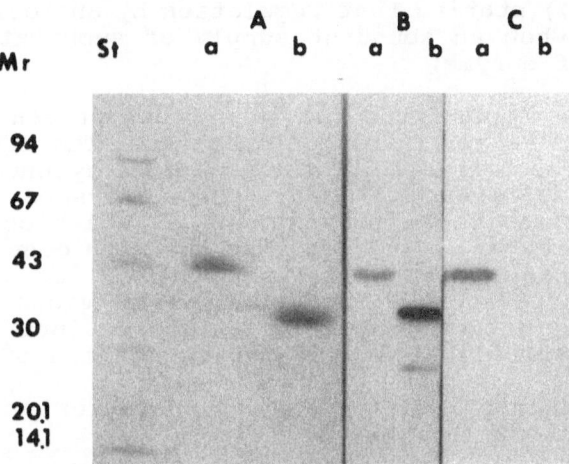

Fig. 2. Western blot analysis of pure xylanase A (**a**) and B (**b**).
A. SDS-PAGE stained with Coomassie Blue. **B.** Immunoblot with
antixylanase B. **C.** Immunoblot with antixylanase A.

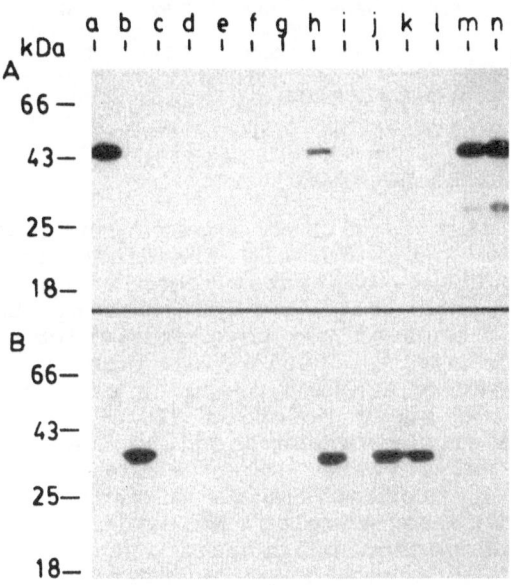

Fig. 3. Comparison by Western blot analysis of culture filtrates
of clones *S. lividans* IAF18 and IAF42 with those of wild-type
strain 3131 and mutant 10-164. Lanes: **a**, xylanase A; **b**, xylanase
B; **c**, 10-164 grown on glucose; **d**, on xylose; **e**, on xylan; **f**,
3131 grown on glucose; **g**, on xylose; **h**, on xylan; **i**, IAF42 grown
on glucose; **j**, on xylose; **k**, on xylan; **l**, IAF18 grown on
glucose; **m**, on xylose; **n**, on xylan.
A. Immunoblot with antixylanase A. **B.** Immunoblot with
antixylanase B.

xylan, an activator-type mechanism for the regulation of xylanases could be predicted according the hypothesis of Savageau (1977) stating that regulation by an activator protein is indicated when an abundant supply of substrate justifies a high demand of enzyme.

In nature xylan forms the interface between cellulose and lignin and is a rather insoluble substrate. Therefore, the xylan breakdown may be initiated by the action of xylanase B releasing soluble oligoxylosides which are then degraded by xylanase A into easily assimilable sugars such as xylobiose and xylose. Such sequential events suggest that the xlnB gene is coding for the key enzyme in the hydrolysis pathway of xylan and the presumptive activator should be in its vicinity (Beck and Warren, 1988). Further studies on the transcriptional level will be needed to establish clearly the regulation of the xylanase genes.

A comparison of the biochemical characteristics of the two xylanases is shown in Table 2.

Table 2. Characteristics of xylanase A and B produced by *Streptomyces lividans* IAF18 and IAF42

Enzyme	M_r	pI	pH	T°	K_m^*	V_{max}^*	Sp.Ac.*	Glycosyl.
Xylanase A	43 kDa	5.2	6.0	60°C	2.59	5.57	733	+
Xylanase B	31 kDa	8.4	6.5	55°C	3.71	1.96	1024	+

*on soluble oat spelts xylan

Enzymatic hydrolysis of xylan

Hydrolysis studies with xylanase A and B were carried out on water soluble and insoluble fractions of xylan from oat spelts. The soluble substrate, consisting of the shorter xyloside chains, was degraded rapidly by both enzymes to oligoxylosides of a DP of two to eight (which was the limit of detection). Xylanase A yielded primarely xylotriose and xylobiose, whereas xylanase B produced a variety of xylosides with five or more sugar moieties (Fig. 4a). The hydrolysis pattern obtained with xylanase A on the water-insoluble xylan fraction showed very little, if any, degradation. On the other hand, xylanase B produced large amounts of oligoxylosides ranging from one to eight xylose molecules (Fig. 4b).

These patterns seem to confirm the previously mentioned hypothesis that xylanase B is necessary to initiate the hydrolysis of xylan associated with the cellulose and lignin in lignocellulosic biomass.

The sequential action of xylanase A and B was studied on total oat spelts xylan and gave some insight on the manner in which the two enzymes acted on the substrate. The hydrolysis pattern obtained with xylanase A remained essentially unchanged when incubated with xylanase B (Fig. 5a), indicating that xylobiose and (or) xylotriose the main products of the former inhibit the latter. Inverting the order of the enzymes resulted in a very rapid and effective hydrolysis, accumulating large quantities of xylotriose, xylobiose and xylose as end products (Fig.5b).

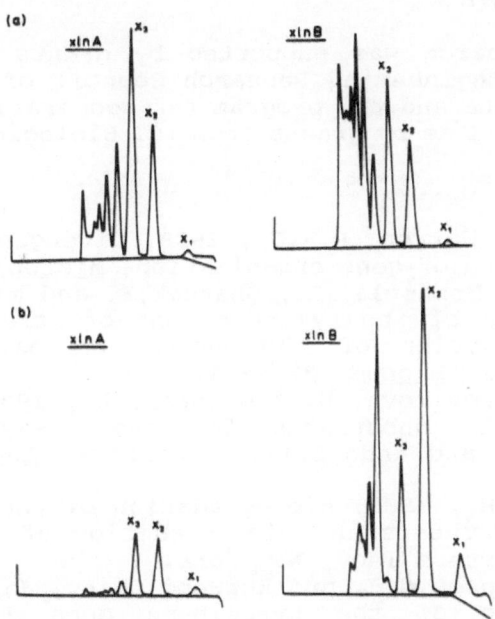

Fig. 4. Comparison of hydrolysis patterns obtained with xylanase A and B on (**a**) water soluble xylan; (**b**) water insoluble xylan (x_1, xylose; x_2, xylobiose; x_3, xylotriose).

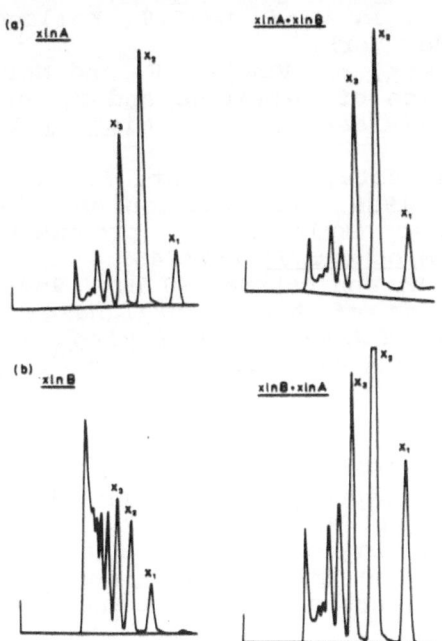

Fig. 5. Sequential hydrolysis of total oat spelts xylan by (**a**) xylanase A followed by xylanase B; (**b**) xylanase B followed by xylanase A (x_1, xylose; x_2, xylobiose; x_3 xylotriose).

Further studies of the two xylanases will be necessary in order to understand fully their mode of action. It will be specially important to carry out such investigations with well defined substrates such as pure oligoxylosides of different degrees of polymerization.

ACKNOWLEDGEMENTS

This research was supported by grants from the Natural Sciences and Engineering Research Council of Canada under the strategic grants and the program for cooperation university and industry as well as by grants from ICI Biological Products NATC.

REFERENCES

Beck, C.F. and Warren, R.A.J., 1988, Divergent promotors, a common form of gene organization. Microbiol. Rev. 52: 318.

Bertrand, J.L, Morosoli, R., Shareck,F. and Kluepfel, D., 1989, Expression of the xylanase gene of *Streptomyces lividans* and production of the enzyme on natural substrates. Biotechnol. Bioeng. 33: 791.

Biely, P., Mislovicova, D. and Toman, R., 1985, Soluble chromogenic substrates for the assay of endo-1,4-ß-xylanases and endo-1,4-ß-glucanases. Anal. Biochem. 144: 142.

Dekker, R. F. H., 1985, Biodegradation of the hemicelluloses, in "Biosynthesis and biodegradation of wood components", Academic Press Inc., New York.

Katz, E., Thompson,C.J. and Hopwood, D.A., (1983), Cloning and expression of the tyrosinase gene from *Streptomyces antibioticus* in *Streptomyces lividans*. J. Gen. Microbiol. 129: 2703.

Kluepfel, D., 1988, Screening of prokaryotes for cellulose- and hemicellulose-degrading enzymes, in "Methods of Enzymology", W. A. Wood and S.T. Kellogg, eds., Academic Press Inc., New York.

Kluepfel, D., Shareck, F., Mondou, F. and Morosoli, R., 1986, Characterization of cellulase and xylanase activities of *Streptomyces lividans* . Appl. Microbiol. Biotechnol. 24: 230.

Kluepfel, D., Vats-Mehta, S., Aumont, F., Shareck, F. and Morosoli, R., 1990, Purification and characterization of a new xylanase (xylanase B) produced by *Streptomyces lividans* 66. Biochem. J. 267: 45.

Lee, S., Forsberg, W. and Rattray, J.B., 1987, Purification and characterization of two endoxylanases from *Clostridium acetobutylicum* ATCC 824. Appl. Environ. Microbiol. 53: 644.

Mondou,F., Shareck, F., Morosoli, R. and Kluepfel, D., 1986, Cloning of the xylanase gene of *Streptomyces lividans* . Gene 49: 323.

Morosoli, R. Bertrand, J.L., Mondou, F., Shareck, F. and Kluepfel, D., 1986, Purification and properties of a xylanase from *Streptomyces lividans* . Biochem. J. 239: 587.

Paice,M.G., Bernier Jr, R. and Jurasek, L. (1988), Viscosity-enhancing bleaching of hardwood Kraft pulp with xylanase from a cloned gene. Biotechnol.Bioeng., 32: 235.

Reilly, P., 1981, Xylanases: structure and function. Basic Life-Sci. 18: 111.

Savageau, M. A., 1977, Design of molecular control mechanisms and the demand for gene expression, Proc. Natl. Acad. Sci. USA 74: 5647.

Vats-Mehta, S., Bouvrette, P., Shareck, F., Morosoli, R. and Kluepfel, D., 1990, Cloning of a second xylanase-encoding gene of *Streptomyces lividans* 66. Gene, 86: 119.

Wong, K.K.Y., Tan, L.U.L. and Saddler, J.N., 1988, Multiplicity of ß-1,4-xylanase in microorganisms: functions and applications. Microbiol. Rev. 52: 305.

PULLULAN-HYDROLYZING ENZYMES OF STREPTOMYCETES: EFFECT OF
CARBON SOURCE ON THEIR PRODUCTION AND CHARACTERIZATION OF
A PULLULANASE-NEGATIVE MUTANT

Frank Hänel, Hans Krügel and Theo Peschke

Central Institute of Microbiology
and Experimental Therapy, GDR Academy of Sciences
P.O. Box 73, 6900 Jena, GDR

INTRODUCTION

Streptomycetes are Gram-positive mycelial soil bacteria
that obtain nutrients from macromolecular organic debris by
using a wide range of hydrolytic enzymes. These enzymes
include proteases, nucleases, lipases, and a variety of
enzymes that hydrolyze different types of often quite complex
polysaccharides (Williams et al., 1983). These include
α-amylases (1,4-α-D-glucan-4-glucanohydrolase, EC 3.2.1.1)
that are responsible for the breakdown of starch by the
endoglucanic cleavage of 1,4-α-linked glucose residues
(Guilbot and Mercier, 1985) and pullulanases (α-dextrin
endo-1,6-α-glucosidase, EC 3.2.1.41), which hydrolyze
α-(1$\rightarrow$6)-glucosidic linkages of starch and pullulan (Bender
and Wallenfels, 1961). α-Amylases of Streptomyces have
received considerable recent attention (Gräfe et al., 1986;
McKillop et al., 1986; Hoshiko et al., 1987; Virolle and
Bibb, 1988; Virolle et al., 1988). In contrast, there are only
a few reports on pullulanases of streptomycetes (Ueda et al.,
1971; Yagisawa et al., 1972). In addition to their importance
as industrial enzymes, pullulanases are also of interest to
basic research. For example, no data are available concerning
a possible induction of pullulanase genes by small oligo-
saccharides derived from starch or pullulan (e.g. maltotriose) or repression by more readily assimilated sugars
(e.g. glucose) in streptomycetes.

To study such regulatory mechanisms we screened a variety
of <u>Streptomyces</u> species for pullulan-hydrolyzing enzyme
activities. In this paper the production of pullulanase,
α-amylase and α-glucosidase (EC 3.2.1.20) activities by
<u>Streptomyces</u> <u>griseus</u> K942 and <u>Streptomyces</u> <u>lividans</u> TK24
is reported and the effect of the carbon source on the syn-
thesis of these enzymes is examined. Furthermore, a pullulanase-negative mutant of <u>S</u>. <u>lividans</u> TK24 with a changed
differentiation behaviour was derived and characterized.

RESULTS

<u>Distribution of pullulan-hydrolyzing enzymes among strepto-</u>
<u>mycetes</u>

Among 39 <u>Streptomyces</u> strains tested, 16 strains (41 %)
showed pullulan-hydrolyzing activity. This is a similar
distribution to another Gram-positive genus <u>Bacillus</u>, where
of 297 strains representing 26 species of <u>Bacillus</u>, 138
(46 %) degraded pullulan in pullulan agar (Morgan et al.,
1979). So pullulan-hydrolyzing enzymes are far more prevalent
among streptomycetes than has been suspected hitherto (Ueda
et al., 1971; Yagisawa et al., 1972). Two pullulanase-positive
representives, one fermenting, one not fermenting maltose
were further investigated, because in other pullulan-degra-
ding microorganisms like <u>Klebsiella</u> <u>pneumoniae</u> the pullulanase
structural and secretion genes are components of the maltose
regulon (d'Enfert et al., 1987).

<u>Pullulanase, α-amylase and α-glucosidase of <u>S</u>. <u>griseus</u> K942</u>

<u>S</u>. <u>griseus</u> K942 did not grow on minimal agar medium
supplemented with maltose, but showed good growth with either
pullulan or Zulkowsky starch as sole carbon source (Table 1).
This is probably due to a defect in the maltose uptake system
since cell-bound α-glucosidase activity could be detected in
ultrasonic extracts of the strain(Table 2). The inability
of <u>S</u>. <u>griseus</u> K942 to take up maltose raises a question
concerning the action pattern of the pullulanase of this
strain. Fig. 1 shows the thin-layer chromatogram of the
products produced through the action of the crude enzyme

216

Table 1. Growth of S. griseus K942, S. lividans TK24 and
S. lividans 849 on minimal agar medium supplemented
in each case with one carbon source as mentioned.

Carbon source	Growth		
	K942	TK24	849
Pullulan	+	+	−
Starch (Zulkowsky)	+	+	+
Maltose	−	+	+

.

Table 2. Cellular α-glucosidase activities of S. griseus
K942 and S. lividans TK24 grown for 24 h at 28 °C
on a chemically defined medium supplemented with
2 % (w/v) of various carbon sources. Cells from
4 separately grown cultivation flasks were pooled
and disrupted by sonication.

Carbon source	Enzyme activity (U mg^{-1} protein)	
	K942	TK24
Starch (soluble)	3.41	0.85
Maltose	0	2.40
Glucose	0.16	0.19
Glycerol	0	0

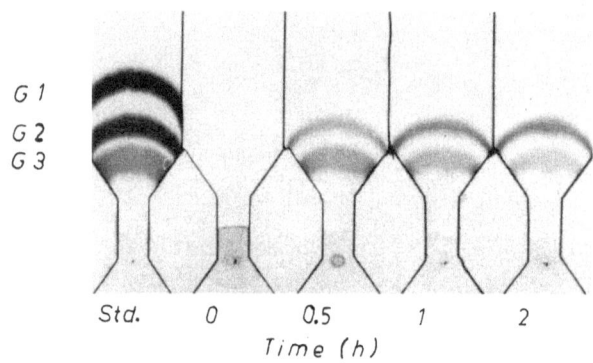

G 1
G 2
G 3

Std.　　0　　0.5　　1　　2

Time (h)

Fig. 1. A thin-layer chromatogram showing the products
released from pullulan by the action of
<u>S. griseus</u> K942 pullulanase during hydrolysis
at 50 °C. 0.5 ml sodium acetate buffer (50 mM,
pH 6.5) containing 0.5 % pullulan and 0.5 ml
crude enzyme mixture were incubated at 50 °C;
10 µl of each sample was applied to TLC.
Std, standards: glucose, maltose and maltotriose (G1-G3).

mixture on pullulan. Both maltotriose and maltose were formed
as the main products. During the time course of hydrolysis of
pullulan maltotriose gradually decreased and maltose was formed
without the production of an appreciable amount of glucose.
Nevertheless, this small, but direct, production of glucose
from pullulan should enable <u>S. griseus</u> K942 to grow on pullulan as sole carbon source.

To assess the effects of different carbon sources on
pullulanase and α-amylase activities of <u>S. griseus</u> K942
grown on each of the carbon sources mentioned on Fig. 2, the
amount of pullulanase present in the cells and in the culture
supernatants and of α-amylase only in the supernatants were
determined after 24 and 48 hours of growth. In the presence
of readily soluble small glucose polymers (Zulkowsky starch,
dextrin) a low cell-bound activity of pullulanase could be
detected, while the highest extracellular enzyme activity per
mg whole cell protein was found on Zulkowsky starch and
glucose. α-Amylase activity was highest on Zulkowsky starch
and dextrin and could not be estimated on glucose. Both

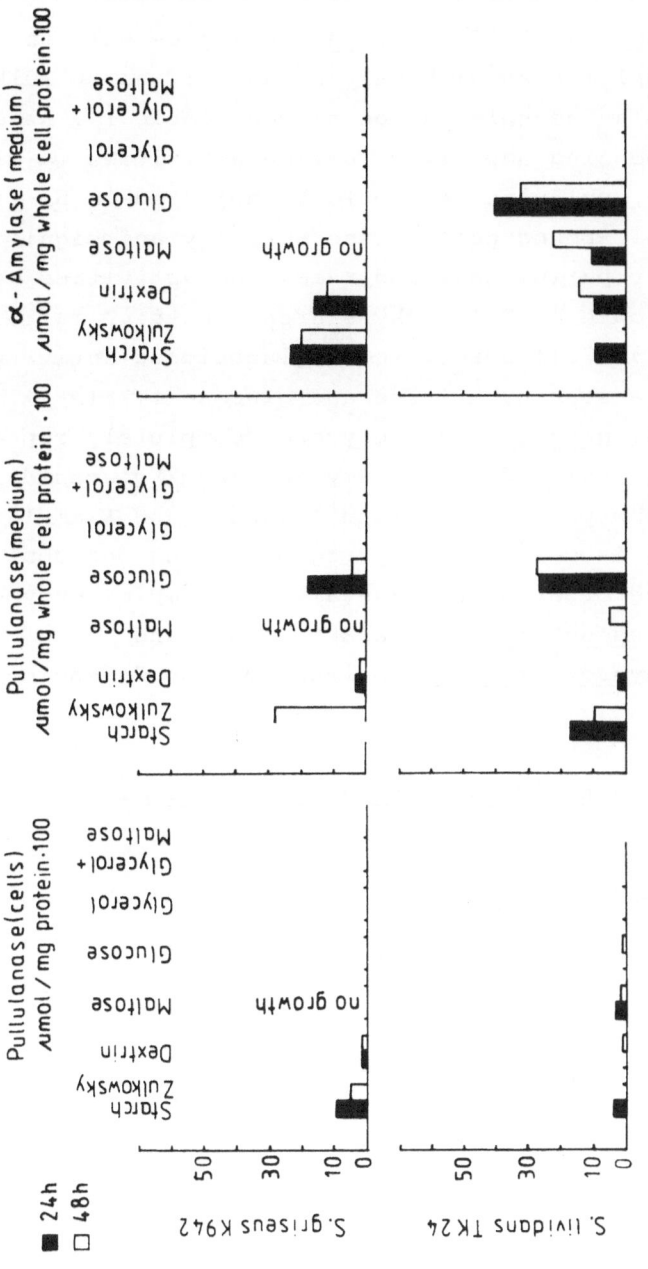

Fig. 2. Effect of different carbon sources on the production of pullulanase and α-amylase activities by S. griseus K942 and S. lividians TK24 and their localization (medium vs washed cells). Both strains were grown on a chemically defined medium for 24 and 48 h 2 % carbon source.

enzymes could not be detected on glycerol. α-Glucosidase activity was found in cells grown on soluble starch and glucose (Table 2).

Pullulanase, α-amylase and α-glucosidase of S. lividans TK24

Unlike S. griseus K942, S. lividans TK24 grew well on minimal agar medium supplemented with maltose as well as on Zulkowsky starch or pullulan as sole carbon source (Table 1). As expected, the corresponding amylolytic enzyme activities (α-glucosidase, α-amylase, pullulanase) could be detected in S. lividans TK24. A low cell-bound pullulanase activity and significant higher extracellular pullulanase and α-amylase activities were estimated during growth on Zulkowsky starch, dextrin, maltose and glucose (Fig. 2). Cell-associated α-glucosidase activity was found on soluble starch, maltose and glucose (Table 2). As in the case of S. griseus K942, glycerol completely repressed the synthesis of pullulanase, α-amylase and α-glucosidase. So it seemed that the production of pullulanase and α-amylase in S. lividans TK24 is co-ordinately regulated and one could assume the existence of a pullulanase-amylase complex enzyme as found for Bacillus subtilis (Takasaki, 1987) and Clostridium thermohydrosulfuricum (Melasniemi and Paloheimo, 1989).

Characterization of a pullulanase-negative mutant of S. lividans TK24

To prove the above-mentioned hypothesis, we looked for pullulanase-negative mutants which still exhibited α-amylase activity after mutagenic treatment of spores of S. lividans TK24 with N-methyl-N'-nitro-N-nitrosoguanidine. Among 1300 mutants tested we found one (S. lividans 849) which did not degrade pullulan in pullulan agar but still hydrolyzed starch (Fig. 3a, b). Sodium dodecyl sulphate polyacrylamide gel electrophoresis of the supernatants derived from S. lividans 849 and its parental strain revealed that besides the pullulanase a series of other proteins detectable in the supernatant were not secreted by the mutant (Fig. 3c). S. lividans 849 did

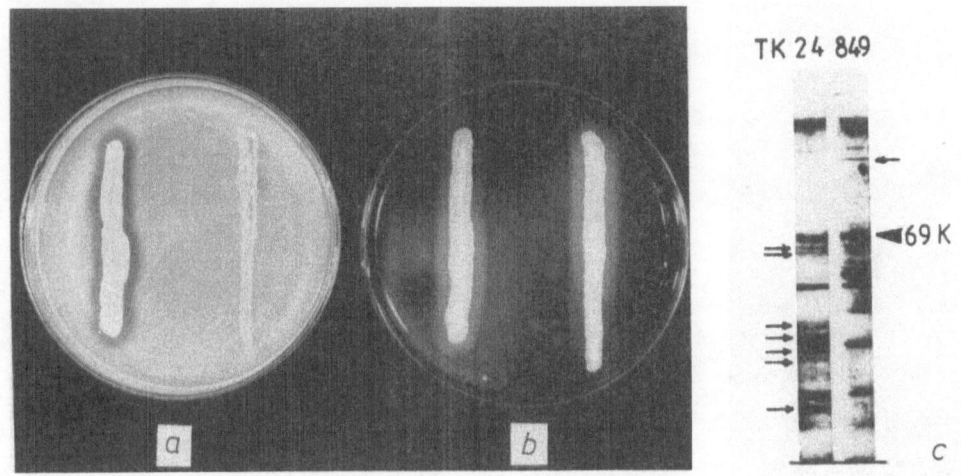

Fig. 3. Pullulan- and starch-degrading activities of
S. lividans TK24 (left on each plate) and S. lividans
849 on pullulan agar (a) and yeast-starch agar (b).
Sodium dodecyl sulphate polyacrylamide gel electrophoresis pattern of the supernatants (c).

not grow on minimal agar medium supplemented with pullulan
as sole carbon source but grew well on Zulkowsky starch and
maltose (Table 1). Starting from results concerning the role
of nutrient limitation and intracellular degradation of
glycogen in the regulation of differentiation processes in
streptomycetes (Ochi, 1987; Chater, 1989) we tested the
differentiation behaviour of S. lividans TK24 and 849 on two
agar media. On yeast-starch agar (gl^{-1}: peptone, 1; baker's
yeast, 15; potato starch, 10; NaCl, 5; agar, 15) both strains
developed aerial mycelium and spores (Fig. 4a, b). Quite
another picture we got on a minimal agar medium (gl^{-1}:
$(NH_4)SO_4$, 2; peptone, 0.5; yeast extract 0.5; maltose or
glucose, 5; trace elements; agar, 15; Na-phosphate buffer,
0.05 M). S. lividans TK24 did not form aerial mycelium and
spores on both carbon sources while its mutant 849 produced
aerial hyphae and spores on maltose (Fig. 4c, d). On glucose
aerial mycelium development and sporulation did not occur
as found for the parental strain.

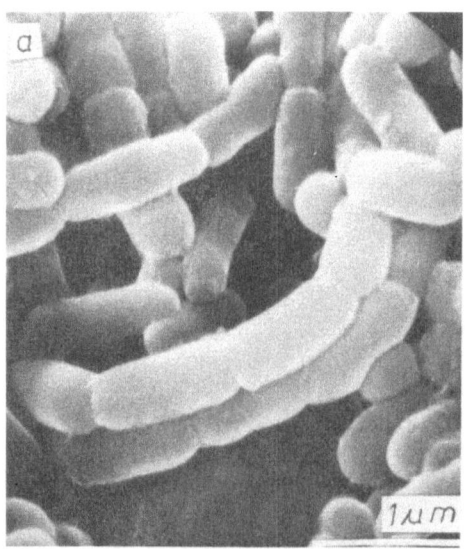
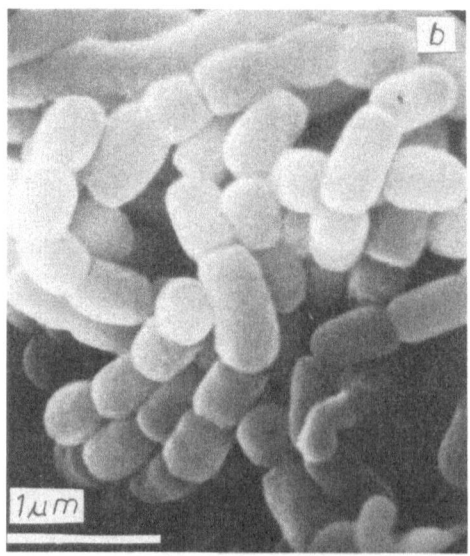
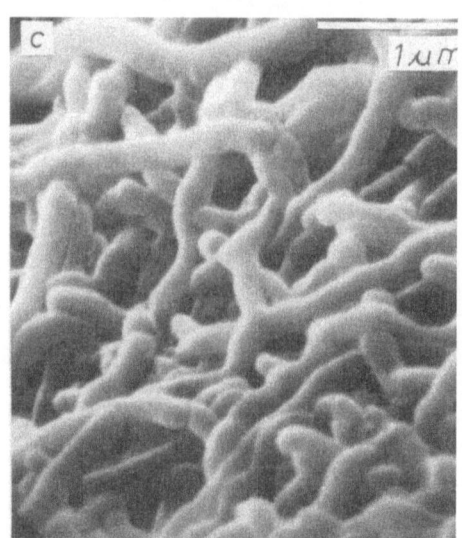
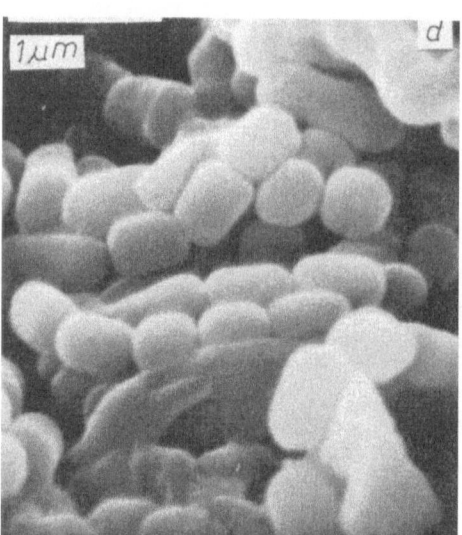

Fig. 4. Scanning electron micrographs of S. lividans TK24
(a, c) and 849 (b, d) cultivated on yeast-starch
agar (a, b) and synthetic agar (c, d) for 7 days.

DISCUSSION

In S. griseus K942, pullulanase and α-amylase showed
different pattern of carbon source regulation. While the
synthesis of α-amylase was repressed by glucose, the pullulanase synthesis was not. Synthesis of all pullulanases described till now, e.g. in Klebsiella aerogenes (Takizawa and
Murooka, 1985) and Clostridium thermohydrosulfuricum (Melasniemi, 1987) is repressed by glucose. In S. lividans TK24 both
enzymes were co-ordinately regulated and could be detected
during cultivation of this strain on glucose medium. On the
other hand, a pullulanase-negative mutant was derived, which
still showed starch-degrading activity. So it seems that
pullulan- and starch-hydrolyzing activities of both strains
represent two distinct enzymes. This is supported by a recent
report on the molecular cloning of an α-amylase gene from S.
lividans TK24 (Lin et al., 1988). When the α-amylase gene of
S. limosus was subcloned on a low copy plasmid in S. lividans
TK24 and Streptomyces coelicolor 1508, synthesis of the enzyme
was repressed by glucose. Interestingly, this was not the case
in the host strain, where α-amylase activity was detected in
the presence of glucose. Glucose repression of aml expression in
S. coelicolor was relieved in mutants deficient in ATP-dependent
glucose kinase (Virolle and Bibb, 1988). The lack of glucose
repression of pullulanase synthesis in S. griseus K942 and
S. lividans TK24 (both strains utilize glucose) and of
α-amylase synthesis in S. lividans TK24 supports the hypothesis
that possible regulatory properties of glucose kinase in relation to carbon catabolite repression are specific for S.
coelicolor and cannot be generalized for all Streptomyces
species (Virolle and Bibb, 1988).

The pullulanase-negative mutant S. lividans 849 did not
grow on minimal agar medium supplemented with pullulan as sole
carbon source. This means that the α-amylase of this strain
cannot hydrolyze pullulan. Assuming similar properties of the
α-amylase of S. griseus K942 and considering the possible defect in the maltose uptake system, maltotriose, maltose and
glucose may be the reaction products of the pullulan-hydrolyzing activity in this strain. Four types of pullulan-hydrolyzing enzymes have been reported: (i) glucoamylase (glucan
1,4-α-glucosidase; EC 3.2.1.3) (Ueda et al., 1963), which

hydrolyzes pullulan from nonreducing ends to produce glucose;
(ii) pullulanase (α-dextrin endo-1,6-α-glucosidase) (Bender
and Wallenfels, 1961), from Klebsiella pneumoniae, which
hydrolyzes α-(1→6)-glucosidic linkages of pullulan to produce
maltotriose; (iii) isopullulanase (EC 3.2.1.57) (Sakano et al.,
1971), from Aspergillus niger, which hydrolyzes α-(1→4)-
glucosidic linkages of pullulan to produce isopanose (6-0-α-
maltosyl-glucose); and (iv) neopullulanase (Kuriki et al.,
1988), from Bacillus stearothermophilus, which hydrolyzes
α-(1→4)-glucosidic linkages of pullulan to produce panose
(6^2-0-α-glucosyl-maltose). Even if the occurrence of a glucoamylase in S. griseus K942 cannot be excluded, the pullulanase
of this strain seems not to belong to one of the types of
pullulan-hydrolyzing enzymes described above.

From the inability of S. griseus K942 to utilize maltose
as a degradation product of pullulan and the changed differentiation behaviour of the pullulanase-negative mutant S. lividans 849, one might expect a role for pullulanase in Streptomyces physiology other than substrate degrading activity.
Since nitrogen limitation is a known prerequisite for differentiation of streptomycetes (Ochi, 1987), the inability of
S. lividans TK24 to form aerial mycelium and spores on the
synthetic agar medium may be caused by the relatively high
concentration of $(NH_4)_2SO_4$. In S. lividans 849 this metabolic
control of differentiation is possibly changed. On the other
hand, glucose as carbon source prevented differentiation in
both parental strain and its mutant. This is reminiscent of
bld mutants of S. coelicolor, which produce normal sporulating
aerial mycelium when grown on mannitol (instead of the normal
carbon source, glucose) as the sole carbon source in minimal
medium (Chater, 1989).

ACKNOWLEDGMENTS

We thank D. Noack and M. Roth for generous provision of
strains; U. Gräfe for provision of pullulan; Ch. Freysoldt
for carrying out the mutagenesis; C. Franke for her technical
assistance; A. Junghans for typing the manuscript.

REFERENCES

Bender, H., and Wallenfels, K., 1961, Untersuchungen an Pullulan. II. Spezifischer Abbau durch ein bakterielles Enzym, Biochem. Z., 334: 79-95.

Chater, K. F., 1989, Multilevel regulation of Streptomyces differentiation, TIG, 5: 372-377.

D'Enfert, C., Chapon, C., and Pugsley, A. P., 1987, Export and secretion of the lipoprotein pullulanase by Klebsiella pneumoniae, Mol. Microbiol., 1: 107-116.

Gräfe, U., Bormann, E. J., Roth, M., and Neigenfind, M., 1986, Mutants of Streptomyces hygroscopicus deregulated in amylase and α-glucosidase formation, Biotechnol. Lett., 8: 615-620.

Guilbot, A., and Mercier, C., 1985, Starch, in: "The Polysaccharides", vol. 3. pp. 209-282. G. O. Aspinall, ed., Academic Press, New York.

Hoshiko, S., Makabe, O., Nojiri, C., Katsumata, K., Satoh, E., and Nagaoka, K., 1987, Molecular cloning and characterization of the Streptomyces hygroscopicus α-amylase gene, J. Bacteriol., 169: 1029-1036.

Kuriki, T., Okada, S., and Imanaka, T., 1988, New type of pullulanase from Bacillus stearothermophilus and molecular cloning and expression of the gene in Bacillus subtilis, J. Bacteriol., 170: 1554-1559.

Lin, Z. L., Chen, J. C., Lee, C. F., Wang, T. T., Chen, Y. L., and Hsu, W. H., 1988, Molecular cloning of an alpha-amylase gene from Streptomyces lividans TK24, in: "Abstract Book, International Symposium on Biology of Actinomycetes" (Tokyo, Japan, May 1988) P3-1.

McKillop, C., Elvin, P., and Kenten, J., 1986, Cloning and expression of an extracellular α-amylase gene from Streptomyces hygroscopicus in Streptomyces lividans 66, FEMS Microbiol. Lett., 36: 3-7.

Melasniemi, H., 1987, Effect of carbon source on production of thermostable α-amylase, pullulanase and α-glucosidase by Clostridium thermohydrosulfuricum, J. Gen. Microbiol., 133: 883-890.

Melasniemi, H., and Paloheimo, M., 1989, Cloning and expres-
    sion of the Clostridium thermohydrosulfuricum α-amylase-
    pullulanase gene in Escherichia coli, J. Gen. Microbiol.,
    135: 1755-1762.

Morgan, F. J., Adams, K. R., and Priest, F. G., 1979, A cul-
    tural method for the detection of pullulan-degrading en-
    zymes in bacteria and its application to the genus Bacil-
    lus, J. Appl. Bacteriol., 46: 291-294.

Ochi, K., 1987, Metabolic initiation of differentiation and
    secondary metabolism by Streptomyces griseus: signifi-
    cance of the stringend response (ppGpp) and GTP content
    in relation to A factor, J. Bacteriol., 169: 3608-3616.

Sakano, Y., Masuda, N., and Kobayashi, T., 1971, Hydrolysis
    of pullulan by a novel enzyme from Aspergillus niger,
    Agric. Biol. Chem., 35: 971-973.

Takizawa, N., and Murooka, Y., 1985. Cloning of the pullula-
    nase gene and overproduction of pullulanase in Escheri-
    chia coli and Klebsiella aerogenes, Appl. Environm.
    Microbiol., 49: 294-298.

Takasaki, Y., 1987, Pullulanase-amylase complex enzyme from
    Bacillus subtilis, Agric. Biol. Chem., 51: 9-16.

Ueda, S., Fujita, K., Komatsu, K., and Nakashima, N., 1963,
    Polysaccharide produced by the genus Pullularia. I. Pro-
    duction of polysaccharide by growing cells. Appl.
    Microbiol., 11: 211-215.

Ueda, S., Yagisawa, M., and Sato, Y., 1971, Production of iso-
    amylase by Streptomyces sp. No. 28, J. Ferment. Technol.,
    49: 552-558.

Virolle, M.-J., and Bibb, M. J., 1988, Cloning, characteriza-
    tion and regulation of an α-amylase gene from Strepto-
    myces limosus, Mol. Microbiol., 2: 197-208.

Virolle, M.-J., Long, C. M., Chang, S., and Bibb, M. J., 1988,
    Cloning, characterization and regulation of an α-amylase
    gene from Streptomyces venezuelae, Gene, 74: 321-334.

Williams, S. T., Goodfellow, M., Alderson, G., Wellington,
    E. M. H., Sneath, P. H. A., and Sackin, M. J., 1983,
    Numerical classification of Streptomyces and related
    genera, J. Gen. Microbiol., 129: 1743-1813.

Yagisawa, M., Kato, K., Koba, Y., and Ueda, S., 1972, Pullu-
    lanase of Streptomyces sp. No. 280, J. Ferment. Technol.,
    50: 572-579.

THE TENDAMISTAT EXPRESSION-SECRETION SYSTEM: SYNTHESIS OF PROINSULIN

FUSION PROTEINS WITH STREPTOMYCES LIVIDANS

Klaus-P. Koller, Günther Rieß, Klaus Sauber,
László Vértesy, Eugen Uhlmann
and Holger Wallmeier

HOECHST AG
6230 Frankfurt/Main 80
FRG

INTRODUCTION

Tendamistat (HOE 467), a potent inhibitor of the human pancreatic α-amylase, is an acidic protein of 74 amino acids (Vértesy et al., 1984). The inhibitor gene was cloned from an amplified genomic sequence of an over-producing strain of Streptomyces tendae and further characterized. By expression of the gene in the heterologous host Streptomyces lividans we demonstrated that secretion of this protein was mediated by a signal peptide dependent mechanism (Koller & Rieß, 1989). We have also evaluated the tendamistat-based secretion system for a number of foreign proteins, for example interleukin II and proinsulin (Bender et al., 1990, Koller et al., 1989). Our main interest focussed on the formation of disulphide bonds, stability and activity of secreted proteins.

Proinsulin is particularly well suited for these experiments because of its complex folding pathway, which requires the formation of three disulphide bridges, and because it is a priori a secreted protein. Recently, we showed the successful secretion of a tendamistat-monkey proinsulin fusion by recombinant Streptomyces. However, formation of disulphide linkages was incorrect (Koller et al., 1989). We now report on the secretory synthesis of a soluble proinsulin fusion protein with correct disulphide bridges obtained after modification of the C-peptide.

RESULTS AND DISCUSSION

To study secretion of proinsulin fusion proteins gene fusions were made between the tendamistat gene and the monkey proinsulin gene. The proinsulin gene codes for a single chain precursor consisting of B-chain (30 amino acids), C-peptide (35 amino acids) and A-chain (21 amino acids). The B- and A -chain are identical to human sequences (Fig. 1). Therefore, human insulin can be prepared from a correctly folded proinsulin precursor e.g. by cleaving the C-peptide with trypsin leaving for example des-Thr(B30) insulin which can be converted into human insulin by

To Prof. W. Wehrmeyer on the occasion of his 60th birthday

transpeptidation in presence of threonine ester or Arg(B31) insulin from which human insulin is easily obtained by the action of carboxypeptidase B (Thim et al., 1989).

```
                                                                      ↓B1
NH₂- Tendamistat - │PHE ASN ALA MET ALA THR GLY ASN SER ALA ARG│PHE VAL

ASN GLN HIS LEU CYS GLY SER HIS LEU VAL GLU ALA LEU TYR LEU VAL CYS GLY
                              ↓B30 C1 ↓
GLU ARG GLY PHE PHE TYR THR PRO LYS THR ARG ARG GLU ALA GLU ASP PRO GLN

VAL GLY GLN VAL GLU LEU GLY GLY GLY PRO GLY ALA GLY SER LEU GLN PRO LEU
                         C35│A1
                            ↓
ALA LEU GLU GLY SER LEU GLN LYS ARG GLY ILE VAL GLU GLN CYS CYS THR SER
                                       A21
ILE CYS SER LEU TYR GLN LEU GLU ASN TYR CYS ASN-COOH
```

Fig. 1. Amino acid sequence of the monkey proinsulin part of a fusion
         protein secreted by recombinant Streptomyces lividans.
         B1-B30 B-chain, C1-C35 connecting (C-)peptide, and A1-A21
         A-chain of proinsulin, respectively. Boxed is the spacer peptide.
         Preferential cleavage sites for trypsin are indicated by arrows.

To establish gene fusions between the tendamistat and the proinsulin gene, first a unique Kpn I restriction site was introduced into the tendamistat coding sequence by site –directed mutagenesis 18bp upstream from the termination codon. Using this restriction site an oligonucleotide linker was designed to connect both genes. It restored the sequence coding for the carboxy terminal end of tendamistat and introduced an 11 amino acid spacer peptide between tendamistat and proinsulin (Fig.1, Fig 2A). The spacer peptide was optimized by a molecular dynamics computer simulation to allow independent folding of the fusion partners (Koller et al., 1989). Additionally, a codon for methionine was introduced to allow chemical cleavage by CNBr to release proinsulin for further processing. Alternatively, enzymatic cleavage of the fusion protein with trypsin or lysylendopeptidase (LEP) for the direct release of insulin derivatives could be applied. Both enzymes do not act on native tendamistat.

Expression of the fusion protein was achieved after cloning the fused genes into the Streptomyces plasmid pIJ702 generating the expression vector pKK500 (Fig. 3). The expression of the gene fusion is driven by a tandem promoter arrangement combining the tyrosinase (mel) and tendamistat promoters which was shown to strongly increase the yields in tendamistat secretion (Koller & Rieß, 1989). Using this construction, the tendamistat – proinsulin fusion protein is secreted in amounts between 20 – 100 mg/l, depending on the culture medium and the fermentation conditions.

228

With longer fermentation times proteolytic digestion of the fusion protein occured. We therefore analyzed the spectrum of secreted proteases, and found that a trypsin and a chymotrypsin-like protease specifically attack the fusion protein. The trypsin-like activity is inhibited by the addition of zinc to the culture medium (Aretz et al., 1989). To inhibit the action of the chymotrypsin-like activity, a different approach was chosen. Since especially the amino acids phenylalanine and leucine are good substrates for chymotrypsin-like endoproteolytic activities, a novel oligonucleotide linker was synthesized in which the codons for phenylalanine and leucine were replaced by the codons for alanine and glycine, respectively. In addition, the codon for methionine was changed to the codon for glutamic acid to avoid restart of protein translation. The spacer peptide sequence is given in Fig. 2B. This modified fusion protein was also secreted and shown to be significantly more stable during long term fermentations.

```
        T74                                                        B1
A)  Cys LEU PHE ASN ALA MET ALA THR GLY ASN SER ALA ARG Phe

        T74                                                        B1
B)  Cys ALA GLY ASN ALA GLN ALA THR GLY ASN SER ALA ARG Phe
```

Fig. 2. Modification of the spacer peptide in tendamistat-proinsulin fusion proteins to reduce the degradation of the fusion protein by a chymotrypsin-like activity. A) original, and B) modified spacer sequence, respectively. Boxed are amino acid substitutions.

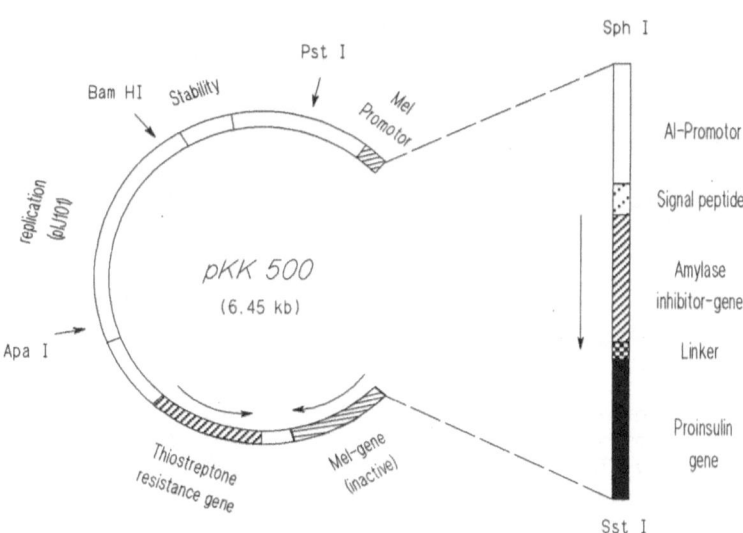

Fig. 3. Vector for expression of tendamistat-proinsulin fusion proteins in Streptomyces lividans TK24.

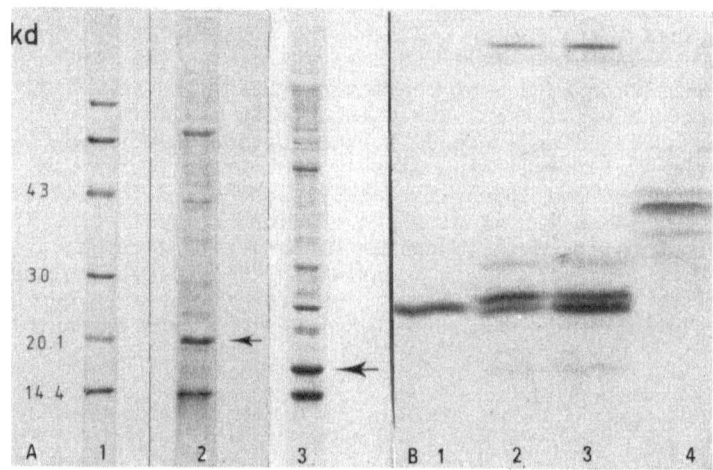

Fig. 4 A. Secretion of tendamistat-monkey proinsulin (2) and
tendamistat-Insu-Lys (3) fusion proteins. Analysis of the
culture supernatant was done on 10 - 17.5% SDS PAGE.
B. Enzymatic cleavage of purified tendamistat-Insu-Lys fusion
protein (4) by LEP. Des-Thr(B30) insulin (1), LEP digestion of
fusion protein after 1 hr (2) and overnight (3).
Analysis was carried out on native 15% PAA gels.

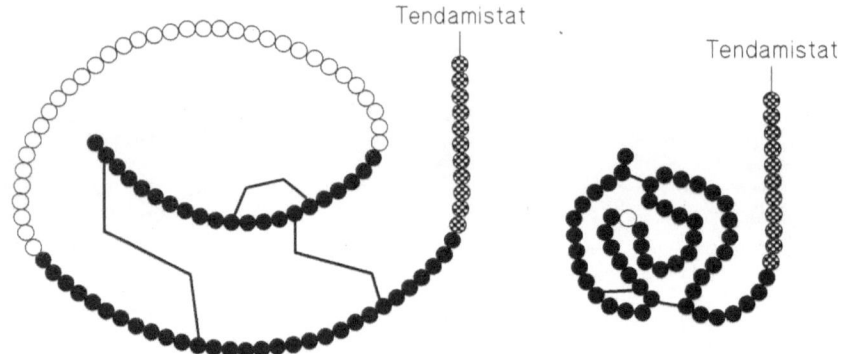

Fig. 5. Comparison of the secondary structure of (A) monkey - and (B)
Insu-Lys-proinsulin fusion proteins. Amino acids composing spacer
peptide (✪), B- and A- chain of human insulin (●), and
C-peptides (○), respectively. (B) adapted from Thim et al.,
1989.

230

The secreted fusion protein was purified from the culture broth by acid precipitation at pH 4.0 and cationic chromatography on Mono S$^{(R)}$. However, cleavage of the protein with trypsin did not yield insulin precursors, for example des-Thr(B30) insulin. Only after reduction, and re-folding of the fusion protein, followed by trypsin digestion, fully active des-Thr(B30) insulin could be obtained (Koller et al., 1989).

Conformational studies on disulphide bridges in the tendamistat-proinsulin fusions as well as analyses of the enzymatic cleavage products suggest that the tendamistat part is correctly disulphide-linked, whereas in the proinsulin part disulphide bridges are inadaequately formed (Vertesy et al., in press). The correctly folded tendamistat part of the fusion protein retains its inhibitory activity. This offers a significant advantage in strain improvement experiments, since the size of the blue halos in the α-amylase inhibitor plate test (Koller et al., 1984) is corresponding to the amount of secreted fusion protein. To address the problem of incorrect disulphide bridge formation in the proinsulin part we examined the influence of the C-peptide on the folding of the precursor. Thim et al. (1986) had shown for the yeast secretory synthesis of single chain proinsulin precursors that reducing the number of amino acids of the C-peptide did not influence the correct formation of disulphide bonds. The lower limit was two basic amino acids substituting for the 35 residues of the original C-peptide.

We decided to replace the C-peptide by one amino acid, lysine, and also placing a lysine in front of the first amino acid of the B chain, phenylalanine. Enzymatic cleavage of such a precursor with LEP, a specific endoproteinase cutting at the carboxy terminal side of lysine would release des-Thr(B30)in a single reaction after complete digestion, which is a fully active insulin derivative (Koller et al., 1989). As tendamistat, although containing one lysine is not cleaved by this enzyme, the two main cleavage products could easily be analyzed on native gels.

To prove this concept a gene was synthesized coding for such a proinsulin precursor and cloned into pKK 500 to replace the proinsulin gene (Koller et al., submitted). This plasmid was transformed into Streptomyces lividans TK 24. After three days of fermentation of the recombinant Streptomyces strain, supernatants of culture broths were analyzed for α-amylase inhibitor activity and proteins reacting with insulin antibodies in immunoblots. Fig. 4A shows the analysis of the supernatant by SDS gel electrophoresis. A major protein of an apparent molecular weight of 16 kD which uniquely reacts with insulin antibodies is newly synthesized in similar yields we had obtained with the tendamistat-proinsulin fusion protein. After purification of the protein as described above, complete digestion with LEP clearly released des-Thr(B30) insulin as shown by analytical gel electrophoresis in native 15% PAA-gels (Fig. 4B). We therefore conclude, that the formation of all three disulphide bridges is correct in the biosynthetic proinsulin precursor which we named Insu-Lys fusion protein. Additionally, the tendamistat part retains its inhibitory activity offering the advantages for strain improvement as described. Models of the secondary structures for both the tendamistat-proinsulin and the tendamistat-Insu-Lys fusion proteins are given in Fig. 5.

Recently, Chang and Chang (1988), Steiert et al. (1989), Bender et al. (1990),and Brawner et al. (1990) showed secretory synthesis of quite different heterologous proteins like tumor necrosis factor, a membrane-bound hydrolase from Flavobacterium, interleukin II, and soluble CD4 receptor domain, respectively. The successful secretion of single chain proinsulin precursors is another example for the high secretion potential of Streptomyces.

On the molar basis, about 38% of Insu-Lys fusion is proinsulin. This corresponds to yields of about 30 - 40 mg insulin/l. Thim et al. (1989) reported expression levels of up to 10.5 mg insulin/l for single chain precursors with shortened C-peptides in the yeast secretion system. Thus yields obtained with recombinant Streptomyces appear to be superior to yeast, not taking into account strain improvement and optimization of culture media and fermentation conditions. Therefore, we regard Streptomyces an attractive alternative to other expression systems for production of human insulin.

ACKNOWLEDGMENTS

We thank Prof. E.-G. Afting and Dr. H. Dornauer for continuous support and U. Geckeler, A. Janssen, W. Klingenberg, D. Rabold, and A. Schwarz for expert technical assistance.

LITERATURE

Aretz, W., Koller, K.-P., and Riess, G., 1989. Proteolytic enzymes from recombinant Streptomyces lividans TK24. 1989. FEMS Microbiology Letters 65:31

Bender, E., Koller, K.-P., and Engels, J.W., 1990. Secretory synthesis of human interleukin-2 by Streptomyces lividans. Gene, 86:227

Brawner, M., Taylor, D., and Fornwald, J. 1990. Expression of the soluble CD-4 receptor in Streptomyces. J. Cellular Biochemistry, Supplement 14A:103

Chang, S.-Y., and Chang, S. 1988. Secretion of heterologous proteins in Streptomyces lividans. Seventh Int. Symp. on Biology of Actinomycetes. Y. Okami, T. Beppu, H. Ogawara (Eds). Japan Scientific Societies Press, Tokyo. p. 103

Koller, K.-P., Engels, J.W., and Uhlmann, E. 1984. Gene amplification and over-production of the α-amylase inhibitor (Hoe 467, Tendamistat) in Streptomyces tendae. Third European Congress on Biotechnology, Vol.III:273

Koller, K.-P., and Riess, G., 1989. Heterologous expression of the α-amylase inhibitor gene cloned from an amplified genomic sequence of Streptomyces tendae. J.Bacteriol. Vol. 171, No.9:4953

Koller, K.-P., Riess, G., Sauber, K., Uhlmann, E., and Wallmeier, H., 1989. Recombinant Streptomyces lividans secretes a fusion protein of tendamistat and proinsulin. Bio/Technology Vol. 7:1055

Steiert, J.G., Pogell, B.M., Speedie, M.K., and Laredo, J. 1989. A gene coding for a membrane-bound hydrolase is expressed as a secreted, soluble enzyme in <u>Streptomyces lividans</u>. <u>Bio/Technology</u>, 7:65

Thim, L., Hansen, M.T., Norris, K., Hoegh,I., Boel, E., Forstrom, J., Ammerer, G., and Fiil, N.P. 1986. Secretion and processing of insulin precursors in yeast. <u>Proc</u> . <u>Natl</u>. <u>Acad</u>. <u>Sci</u>. USA 83:6766

Thim, L., Snel, L., Norris, K., and Hansen, M.T. 1989. Insulin and insulin analogs from <u>Saccharomyces cerevisiae</u>. In: Genetics and Molecular Biology of <u>Industrial Microorganisms</u>. C.L. Hershberger, S.W. Queener, G. Hegemann (Eds).American Scociety for Microbiology, Washington p. 322

Vértesy,L., Oeding,V., Bender,R., Zepf,K. and Nesemann,G. 1984. Tendamistat (HOE 467), a tight binding α–amylase inhibitor from <u>Streptomyces tendae</u> 4158. Isolation and biochemical properties. <u>Eur</u>. <u>J</u>. <u>Biochem</u>. 141:505

Vértesy,L., Tripier,D., Koller,K.-P. and Riess,G. 1991. Disulphide bridge formation of proinsulin fusion proteins during secretion in <u>Streptomyces</u>. <u>Biol</u>. <u>Chem</u>. <u>Hoppe-Seyler</u> 372: in press

CONSTRUCTION AND USE OF A SECRETION VECTOR IN *STREPTOMYCES*

Claudine Piron-Fraipont, Mauro V. Lenzini and Jean Dusart

Fundamental Bacteriology Unit, Department of Microbiology
University of Liège
Liège, Belgium

INTRODUCTION

The main research project of the Fundamental Bacteriology Unit in Liège is the unravelling, at the molecular level, of the mode of action of β-lactam antibiotics. For this purpose, most of the model target enzymes have been selected in the genus *Streptomyces* (Frère and Joris, 1985; Ghuysen *et al.*, 1989). In the frame of this program, we were brought along to overproduce several of these model enzymes as well as several β-lactamases. This was carried out by cloning the corresponding genes in *Streptomyces* multicopy plasmids as pIJ702 or related vectors. This part of the work is now over, seven genes (three β-lactamases, and four penicillin-sensitive DD-peptidases) have been cloned, and sequenced and their products were overexpressed in *Streptomyces lividans* (Dehottay *et al.*, 1986, 1987; Duez *et al.*, 1987 and unpublished results; Lenzini *et al.*, 1987, 1988; Piron-Fraipont *et al.*, 1989; Houba *et al.*, 1989).

In the meantime, several papers have been published, emphasizing the possible role of *Streptomyces* as a host for the production of extracellular heterologous proteins (Chang and Chang, 1988; Nagashima *et al.*, 1989). We have thus thought about the construction of a *Streptomyces* secretion vector, using the transcription and secretory signals from one of our genes.

RESULTS AND DISCUSSION

Transcriptional analysis of the *dac* gene

The gene encoding the extracellular DD-carboxypeptidase of *Streptomyces* R61 (Duez *et al.*, 1987) — the *dac* gene — was chosen for this purpose. The presence of a unique *Pst*I site 11 codons downstream of the processing site allowed the construction of fused genes without altering this processing site (see Fig. 1).

The region upstream of the *dac* structural gene was analysed in promoter probing experiments, using the promoter-probe vector pIJ424, the reporter gene of which codes for kanamycin resistance (Hopwood *et al.*, 1985). Two different fragments shown in Fig. 2 were tested. The resistance towards kanamycin induced in *S. lividans* by pIJ424 containing the 1565-bp *Bgl*II-*Bcl*I segment or the 631-bp *Sau*3A-*Bcl*I segment, came to 500 and 800 μg of

SalI
GTCGACACCCTGCTCACCACCGCGGCCCAGGCGCACAGCGACGACATGGTCGCCCGCGCCCTTCTACTCCCACACCTCCCC   80

CGACGGCAGCCAGTGCCGGGGACCGGGCCGCCGCCGCGGCTCCGAGCGCCGCACCATCGGGGAGAACATCGCCTGCGGCC   160

AGCGCTCCCCCGCCGAGGTCGTCCGCGCGTGGATGAACAGCCCGGGCCACCGCGCCAACATCCTCAAGCCCGATTTCACC   240
                                                                          Sau3A
CACATAGCCATCGGCTTCGCGGGCGGCGGCTCGGCCGGGCCGACGTACTGGACGCAGCTTTTCGGCCCGCTGAAGGATCT   320
            -35   -10   ** **
TTTCCGCTTTCCTTCGCGTGACATGCAACCCATCTGCCCCTCCTGCGCGTAGAGATGGTGCCGGCGTTCGCATCCGCGGC   400

                                        -31 Met Val Ser Gly Thr Val Gly Arg Gly Thr Ala
                         SD
GCGGTACACAGAGGGAACTCGGGAGAAGAATCAG   ATG GTC TCA GGA ACG GTG GGC AGA GGT ACG GCG   467

 20   -10   NotI ↓
Leu Gly Ala Val Leu Leu Ala Leu Leu Ala Val Pro Ala Gln Ala Gly Thr Ala Ala Ala
CTG GGC GCG GTG CTG TTG GCC CTC CTC GCA GTC CCC GCA CAG GCC GGC ACC GCC GCG GCC   527

  1 Sau3A   10 PstI   20
Ala Asp Leu Pro Ala Pro Asp Asp Thr Gly Leu Gln Ala Val Leu His Thr Ala Leu Ser
GCG GAT CTG CCG GCA CCC GAC GAC ACC GGT CTG CAG GCG GTG CTG CAC ACG GCC CTT TCC   587

                                    30 SalI   40
Gln Gly Ala Pro Gly Ala Met Val Arg Val Asp Asp Asn Gly Thr Ile His Gln Leu Ser
CAG GGA GCC CCC GGT GCG ATG GTG CGG GTC GAC GAC AAC GGC ACG ATC CAC CAG TTG TCG   647

                        50   60
Glu Gly Val Ala Asp Arg Ala Thr Gly Arg Ala Ile Thr Thr Thr Asp Arg Phe Arg Val
GAG GGA GTC GCC GAC CGG GCC ACC GGG CGT GCG ATC ACC ACG ACC GAC CGG TTC CGC GTC   707

- 140 BclI
                                              Leu Ile Thr
- CTG ATC ACC

Figure 1. Sequence of the *Streptomyces* R61 *dac* gene and position of the
         restriction sites of interest. The Shine-Dalgarno sequence (SD),
         transcriptional starts (asterisks) and processing site (vertical
         arrow) are also shown.

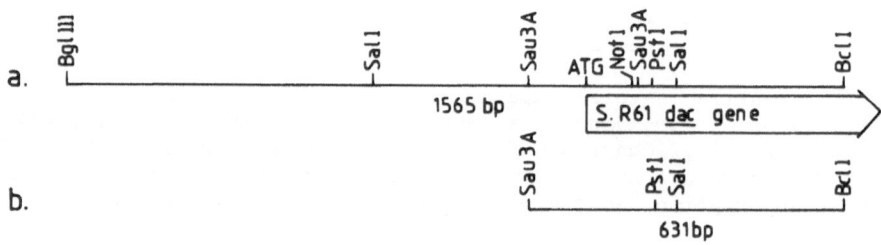

Figure 2. DNA segments used in promoter-probing assays.

kanamycin per ml, respectively. This result suggests a negative effect of
the *Bgl*II-*Sau*3A stretch on the activity of the downstream promoter.

    S1 mapping analysis was carried out using as a probe the *Sal*I-*Not*I
segment (see Fig. 1). Four major and one minor hybrids were observed,
around 55-bp upstream of the ATG initiator codon (see Fig. 1). Hexamer
consensus - 10 and - 35 sequences were tentatively identified although the
spacing exceeds the usual 17-18 bp length.

Construction of the secretion vector : pDML116

The *Sau*3A-*Pst*I segment (Fig. 1) was isolated and inserted in pIJ702 giving rise to pDML116 (see Fig. 3) where a foreign gene can be inserted in the unique *Pst*I site. It should be noticed that the junction between *Sau*3A (from the insert) and *Bgl*II (from the vector) has regenerated a *Bgl*II site which is unique in the construction.

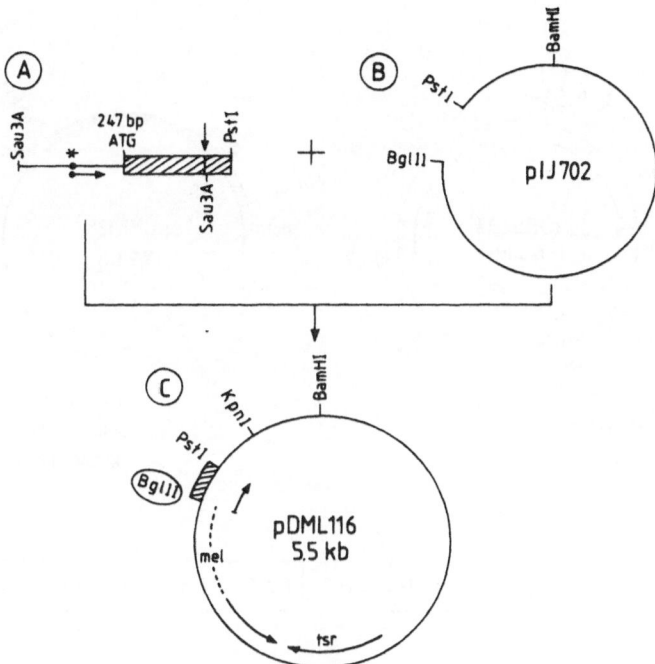

Figure 3. Construction of pDML116.

As a first trivial test, the original R61 *dac* gene was reinserted behind its signal sequence. When the plasmid (pDML120) containing this reconstituted *dac* gene was introduced in *S. lividans*, it induced the synthesis of 30-40 mg of enzyme per litre of culture, instead of the 15-20 mg of enzyme produced by the primary clone *S. lividans*/pDML110 (Duez et al., 1987). The difference at the gene level between pDML120 and pDML110 is the presence in the latter of the *Bgl*II-*Sau*3A upstream segment, which was shown hereabove to exert a negative effect on the expression of the reporter gene (*in vivo* promoter probing assays). A possible explanation of this negative effect could be the occurrence in this upstream region of an open reading frame which overlaps with the promoter of the *dac* gene, the TAG terminator codon being part of the - 10 consensus sequence (see Fig. 1).

Construction of pDML128 : expression-secretion of the R-TEM β-lactamase (Fig. 4)

pDML116 has been tested with an easily detectable *Escherichia coli* gene, the R-TEM β-lactamase gene form pBR322. The gene was isolated from form pJBS633, where it occurs devoid of its own signal sequence (Broome-Smith and Spratt, 1986). pDML116 was linearized with *Pst*I and the *Pst*I ends were blunted before ligation with the *Pvu*II-*Nru*I fragment form pJBS633. *S. lividans* was transformed with the ligation mixture and one single β-lactamase-producing clone was found. The nucleotide sequence of the junction has been determined and is given in Fig. 4. The fused protein thus consists of the entire R-TEM β-lactamase added with 11 N-terminal residues deriving from the Dac protein.

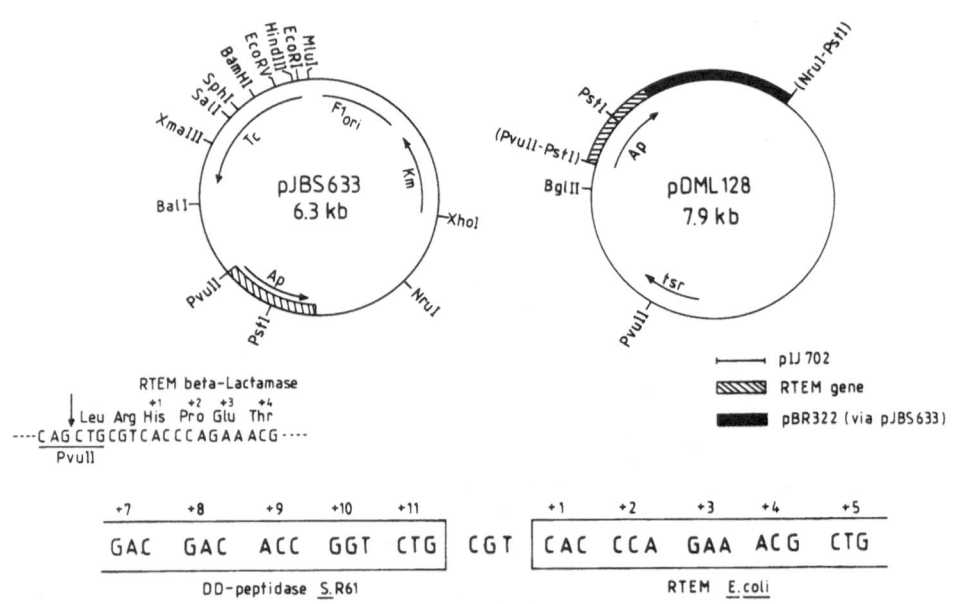

Figure 4. Construction of pDML128.

Under optimal conditions of growth, the yield of β-lactamase obtained from *S. lividans*/pDML128 was by far lower the yield of DD-carboxypeptidase from *S. lividans*/pDML120. Only 0.3 mg of R-TEM β-lactamase was produced per litre of culture. The enzyme was almost entirely found in the culture fluid, indicating that the *Streptomyces* signal sequence is efficient for the export of the *E. coli* protein from *Streptomyces* cells.

When the R-TEM gene was present in *S. lividans* under the control of its own *E. coli* promoter and signal sequences (*S. lividans*/pDML129, see Fig. 5), the yield of β-lactamase is still lower than that obtained with pDML128. The most striking observation is that one-third of the total activity is found in the cytoplasmic fraction. This suggests that this *E. coli* signal sequence does not operate properly in *Streptomyces*. The same observation has been made in our laboratory with the *amp*C β-lactamase gene from the *Enterobacter cloacae* chromosome (unpublished result). On the contrary, the signal sequence of the *Salmonella typhimurium* oxa-2 β-lactamase has been reported to be fully efficient in *S. lividans* for exporting the Oxa-2 protein from *S. lividans* (Ali and Dale, 1986).

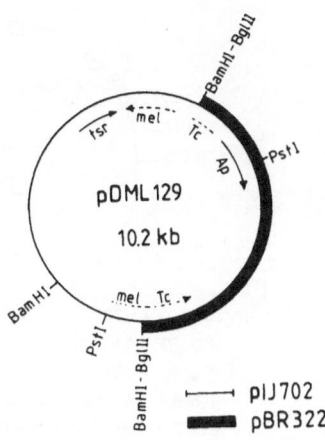

Figure 5. Plasmid pDML129 was obtained by direct
fusion between pBR322 and pIJ702.

Use of strong promoters in pDML116

As noticed above, pDML116 has a unique *Bgl*II site upstream of the *dac*
promoter region (Fig. 3,C). Promoter sequences known for their high promo-
ting strength were inserted in this *Bgl*II site. We expected to see them
taking the control of the gene and determining the level of expression of
the downstream structural gene.

Two promoter regions were selected which were available as short *Bgl*II
segments. The first one has been isolated in our laboratory from the genomic DNA of *Bacillus licheniformis* and has been shown to activate the kanamycin resistance reporter gene of pIJ424, allowing the *S. lividans* host
cells to grow in the presence of 800-900 µg of kanamycin per ml (unpublished results). The second one was the *erm*E promoter from *Saccharopolyspora
erythraea* (Bibb *et al.*, 1986), the activity of which appeared to be still
higher when tested in pIJ424 (1000 µg kanamycin per ml).

Plasmids pDML119 and pDML122 were constructed (Fig. 6). They contain
the *B. licheniformis* or the *S. erythraea* promoter region properly oriented
immediately upstream of the entire *dac* gene.

S1 mapping analysis showed that in both constructs only the added promoter is active. As hoped, it has taken precedence over the *dac* promoter
but, to our disappointment, the efficiency was not the expected one. The
yields of DD-peptidase from *S. lividans* transformed with pDML119 or pDML122
were 9 and 7 mg of enzyme per litre, respectively, instead of 30-40 mg when
only the complete *dac* gene is present (pDML110).

It has been proposed by several authors (Iserentant and Fiers, 1980;
Hall *et al.*, 1982) that the expression of a gene is maximized when the secondary structure of the mRNA is such that the initiator codon is accessible.

Predictions of secondary structure of the RNA transcripts expected
from pDML110, pDML119 and pDML122, were established using the Fold and
Squiggles program from the Wisconsin group (Freier *et al.*, 1986). In the
first case, the ribosome-binding site and the AUG initiator codon of the
dac gene are always found at the end of a hairpin structure. The AUG codon
is located entirely or partially (A being the only paired base) in a loop.
This situation is very rarely observed in the transcripts from pDML119 and
pDML122. These considerations could explain the low yields of DD-peptidase

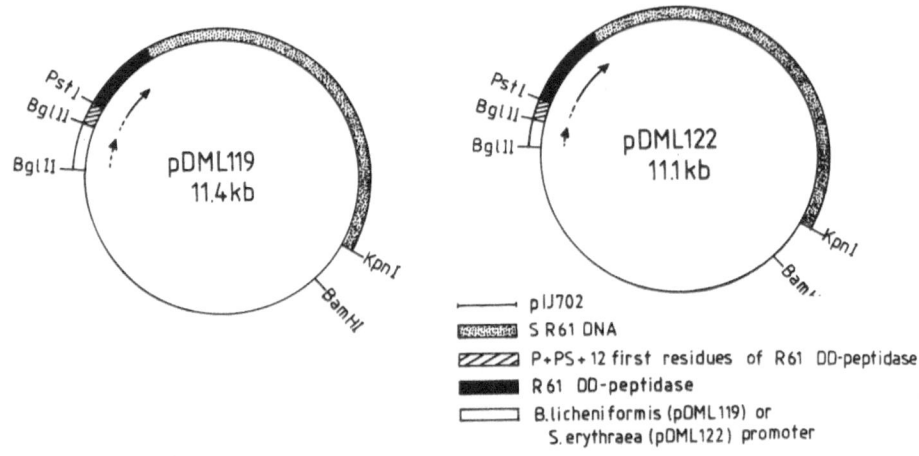

Figure 6. Structure of pDML119 and pDML122.

obtained from these constructs. However, it is to be kept in mind that se-condary structures are predicted with a probability of 82 % only and that several other factors could influence the expression of a gene.

REFERENCES

Ali, N.A., and Dale, J.W., 1986, Secretion by *Streptomyces lividans* of a cloned Gram-negative β-lactamase, *FEMS Microbiol. Lett.*, 33:277.

Bibb, M.J., Janssen, G.R., and Ward, J.M., 1986, Cloning and analysis of the promoter region of the erythromycin resistance gene (*ermE*) of *Streptomyces erythraeus*, *Gene*, 41:E357.

Broome-Smith, J.K., and Spratt, B.G., 1986, A vector for the construction of translational fusions to TEM β-lactamase and the analysis of protein export signals and membrane protein topology, *Gene*, 49:341.

Chang, S.Y., and Chang, S., 1988, Secretion of heterologous protein in *Streptomyces lividans*, in : "Biology of Actinomyces '88", Y. Okami, T. Beppu, and H. Ogawara, eds, Japan Scientific Societies Press, Tokyo.

Dehottay, P., Dusart, J., De Meester, F., Joris, B., Van Beeumen, J., Erpi-cum, T., Frère, J.M., and Ghuysen, J.M., 1987, Nucleotide sequence of the gene encoding the *Streptomyces albus* G β-lactamase precursor, *Eur. J. Biochem.*, 166:345.

Dehottay, P., Dusart, J., Duez, C., Lenzini, M.V., Martial, J.A., Frère, J.M., Ghuysen, J.M., and Kieser, T., 1986, Cloning and amplified expression in *Streptomyces lividans* of a gene encoding extracellular β-lactamase from *Streptomyces albus* G, *Gene*, 42:31.

Duez, C., Piron-Fraipont, C., Joris, B., Dusart, J., Urdea, M.S., Martial, J.A., Frère, J.M., and Ghuysen, J.M., 1987, Primary structure of the *Streptomyces* R61 extracellular DD-peptidase. 1. Cloning into *Strep-tomyces lividans* and nucleotide sequence of the gene, *Eur. J. Bio-chem.*, 162:509.

Freier, S.M., Kierzek, R., Jaeger, J.A., Sugimoto, N., Caruthers, M.H., Neilson, T., and Turner, D.H., 1986, Improved free-energy parame-ters for predictions of RNA duplex stability, *Proc. Natl. Acad. Sci. USA*, 83:9373.

Frère, J.M., and Joris, B., 1985, Penicillin-sensitive enzymes in peptido-glycan biosynthesis, *Crit. Rev. Microbiol.*, 11:299.

Ghuysen, J.M., Frère, J.M., Joris, B., Dusart, J., Duez, C., Leyh-Bouille, M., Nguyen-Distèche, M., Coyette, J., Dideberg, O., Charlier, P., Dive, G., and Lamotte-Brasseur, J., 1989, Inhibition of enzymes involved in bacterial cell wall synthesis, <u>in</u> : "Design of enzyme inhibitors as drugs", M. Sandler, and H.J. Smith, eds, Oxford University Press, Oxford.

Hall, M.N., Gabay, J., Debarbouillé, M., and Schwartz, M., 1982, A role for mRNA secondary structure in the control of translation initiation, *Nature*, 295:616.

Hopwood, D.A., Bibb, M.J., Chater, K.F., Kieser, T., Bruton, C.J., Kieser, H.M., Lydiate, D.J., Smith, C.P., Ward, J.M., and Schrempf, H., 1985, "Genetic manipulation of *Streptomyces*. A laboratory manual", The John Innes Foundation, Norwich.

Houba, S., Willem, S., Duez, C., Molitor, C., Dusart, J., Frère, J.M., and Ghuysen, J.M., 1989, Nucleotide sequence of the gene encoding the active-site serine β-lactamase from *Actinomadura* R39, *FEMS Microbiol. Lett.*, 65:241.

Iserentant, D., and Fiers, W., 1980, Secondary structure of mRNA and efficiency of translation initiation, *Gene*, 9:1.

Lenzini, M.V., Ishihara, H., Dusart, J., Ogawara, H., Joris, B., Van Beeumen, J., Frère, J.M., and Ghuysen, J.M., 1988, Nucleotide sequence of the gene encoding the active-site serine β-lactamase from *Streptomyces cacaoi*, *FEMS Microbiol. Lett.*, 49:371.

Lenzini, M.V., Nojima, S., Dusart, J., Ogawara, H., Dehottay, P., Frère, J.M., and Ghuysen, J.M., 1987, Cloning and amplified expression in *Streptomyces lividans* of the gene encoding the extracellular β-lactamase from *Streptomyces cacaoi*, *J. Gen. Microbiol.*, 133:2915.

Nagashima, M., Okumoto, Y., and Okamishi, M., 1989, Nucleotide sequence of extracellular xylanase in *Streptomyces* species n°36a and construction of secretion vectors using xylanase gene, <u>in</u> : "Trends in Actinomycetology in Japan", Y. Koyama, ed., Society for Actinomycetes, Tokyo.

Piron-Fraipont, C., Duez, C., Matagne, A., Molitor, C., Dusart, J., Frère, J.M., and Ghuysen, J.M., 1989, Cloning and amplified expression in *Streptomyces lividans* of the gene encoding the extracellular β-lactamase of *Actinomadura* R39, *Biochem. J.*, 262:849.

GENETIC INSTABILITY

H. Schrempf

F B Biologie/ Chemie
Universitat Osnabruck
F R G

Streptomyces strains show a high degree of instability. This manifests itself in the occurrence of variants which have lost the ability to produce aerial mycelium and spores, pigments, antibiotics, to form argininosuccinate synthetase or to utilise certain amino acids as a nitrogen source. Variants which are sensitive or resistant to various antibiotics can be easily isolated.

Using chromosome-walking studies and pulsed-field electrophoresis, it could be demonstrated that genes encoding variable traits are located within the chromosome, of which continuous stretches of up to several hundred kb of DNA can be lost at a high frequency.

Concomitant with the deletions, amplification of various amplifiable DNA units are likely to occur. The functions of most amplifiable DNA units are so far unknown but in some cases it was possible to select for amplification of antibiotic resistance determinants.

Using continuous culture conditions which have been optimised for vector-containing strains, it may be possible to select stable strains of Streptomyces.

The following contributions report recent progress towards an understanding of plasmid and genome plasticity.

GENETIC INSTABILITY IN STREPTOMYCES

H. Schrempf

FB Biologie/Chemie
Universität Osnabrück, FRG

INTRODUCTION

Streptomycetes are Gram-positive soil microorganisms which differentiate into substrate mycelium and spores. They produce a broad range of different metabolites, such as extracellular enzymes, enzyme inhibitors, pigments, cytostatics and antibiotics. Results from several researchers (1-7,10) have demonstrated that Streptomyces strains are often genetically unstable.

UNSTABLE CHARACTERISTICS

Typically, Streptomyces strains lose characteristics such as aerial mycelium and spore formation, antibiotic, pigment, A-factor, or extracellular enzyme production spontaneously or after treatment with plasmid-curing agents at frequencies of 10^{-2} to 10^{-3}.

In addition, many characteristics used in numerical taxonomy differ significantly among several variants and their corresponding wild-type strains. Thus, the isolated variants may have to be grouped in a cluster different from that of the wild-type strain. Also, new secondary metabolites have been identified within variants (16).

SEGREGATION PATHWAYS

To determine whether certain functions of Streptomyces strains are being altered in a predictable manner, segregation patterns of different wild-type strains were studied. Analyses of these data revealed that variants no longer differentiating aerial mycelium or spores or both, frequently lose their ability to produce melanin or antibiotics (e.g. Streptomyces reticuli, Streptomyces scabies and several others).

Morphological variants of Streptomyces lividans give rise to chloramphenicol-sensitive (Cml^{S}) variants which then segregate further to argG (argininosuccinate synthetase)-negative variants (1,2,7). If, however, selection is kept for chloramphenicol resistance, tetracycline-sensitive (Tet^S) strains can be isolated. They segregate further at a high frequency (10%) to Ntr variants, which grow well only on minimal medium plates containing glutamate or NH_4Cl in place of asparagine as a nitrogen source (6). These variants still contain glutamine synthetase, which can be modified by adenylation but which is no longer positively regulated. If selection for chloramphenicol resistance is released, these strains are able to segregate further to Tet^S Ntr Cml^S Arg$^-$ strains. Similarly Cml^S Arg$^-$ strains can acquire the Tet^S and Ntr characteristics. Recently we demonstrated that variants of any type can segregate further to variants which are also sensitive to $HgCl_2$ or fusidic acid or both.

CHARACTERISTICS OF UNSTABLE GENES

To determine why certain functions are not expressed within different variants, unstable genes were cloned and characterised.

The argG gene of S. lividans encodes argininosuccinate synthetase and has been cloned and expressed in Escherichia coli (2). The chloramphenicol resistance gene has been obtained by "shotgun" cloning Sau3A fragments of S. lividans wild-type DNA in the low-copy-number vector pGM2 (17) and transformation of the S. lividans variant N1 (Cml^S Tet^S Ntr). The mechanism of chloramphenicol resistance is so far unknown, but we could exclude acetylation of the antibiotic and the action of a hydrolase as shown for S. venezuelae (13). The structural gene encoding tetracycline resistance has also been obtained by "shotgun" cloning, using the N1 strain as host. We could demonstrate that it contains sequences which are common within GTP binding proteins (Dittrich, Schrempf, to be published) and shares high homology with the octI gene recently identified in Streptomyces rimosus (11).

DELETION AND AMPLIFICATION

In S. lividans and several other Streptomyces strains the gene encoding arginosuccinate synthetase (argG), which has been cloned by different groups (2,10,12) is deleted at a high frequency in spontaneously occuring Arg$^-$ mutants.

The profile of genetic instability in S. lividans is stepwise. First, there is a loss of the chloramphenicol resistance determinant (cml^r), followed by deletion of the argG gene (2).

Chromosome walking studies indicated that the chloramphenicol resistance determinant is located outside the well-characterised region surrounding the argG gene and the occurrence of very large deletions within the DNA of Cml^S Arg$^-$ strains.

Genetic (6) data suggest that the locus encoding chloramphenicol resistance is linked to the argG gene.

246

Interestingly, in S. lividans both unstable loci, argG and cml^r and the amplifiable DNA unit (AUD) are located within the "silent" region of the S. lividans genome.

By use of gene libraries, DNA regions surrounding the various unstable determinants have been characterised. Comparative hybridisations revealed that most argG variants arise by deletion of at least 73 kilobases (kb), including the argG gene. Chloramphenicol-sensitive, argG variants delete at least an additional 15kb of DNA, including the determinant for chloramphenicol resistance (16).

Two tandem copies of a 6.8kb AUD are located 25kb away from the argG gene. By use of gene libraries, in the wild-type genome a region next to the argG locus has been sequenced which shares homology with part of the amplifiable sequence. It seems likely that recombination among these homologous sequences leads to deletion (Betzler, Schrempf, to be published). Thus, most argG cml^S strains deriving from the wild-type strain contain, in addition to the extensive deletions, 300 to 500 copies of the 6.8kb AUD in tandem array (1,2).

In S. lividans, we demonstrated that amplification occurs in a predictable manner only if two copies of the 6.8kb AUD sequence are present and that deletions extend into the AUD (16). On the other hand, amplification of the 6.8kb AUD does not occur if selection is kept for chloramphenicol. In these strains, however, tetracycline-sensitive variants, which frequently have an Ntr phenotype, arise at high frequencies. In contrast to Cml^S Arg$^-$ strains, only a small portion of Tet^S variants contain ADS with no specific pattern to the type of amplification present. Tetracycline-sensitive variants lose the structural gene encoding tetracycline resistance and at least 13kb of DNA surrounding it (11). Tet^S Ntr strains carry an additional deletion of at least 4.4kb which represents an AUD.

Fusidic acid sensitive variants lack the ability to modify the steroid antibiotic (v.d. Haar, Schrempf, to be published). The unstable locus for fusidic acid resistance is currently under investigation.

CHARACTERISTICS OF AMPLIFICATION

Within S. lividans two small AUDs have been characterised. The 6.8kb AUD is present in one copy in TK23, has a tandem repeated structure in the wild-type genome and is amplified in most Cml^S variants. Both AUDs are flanked by direct repeats. A 4.4kb AUD is present in one copy in the wild-type genome and has been found amplified within DNA of one Tet^S strain. Amplification of an AUD element rarely occurs when it is present in one copy. Thus, amplification of the one 4.4kb AUD copy in the wild-type and the 6.8kb AUD in the TK23 genome is a rare event (16).

To test whether it was necessary to convert the single-copy AUD status to a duplicated form in order to amplify this element, we constructed a recombinant vector in E. coli, unable to replicate in S. lividans containing a copy of the 5.7kb ADS (deriving from the 6.8kb AUD) and two selectable markers (kanamycin resistance and thiostreptone resistance). This plasmid was transformed into TK23 under selection pressure for the selectable markers, permitting the isolation of recombinants in which the plasmid had integrated at the AUD. An analysis of these strains showed that a duplicated structure, containing the selectable markers, was the consequence of integration. A preliminary analysis showed that a cml^s argG variant derived from such a strain contained amplification of the recombinant amplifiable element and that the proximal deletion endpoint was defined by this amplification. Moreover, selection pressure to the original transformants for an increased copy number of the kanamycin resistance gene resulted in an increase of populations containing a reiteration of the recombinant amplifiable sequence.

In most cases studied, amplification of specific sequences appears without apparent selection, and it is noticeable that certain amplification events are observed concomitantly with the appearance of specific deletion events. Thus, many melanin-negative variants which derived by deletion of the corresponding gene from S. reticuli and S. scabies contain ADS (14,15).

ANALYSIS WITH PULSED-FIELD ELECTROPHORESIS

Using pulsed-field electrophoresis we were able to separate high-molecular-weight DNA and giant plasmids of most Streptomyces strains, except from S. lividans 1326 and its derivatives (16). Genes encoding tyrosinase (tyr) and polyketide synthetase (actI) are located on the high-molecular-weight chromosomal DNA of all strains analysed. Likewise, the argG, cml^r, tet^r genes, the 6.8kb and the 4.4kb AUD reside on the chromosomal DNA of S. coelicolor A3(2). Most tetracycline-sensitive variants of S. coelicolor A3(2) delete 100kb DNA including the tet^r gene and the 4.4kb AUD characterised previously in S. lividans (see above). Chloramphenicol-sensitive S. coelicolor A3(2) variants each lack a 400kb fragment of chromosomal DNA on which the cml^r gene is located. Melanine nagative variants of S. reticuli delete up to 400kb of chromosomal DNA and contain about 1500kb of amplified DNA, which can be lost in some variants.

FUNCTION OF ADS

Different classes of ADS have been identified, some of which share homology (16); most of them have been found without apparent selection in strains which contain deleted DNA and encode no known functions. Until now, only in a few cases was it possible to select for amplification of specific genes. In Streptomyces tendae the gene encoding an amylase inhibitor has been shown to be amplified within a large AUD after screening for an overproducer following mutagenic treatment. In Streptomyces achromogenes it was possible to select for variants carrying several hundred copies of spectinomycin-resistance (9). In S. lividans we could select for highly chloramphenicol-resistant strains carrying the resistance determinant and neighbouring DNA regions amplified.

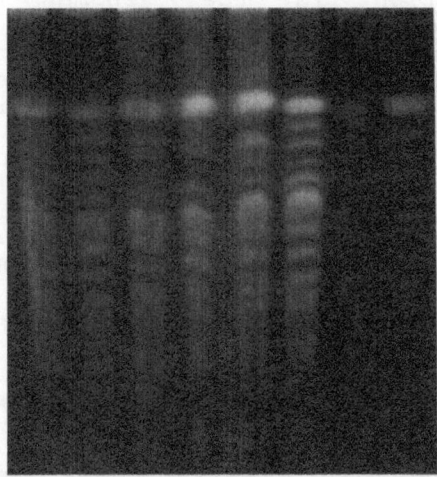

Fig.1. Chromosomal DNAs of <u>S</u>. <u>reticuli</u> (WT) and different variants of <u>S</u>. <u>reticuli</u> (a-e) were cleaved with <u>Bfr</u>I and analysed by pulsed-field electrophoresis. L-multimers of λ, Y-chromosomes of <u>Saccharomyces</u> <u>cerevisiae</u> YNN 295.

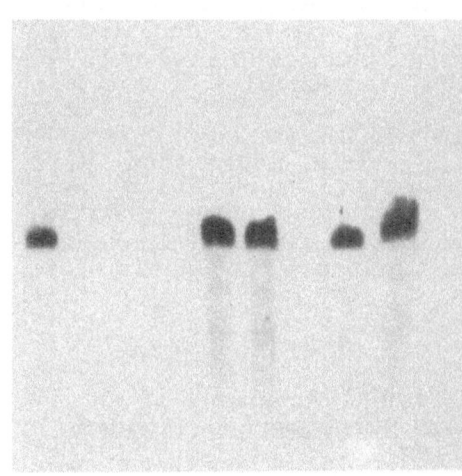
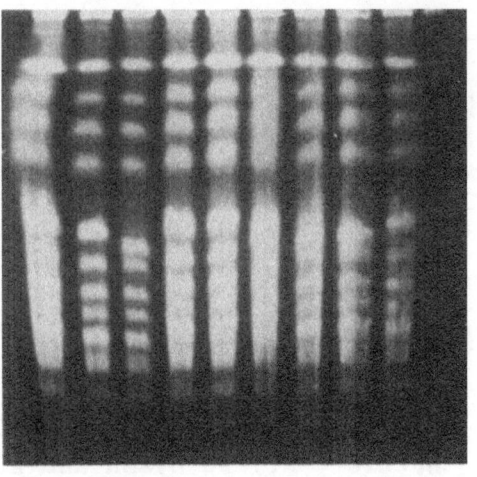

Fig. 2. Chromosomal DNA of <u>S</u>. <u>coelicolor</u> M110 (WT) and different variants (1-8) were cleaved by <u>Ase</u>I (right) and hybridised with the <u>tet</u>^{r} gene cloned from <u>S</u>. <u>lividans</u> (left) Y-chromosomes of <u>Saccharomyces</u> <u>cerevisiae</u> YNN295.

CONCLUSION

These studies have revealed that variants delete large stretches of DNA. Preliminary data suggest different "hot spots" for deletion in the genomes of different _Streptomyces_ species. Thus, in _S. lividans_ (1,2,6), Cml^S Arg$^-$ mutants appear without selection at high frequencies spontaneously, as do Tet^S Ntr strains. In other _Streptomyces_ strains - melanin and macrolide-negative - developmental variants arise frequently (15). Hybridisation experiments in which unstable genes were used as probes (_argG_, _cml_^{r}, _tet_^{r} and _tyr_) revealed that these genes often delete. In _S. lividans_, (2) _S. reticuli_ (14) and _Streptomyces glaucescens_ (3) in addition to genes encoding unstable gene functions, large stretches of DNA of ,as yet, unknown function are lost. As shown recently by chromosomal walking studies (16) and by pulsed-field electrophoresis (Schrempf, Müller, Mers, to be published) up to several hundred kb of chromosomal DNA may be absent in variants of several _Streptomyces_ species. The functions of a great portion of the large, continuous stretches of deletable DNA are unknown. However, we recently noticed that a large portion of about 350 traits used for numerical classification of _Streptomyces_ strains is changed within variants which lack a considerable part of their DNA (Kämpfer, Schrempf, to be published).

The deletable DNA sequences have been localised within several genomes in close proximity to one or two AUDs. Deletable and amplifiable sequences are part of the silent genomic region in _S. lividans_. Structural instability appears to affect certain genes at a high frequency.

The amplification itself is a two-step process. Single-copy amplifiable elements are converted by a rate-limiting, unequal crossing-over event to a duplicated structure. This duplicated structure can then be efficiently amplified, likely with a mechanism involving replication.

Further studies are required to understand the numerous events which are involved in generating new _Streptomyces_ variants. Additional knowledge should greatly enhance our understanding of genome plasticity and increase our ability to control the highly unstable production of secondary metabolites.

REFERENCES

1. **Altenbucher, J. and J. Cullum.** 1985. Structure of an amplifiable DNA sequence in _Streptomyces_ _lividans_ 66. Mol. Gen. Genet. 201:192-197.
2. **Betzler, M., P. Dyson and H. Schrempf.** 1987. Relationship of an unstable argG gene to a 5.7-kilobase amplifiable DNA sequence in _Streptomyces_ _lividans_ 66. J. Bacteriol. 169: 4804-4810.

3. **Birch, A. A. Häusler, M. Vögtli, W. Krek and R. Hütter.** 1989. Extremely large chromosomal deletions are intimately involved in genetic instability and genomic rearrangements in Streptomyces glaucescens. Mol. Gen. Genet. 217:447-458.

4. **Cullum, J., J. Altenbucher, F. Flett, W. Piendl and J. Platt.** 1986. DNA amplification and genetic instability in Streptomyces. Biotech. Genet. Eng. Rev. 4:59-78.

5. **Demuyter, P., P. Leblond, B. Decaris and J. M. Simonet.** 1988. Characterization of two families of spontaneously amplifiable units of DNA in Streptomyces ambofaciens J. Gen. Microbiol. 134:2001-2007.

6. **Dyson, P. and H. Schrempf.** 1987. Genetic instability and DNA amplification in Streptomyces lividans 66. J. Bacteriol. 169:4796-4803.

7. **Häusler, A., A. Birch, W. Krek, J. Piret and R. Hütter.** 1989. Heterogeneous genomic amplification in Streptomyces glaucescens: structure, location and junction sequence analysis. Mol. Gen. Genet. 217:437-446.

8. **Hintermann, G. R. Crameri, M. Vögtli and R. Hütter.** 1984. Streptomycin-sensitivity in Streptomyces glaucescens is due to deletions comprising the structural gene coding for a specific phosphotransferase. Mol. Gen. Genet. 196:513-520.

9. **Hornemann, U., D. J. Otto, C. G. Hoffmann and A. C. Bertinuson.** 1987. Spectinomycin resistance and associated DNA amplification in Streptomyces achromogenes subsp. rubradiris. J. Bacteriol. 169:2360-2366.

10. **Ishihara, H., M. M. Nakano and H. Ogawara.** 1985. Cloning of a gene from Streptomyces species, complementing argG mutations. J. Antibiot. 38:787-794.

11. **Kessler, A., W. Dittrich, M. Betzler and H. Schrempf.** 1989. Cloning and analysis of a deletable tetracycline-resistance determinant of Streptomyces lividans 1326. Mol. Microbiol. 3:1103-1109.

12. **Meade, H.** 1985. Cloning of argG from Streptomyces: loss of gene in Arg mutants of S. cattleya. Bio/Technology 3:917-918.

13. **Mosher, R. H., N. P. Ranade, H. Schrempf and L C. Vining.** Chloramphenicol resistance in Streptomyces: cloning and characterization of a chloramphenicol hydrolase gene from Streptomyces venezuelae. 1990. J. Gen. Microbiol. 136:293-301.

14. **Schrempf, H.** 1983. Deletion and amplification of DNA sequences in melanin-negative variants of Streptomyces reticuli. Mol. Gen. Genet. 189:501-505.

15. **Schrempf, H.** 1985. Genetic instability: amplification, deletion and re-arrangement within <u>Streptomyces</u> DNA, p. 436-440. In L. Leive (ed.), Microbiology - 1985. American Society for Microbiology, Washington, D.C.

16. **Schrempf, H., P. Dyson, W. Dittrich, M. Betzler, C. Habiger, B. Mahro, V. Brönneke, A. Kessler and H. Düvel.** 1988. Genetic instability in <u>Streptomyces</u>. In: Biology of Actinomycetes. pp. 145-150. Okami, Y., T. Beppu and H. Ogawara (eds.). Tokyo: Japan Scientific Press.

17. **Wohlleben, W., G. Muth, E. Birr and A. Pühler.** 1986. A vector system for cloning in <u>Streptomyces</u> and <u>Escherichia coli</u>. In: Biological, biochemical and biomedical aspects of Actinomycetes. pp. 99-101. Szabo, G., S. Biro and M. Goodfellow (eds.). Akademial Kiado, Budapest.

A NEW SYSTEM TO STUDY DNA AMPLIFICATION IN *STREPTOMYCES LIVIDANS*

Josef Altenbuchner and Christa Eichenseer

Institute of Industrial Genetics, University of
Stuttgart, Azenbergstr. 18, 7000 Stuttgart 1, FRG

INTRODUCTION

Streptomycetes are known for an unusual high genetic instability caused by spontaneous deletions of up to 20% of their genome size. Very often the deletions are accompanied by one or more amplifications of DNA sequences on the chromosome[1-5].

In *S.lividans* a spontaneous deletion containing a chloramphenicol resistance (Cml) was observed in around 0.5% of the spores [6]. The first deletion somehow triggers a second deletion in more than 10% of the spores resulting in the loss of a gene encoding argininosuccinate synthetase (*argG*). One endpoint of this second deletion is found within an amplified DNA (ADS1) consisting of a 5.7kb repeated unit. In wildtype strains the amplifiable DNA (AUD1) has a structure of three tandemly repeated 1kb sequences interspersed by two 4.7kb repeated sequences located at a distance of 30kb from the *argG* gene [7,19].

Several models explaining the deletion-amplification mechanism in Streptomycetes were put forward but none of them has been proven so far [8,9]. Besides *Streptomyces*-specific properties like filamentous growth, the genetic and biochemical analysis of these events is complicated by the non-inducibility of the deletion-amplification process in a growing colony and the enormous extent of the DNA rearrangements, which usually occur on chromosomes and so far were not reproduced on plasmids.

Concerning the *S.lividans* system there are additional difficulties which are causing problems in the investigation of the deletion-amplification event:
- The deletion-amplification event has to be followed through two steps, which is more time-consuming and labour intensive compared to a single event.
- The presence of three respectively two repeated sequences in the AUD1 region makes it difficult to manipulate the sequences by mutation.

Nevertheless the *S.lividans* system has the invaluable advantage of being highly reproducible concerning the frequency and structure of the amplified DNA. Furthermore due to the very high transformation rates and the lack of a restriction system it is the only system so far allowing the integration and amplification of foreign genes. In the following we present a new system which overcomes most of the *S.lividans*-specific problems and brings about a more efficient analysis of the deletion-amplification process.

MATERIAL AND METHODS

Bacterial strains, phages and plasmids

The bacterial strains, plasmids and phages used in this work are listed in table I, II and III.

DNA manipulation

Standard procedures such as small scale plasmid preparations, restriction enzyme analysis, DNA ligation, agarose gel

Table I. Bacterial strains

STRAIN	MARKER	REFERENCE
E.coli JM109	*relA, endA, gyrA*96, *thi, hsdR*17, *supE*44, *recA*1, Delta(*lac-pro*), [F',*traD*36, *proAB, lacI*^{q}ZDeltaM15]	[23]
S.lividans 1326	wildtype	[11]
S.lividans TK64	*str-6, pro-2*	[11]
S.lividans AJ100	*str-6, pro-2*, Cml^S, Arg$^-$	this work

Table II. Plasmids

PLASMID	MARKER	DESCRIPTION / REFERENCE
pIC19H, pIC20H	Ap^R, *lacZ*'	[13]
pIJ702	Ts^R, Mel$^+$	[22]
pHP45	Ap^R, Sm^R, Sp^R	[16]
pJOE756	Ap^R	[7]
pJOE803	Ap^R, Ts^R	Ts^R gene cloned into pIC20H
pJOE867	Km^R	Km^R gene of Tn5 ligated to Rts1 replication origin
pJOE883	Ap^R	*SacI* fragment of lambda AJ201 cloned into pIC19H
pJOE882, pJOE884	Km^R	*SacI* fragments from the left resp. right side of the amplifiable region of AJ100 cloned into pJOE867
pJOE818, pJOE818.1, pJOE818.2, pJOE818.3, pJOE819, pJOE820, pJOE820.2, pJOE820.3, pJOE820.4, pJOE820.5, pJOE820.6, pJOE951, pJOE953, pJOE954, pJOE955	Ap^R, Ts^R	derivatives of pJOE803 containing different parts of the AUD1 region of *S.lividans*

254

electrophoresis, nick translation, Southern hybridisation, DNA sequencing, transformation of *E.coli* and *S.lividans* were done as described in Maniatis et al.[21] and Hopwood et al.[11]. For construction of genomic libraries the phage lambdaRES[12] was used instead of lambdaEMBL4 [24].

Table III. Phages

PHAGE	DESCRIPTION	REFERENCE
M13mp18, M13mp19	ssDNA vector for DNA sequencing	[23]
lambdaRES	lambda replacement vector	[12]
lambdaEMBL4	lambda replacement vector	[24]
lambdaMT686	lambda EMBL4 recombinant phage containing the AUD1 region of *S.lividans*TK64	[7]
lambdaAJ201	lambdaRES recombinant phage containing the deletion junction of *S.lividans* AJ100	this work
lambdaAJ202	lambdaRES recombinant phage containing a deletion endpoint from *S.lividans* 1326	this work

Construction of the integration vectors

A 1.05kb *Bcl*I fragment of pIJ702 [22] containing the thiostrepton resistance gene (*tsr*) was first cloned between the two *Bam*HI sites of pHP45 [16] replacing the Omega fragment. The *tsr* gene was recovered as a 1.05kb *Sma*I fragment and cloned into the *Nru*I site of pIC20H to give pJOE803. A small deletion had occured in pJOE803 removing a *Sac*I and a *PaeR*7 site in the multiple cloning site adjacent to *Nru*I. The plasmids pJOE818, pJOE819 and pJOE820 were obtained by cloning a 5.7kb *Bam*HI, a 5.7kb *Bgl*II and a 5.7kb *Sac*I fragment from lambdaMT686 into the *Bam*HI resp. *Sac*I site of pJOE803. Similarly, the plasmids pJOE955 containing the left end of AUD1 and the plasmids pJOE954, pJOE953 and pJOE951 containing differnt amounts of the right side of AUD1 were constructed by cloning a 3.9kb *Eco*RI/*Bgl*II, a 5.3kb *Eco*RI/*Bgl*II, a 3.9kb *Eco*RI/*Sac*I and a 2.6kb *Eco*RI/*Bam*HI fragment into DNA of pJOE803 cleaved by the same enzymes respectively. Deletion of a 3kb *Bam*HI/*Bgl*II and a 4.3kb *Sac*I fragment from pJOE820 and a 4.4kb *Sac*I fragment from pJOE818 gave the plasmids pJOE820.2, pJOE820.6 and pJOE818.2. Cloning of a 2kb and a 1.6kb *Pst*I fragment from pJOE820 and in both orientations a 1.6kb *Pst*I fragment from pJOE953 into the *Pst*I site of pJOE803 gave the plasmids pJOE820.5, pJOE820.4, pJOE953.1 and pJOE953.2. The plasmids pJOE820.3, pJOE818.1 and pJE818.3 were generated by cloning from pJOE820 a 3.9kb *Sac*I fragment and from pJOE818 a 4.15kb *Pst*I fragment and a 3kb *Bgl*II/*Bam*HI fragment into the *Sac*I, the *Pst*I and the *Bam*HI site of pJOE803 respectively.

Construction of pJOE867, pJOE882 and pJOE884

The plasmid pJOE867 consists of a *Hind*III/*Sma*I fragment containing the kanamycin resistance of Tn5 [17], the origin of replication of Rts1 [18] obtained from pTW601-1 as a *Eco*RI/*Hind*III fragment and part of the multiple cloning side of pIC20R [13]. Except for the few basepairs of the multiple cloning side there is no homology between pJOE867 and the pICplasmids. The plasmids pJOE882 and pJOE884 were obtained by cloning the two *SacI* fragments flanking the AUD1 region of AJ100 from lambdaAJ201 into the *SacI* site of pJOE867 (fig.4).

DNA sequencing

For DNA sequencing a *SacI/PaeR7* fragment of lambdaAJ201 containing the deletion junction was cloned into M13mp18 and M13mp19 each, cleaved by *SacI/SalI*. This allowed the DNA sequence determination of both strands at the junction. To sequence the deletion endpoints a *SacI/PaeR7* fragment of lambdaAJ202 and a *BglII/PaeR7* fragment of lambdaMT686 were cloned into M13mp18 cleaved by endoR *SacI/SalI* and *BamHI/SalI* respectively. Hereby the DNA sequence of only one strand of the deletion endpoints was determined.

RESULTS

1.Isolation of a Cml^S Arg$^-$ strain containing only one incomplete copy of AUD1

Previously it has been shown that ligation of foreign genes to a copy of the 5.7kb amplified DNA and integration into the chromosome of *S.lividans* via homologous recombination leads to a coamplification of these genes in corresponding Cml^S Arg$^-$ strains [10]. One such mutant of *S.lividans* TK64 containing a coamplified thiostrepton resistance (Ts^R) and the melanine production genes (Mel) of pIJ702 was used to screen derivatives for loss of the amplified DNA. The strain was grown in YEME-medium and subsequently protoplasted and regenerated on R2-agarplates containing tyrosine and copper sulfate [11]. Around 10% of the regenerated colonies showed less or no production of melanine. An apparently white colony segregating Ts^R Mel$^+$ colonies only at a low level of less than 5% was again protoplasted and regenerated on R2-plates. Eventually a colony completely negative for thiostrepton resistance and melanine production was found (AJ100) and used for further analysis.

2.Characterisation of *S.lividans* AJ100

Southern blot hybridisation to chromosomal DNA of strain AJ100 using radioactive labeled pJOE756-DNA [7] (ADS1-DNA, cloned as a 5.7kb *BamHI* fragment into pBR325) gave a positive signal on the autoradiogram (data not shown). To characterize the remaining parts of AUD1 in the chromosome of AJ100 a genomic library of AJ100 was constructed in lambdaRES [12]. Plaques were screened by plaque-hybridisation using again radioactive labeled DNA of pJOE756. A recombinant phage showing a positive signal (lambdaAJ201) was isolated and characterized by restriction enzymes. A comparison of the restriction map of this phage to the the AUD1 wildtype sequence of *S.lividans* cloned in lambdaEMBL4 (lambdaMT686) revealed that most of the AUD1 region was deleted in AJ100 with only one copy of the 1kb repeat and an incomplete copy of the 4.7kb sequence left (see fig.1).
For further analysis of the deletion junction in AJ100 a *SacI* fragment from lambdaAJ201 was cloned into the *E.coli* plasmid pIC19H [13] to give pJOE883 (see fig.1). DNA of

pJOE883 was then used as a probe to screen a genomic library of *S.lividans* 1326 wildtype [14], again constructed in lambdaRES. The recombinant phage lambdaAJ202 gave a positive hybridisation signal and contained the deletion endpoint of the left junction sequence of AJ100.

Finally the DNA sequences at the deletion junction and the corresponding wildtype regions were determined by the chain termination method of Sanger [15]. The DNA sequences are shown in figure 1. The only obvious similarity between the deletion endpoints is a five basepair directly repeated sequence which seemed to have functioned as a side for intra-molecular recombination leading to the observed deletion.

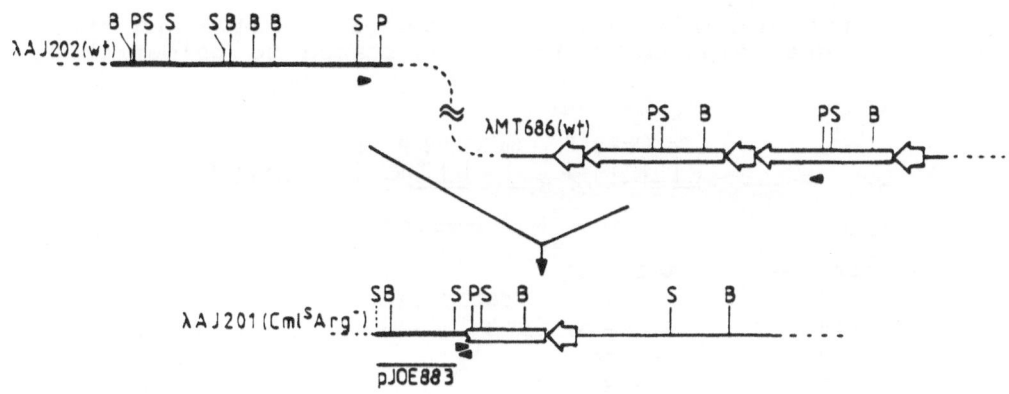

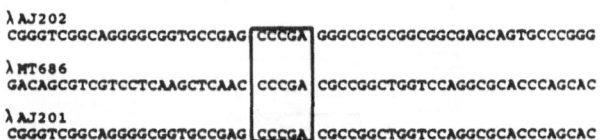

Fig.1. (Top) Restriction endonuclease maps of DNA fragments inserted in the phages lambdaAJ202, AJ201 and MT686. The cleavage sites for the enzymes *Bam*HI (B), *Pae*R7 (P) and *Sac*I (S) are shown. The phages lambda AJ202 and AJ201 contain chromosomal DNA from *S.lividans* 1326 (wildtype) and *S.lividans* AJ100 (Cml^S Arg$^-$) respectively cloned into lambdaRES. LambdaMT686 containing the AUD1 sequence from *S.lividans* TK64 was described previously [7]. The repeated DNA sequences of AUD1 are symbolised by open arrows. The filled in triangles indicate from where and in which direction the DNA sequences shown below were analysed. Also shown is the *Sac*I fragment to the left side of the AUD1 sequence of lambdaAJ201 which was cloned into pIC19H to give pJOE883 (one of the two *Sac*I sites of the fragment is derived from the multiple cloning site of lambda RES).
(Below) DNA sequences of the deletion endpoints and the novel deletion junction of AJ100. The five basepair direct repeated DNA sequences at which the deletion had taken place are shown in a frame.

3.Construction of integration vectors

The strain AJ100 is stable and showed no DNA amplification even though it was kept and propagated for over two years. To find out if a second copy of the 5.7kb repeat or part of it would restore the DNA amplification we constructed a series of integration vectors. All the constructions were done in *E.coli* using the plasmid pIC20H [13]. A copy of the *tsr* gene of pIJ702 [11] was integrated into the multiple cloning site of this plasmid to give pJOE803. Then overlapping fragments from lambdaMT686 covering the complete AUD1 sequence of *S.lividans* wildtype including the flanking sequences were cloned into pJOE803. Finally deletions and subclonings of these plasmids led to the derivatives shown in figure 2. More details of the constructions are given in Material and Methods.

All the plasmids were transformed into AJ100 and transformants were selected by their resistance to thiostrepton.

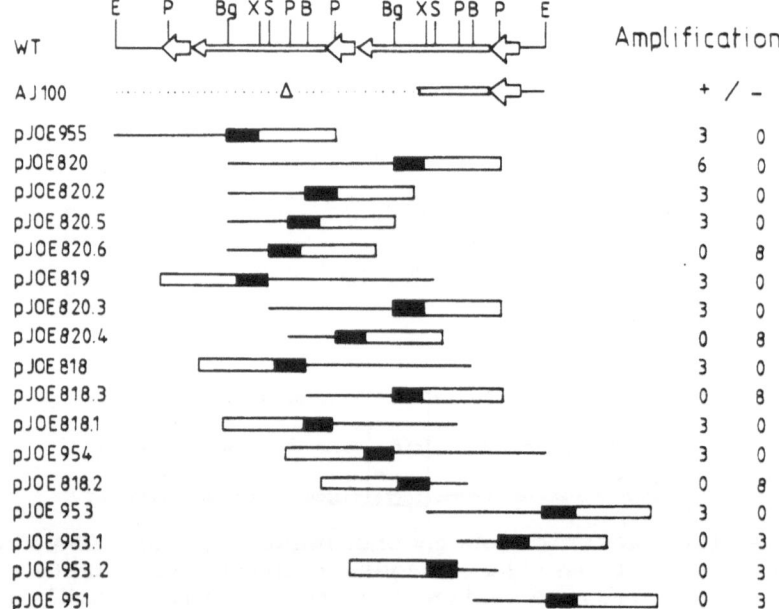

Fig.2. List of integration vectors used for transformation of AJ100. The integration vectors are composed of fragments containing DNA sequences from the AUD1 region (thin line) and a thiostrepton resistance (filled bar) cloned into pIC20H (open bar). The AUD1 derived fragments are drawn according to their origin and extension below a restriction map of the AUD1 region cloned in lambdaMT686. Restriction sites are given for *Bam*HI (B), *Bgl*II (Bg), *Eco*RI (E), *Pst*I (P), *Sac*I (S) and *Pae*R7 (X). The repeated sequences of AUD1 present in *S.lividans* wildtype and the ones left in AJ100 (drawn below the restriction map) are marked by open arrows. The numbers on the right side of the figure indicate how many of the analysed transformants had the respective plasmid amplified (+) or not (-).

At least three colonies from each of the transformed plasmids were inoculated in YEME liquid culture and total DNA was prepared by CsCl density gradient centrifugation [7]. If no amplification of a particular plasmid was detected, DNA from five more transformants was prepared. The results are summarized in fig. 2.

The DNA was prepared at the earliest possible stage after regeneration of the transformants. At this time the DNA amplification had already reached a constant level and further rounds of growth on R2 agarplates or liquid culture did not change the amount of amplified DNA, neither was amplification observed later on in transformants lacking amplification at an early stage of DNA preparation.

In all transformants showing amplification exactly the transformed plasmid was amplified without DNA rearrangemnts like deletions in the *E.coli*-part of the integration vectors. A few examples of amplified plasmids found in transforments of AJ100 are shown in fig. 3.

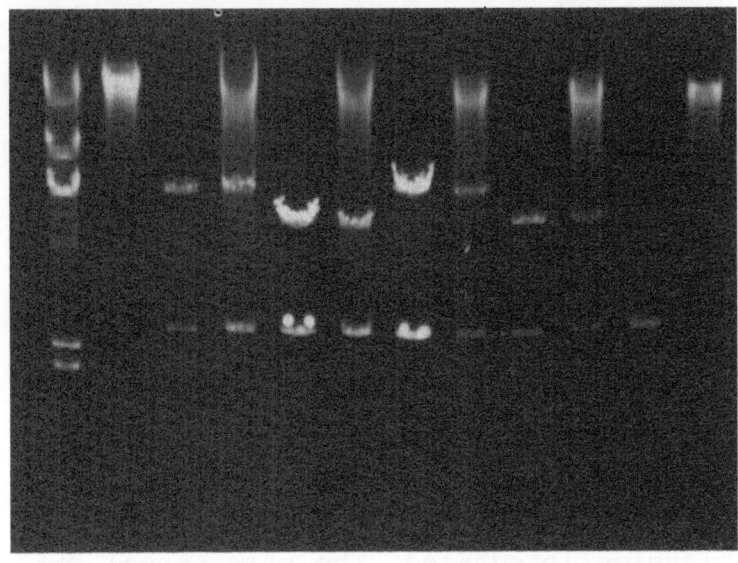

Fig. 3. Agarosegelelectrophoresis of integration vectors and chromosomal DNAs of AJ100 transformed by the integration vectors. All the DNAs were digested by endoR *Hind*III. To the left side of the chromosomal DNAs the corresponding integration vectors are shown. Track1-12: lambda molecularweight standard (1), chromosomal DNA of AJ100 (2), AJ100pJOE818 (4), AJ100pJOE818.1 (6), AJ100 pJOE819 (8), AJ100pJOE820.3 (10), AJ100pJOE820.4 (12) and plasmid DNA isolated from *E.coli* of pJOE818 (3), pJOE881.1 (5), pJOE819 (7), pJOE820.3 (9), pJOE820.4 (11).

4.Amplification of the integration vectors in AJ100 produces further deletions

All strains with DNA amplifications show deletions beginning in or near the amplified DNA [3,5,9]. So far it is unknown if the deletions are a prerequisite or just a result of the amplification process. Therefore it was interesting to see if the amplification of the integration vectors in AJ100 is accompanied by new deletions. Southern blots of chromosomal DNA from several different transformants of AJ100 with and without amplifications were hybridized with radioactive labeled DNA of pJOE882, which contains a DNA fragment from the left side of the deletion junction of AJ100, cloned into pJOE867. The results shown in fig.4 demonstrate that all strains containing a new amplification do not hybridize to pJOE882 and therefore must have a new deletion into the same direction as the original one responsible for the Arg⁻ phenotype. On the other hand there was no change in the hybridisation pattern of the transformants lacking an amplification. As a control DNA from the opposite side of AUD1, ob-

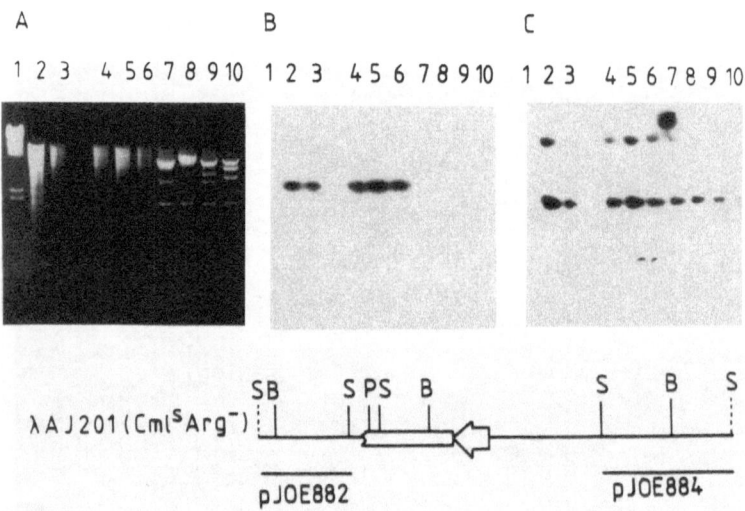

Fig 4. Southern blot hybridisation of chromosomal DNAs of AJ100 transformants containing amplified or nonamplified plasmids integrated. The chromosomal DNAs were digested by endoR *BamHI/SacI*, the fragments electrophoretically separated on an agarosegel (left), blotted on a nitrocellulose membrane and hybridized to DNA of pJOE882 (middle) and pJOE884 (right) which were labelled with [³²P]dCTP in a nicktranslation reaction. The location of the *SacI* fragments flanking the AUD1 sequence of AJ100 and cloned from lambdaAJ201 into pJOE867 to give the plasmids pJOE882 and pJOE884 is indicated (bottom of fig.4). Track 1-10: lambda molecularweight standard (1), TK64 (2), AJ100 (3), AJ100 pJOE820.6 (4), AJ100 pJOE818.2 (5), AJ100 pJOE820.4 (6), AJ100 pJOE820 (7), AJ100 pJOE820.2 (8), AJ100 pJOE820.3(9), AJ100 pJOE818.1 (10). Symbols: B (*BamHI*), S (*SacI*), P (*PaeR7*).

tained as a *Sac*I fragment from lambdaAJ201 and cloned into pJOE867 (pJOE884), was used for the same hybridisation. In all transformants the same bands as in AJ100 were lighting up. This confirms earlier results saying that DNA to the right side of AUD1 is not effected by the amplification process [7,19].

DISCUSSION

Spontaneously arising *S.lividans* derivatives like TK23 have (probably due to homologous recombination) one copy of the 5.7kb repeated sequence deleted and therefore possess only two copies of the 1kb and one copy of the 4.7kb sequence in the AUD1 region [19]. These strains are known to amplify the 5.7kb sequence only at a very low level or not at all. The ability for high amplification can be restored by integration of a second copy of the 5.7kb sequence. In AJ100 the situation is very similar but this strain has the advantage to amplify the integrated vectors immediately without a time-consuming screening for Cml^S Arg^- mutants and the lack of a second copy of the 1kb repeated sequence should help to elucidate the contribution of the 1kb sequence to the amplification process. Furthermore the 4.7kb sequence is diminished, which lowers the size of the amplifiable structure to be investigated for amplification functions.

To restore the amplification process in AJ100 it is necessary to provide a DNA fragment which is homologous to the AUD1 sequence retained in AJ100. Hereby no specific DNA sequence seems to be necessary. This is seen for example at the plasmids pJOE955 and pJOE820.5 which don't share any obvious homology. But the size of the fragment necessary for amplification is to some extent dependent of the region where it comes from. For example, the plasmid pJOE820.5 contains a 2.15kb insert from the AUD1 sequence but because of the deletion in AJ100 there is just 1.27kb of it homologous to the AUD1 sequence in AJ100. Integration of pJOE820.5 into AJ100 leads to amplification of the plasmid, wheras the plasmid pJOE818.2 containing a 1.3kb insertion too and the plasmids pJOE820.4, pJOE953.1 and pJOE953.2 containing even a 1.6kb insertion homologous to the AUD1 sequence of AJ100 don't amplify in AJ100. Even more convincing examples are the plasmids pJOE955, pJOE818.3 and pJOE951 which contain the 1kb repeated sequences from the left, middle and right side of the AUD1 region respectively. The plasmid pJOE955 has a 3.9kb insertion, 2.2kb are derived from the AUD1 sequence. But due to the deletion in AJ100 only the 1kb repeated sequence is homologous to the AUD1 sequence of AJ100. All three transformants had pJOE955 amplified. The plasmid pJOE818.3 has a 3kb insert derived from AUD1 but again due to the deletion of AJ100 only 1.8kb, i.e the middle 1kb repeated sequence and 0.8kb of the left side are homologous to the AUD1 region of AJ100. None of the transformants had pJOE818.3 amplified. Finally the plasmid pJOE951 containing the 1kb repeated sequence from the right side has a 2.6kb insert completely homologous to AJ100, 1.8 kb DNA derived from AUD1. None of the transformants had the plasmid amplified.

The examples demonstrate that it is not sufficient to have just two directly repeated sequences to get amplification as proposed by the model of Young and Cullum [8]. It is more likely that there are functions encoded on the AUD1 sequence, presumably at the right flanking region, which are responsible for the amplification. The lack of amplification of various integration vectors might be due to insertional inactivation of amplification functions, the lack of amplification functions provided on the integration vectors and differences in the repeated DNA sequences as observed for the 1kb repeats by Piendl (unpublished).

In addition, the immediate DNA amplification in all transformants of AJ100 using an appropriate integration vector indicates an AUD1 specific amplification mechanism activated in Cml^S Arg^- mutants. The activation might be due to the deletion of a regulatory gene in Arg^- strains repressing, or to a switch on of an activator gene activating the amplification functions.

The additional deletions seen in new amplified AJ100 strains could be explained by the assumption that deletions are very likely not a prerequisite but an accidental result of the amplification process. The presence of just five basepair direct repeated DNA sequences at the deletion endpoints indicate that the deletions start more or less randomly but in only one direction in the amplifying DNA and end randomly in the flanking chromosomal region. Similar "microhomologies" of several basepairs direct or imperfect inverted repeats have been described for spontaneous deletions in *S.glaucescens* too [3]. The enormous size of the deletions of several hundredthausand basepairs might be explained by the hypothesis that there is not one but a series of deletions accompaning repetitious rounds of amplification of AUD1 from two copies to several hundred copies per chromosome adding up to a very large deletion.

To understand the deletion-amplification process more investigations are necessary. Hereby the strain AJ100 and the integration vectors will certainly be a very valuable tool.

REFERENCES

[1] Cullum,J., Altenbuchner,J., Flett,F., Piendl,W. and Platt,J.(1987). Genetic instability and DNA amplification in *Streptomyces*. In: Genetics of Industrial Microorganisms (eds. M.Alacevic, D.Hranueli, Z.Toman). Ognijen Prica Printing Works, Karlovac.
[2] Hershberger,C.L.(1988). Genetic structure and instability in antibiotic producing Streptomycetes. In: Biology of Actinomycetes'88 (eds. Y.Okami, T.Beppu, H.Ogawara) Japan Scientific Society Press, Tokyo.
[3] Birch,A., Haeusler,A., Voegtli,M., Krek,W. and Huetter,R. (1989). Extremely large chromosomal deletions are involved in genetic instability and genomic rearrangements in *Streptomyces glaucescens*. Mol.Gen.Genet. *217*,447-458.
[4] Leblond,P., Demuyter,P., Moutier,L., Laakel,M., Decaris, B. and Simonet,J.-M.(1989). Hypervariability, a new phenomenon of genetic instability, related to DNA amplification in *Streptomyces ambofaciens*. J.Bacteriol. *171*,419-423.
[5] Hornemann,U., Otto,Ch.J. and Zhang,X.Y. (1989). DNA amplification in *Streptomyces achromogenes* subsp *rubradiris* is acompanied by a deletion, and the amplified sequences are conditionally stable and can be eliminated by two pathways. J.Bacteriol. *171*,5817-5822.
[6] Altenbuchner,J. and Cullum,J.(1984). DNA amplification and an unstable arginine gene in *Streptomyces lividans* 66. Mol.Gen.Genet. *195*,134-138.

[7] Altenbuchner,J. and Cullum,J.(1985). Structure of an amplifiable DNA sequence in *Streptomyces lividans* 66. Mol. Gen. Genet. *201*, 192-197.

[8] Young,M. and Cullum,J.(1987). A plausible mechanism for large-scale chromosomal DNA amplification in Streptomycetes. FEBS Letters *212*,10-14.

[9] Altenbuchner,J., Eichenseer,Ch. and Bruederlein,M.(1988). DNA amplification and deletion in *Streptomyces lividans*. In: Biology of Actinomycetes'88 (eds. Y.Okami, T.Beppu, H.Ogawara). Japan Scientific Society Press, Tokyo.

[10] Altenbuchner,J. and Cullum,J.(1987). Amplification of cloned genes in *Streptomyces*. Biotechnology *5*,1328-1329.

[11] Hopwood,D.A., Bibb,M.J., Chater,K.F., Kieser,T., Bruton C.J., Kieser,H.M., Lydiate,D.J., Smith,C.P., Ward,J.M. and Schrempf,H.(1985) Genetic Manipulation of Streptomyces; A Laboratory Manual. The John Innes Foundation, Norwich.

[12] Altenbuchner,J. and Bruederlein,M.(1987). LambdaRES, a replacement vector with an automated excision mechanism. Biol.Chem. Hoppe-Seyler *386*,1016.

[13] Marsh,J.L., Erfle,M. and Wykes,E.J.(1984). The pIC-plasmid and phage vectors with versatile cloning sites for recombinant selection by insertional inactivation. Gene *32*, 481-485.

[14] Hopwood,D.A., Kieser,T., Wright,H.M. and Bibb,M.(1983). Plasmids, recombination and chromosome mapping in *Streptomyces lividans* 66. J.Gen.Microbiol.*129*,2257-2269.

[15] Sanger,F., Nicklen,S. and Coulson,A.R.(1977). DNA sequencing with chain terminating inhibitors. PNAS *74*,5463-5467.

[16] Frey,J. and Krisch,H.M. (1985). Omega mutagenesis in Gram-negative bacteria: a selectable interposon which is strongly polar in a wide range of bacterial species. Gene *36*, 143-150.

[17] Jorgensen,R.A., Rothstein,S.J. and Reznikoff,W.S.(1979). A restriction enzyme cleavage map of Tn5 and location of a region encoding neomycin resistance. Molec.Gen.Genet.*177*,65-72.

[18] Kamio,Y., Tabuchi,Y., Itoh,Y., Katagiri,H. and Terawaki, Y.(1984). Complete nucleotide sequence of mini-Rts1 and its copy mutant. J.Bacteriol.*158*,1185-1191.

[19] Schrempf,H., Dyson,P., Dittrich,W., Betzler,M., Habiger, C., Mahro,M., Broenneke,V., Kessler,A., and Duevel,H.(1988). Genetic instability in *Streptomyces*. In: Biology of Actinomycetes'88 (eds. Y.Okami, T.Beppu, H. Ogawara). Japan Scientific Press, Tokyo.

[20] Haeusler,A., Birch,A., Krek,W., Piret,J. and Huetter,R. (1989). Heterogeneous genomic amplification in *Streptomyces glaucescens*: Structure, location and junction sequence analysis. Mol.Gen. Genet. *217*,437-446.

[21] Maniatis,T., Fritsch,E.F., and Sambrook,J.(1982). Molecular cloning: A laboratory manual. Cold Spring Harbor Laboratory, Cold Spring Harbor, New York.

[22] Katz,E., Thompson,C.J., and Hopwood,D.A.(1983). Cloning and expression of the tyrosinase gene from *Streptomyces antibioticus* in *Streptomyces lividans*. J.Gen.Microbiol.*129*,2703-2714.

[23] Yanisch-Peron,C., Vieira,J. & Messing,J.(1985). Improved M13 phage cloning vectors and host strains: nucleotide sequences of the M13mp18 and pUC19 vectors. Gene*33*,103-119.

[24] Frischauf,A-M., Lehrach,H., Poustka,A. and Murray,N. (1983). Lambda replacement vectors carrying polylinker sequences. J.MOL.Biol.*170*,827-842.

ANALYSIS OF AMPLIFICATIONS AND DELETIONS

IN *Streptomyces* SPECIES

John Cullum[1], Fiona Flett[2], Birgit Gravius[1],
Daslav Hranueli[3], Kiyotaka Miyashita[4], Jasenka Pigac[3],
Uwe Rauland[1] and Matthias Redenbach[1]

[1] LB Genetik der Universität Kaiserslautern
Paul-Ehrlich-Str. 22
D-6750 Kaiserslautern, F.R.G.
[2] Department of Biochemistry and
Applied Molecular Biology
U.M.I.S.T.
Manchester, U.K.
[3] "PLIVA" Research Institute
Zagreb, Yugoslavia
[4] National Institute of Agro-Environmental Sciences
Tsukuba, Japan

INTRODUCTION

Genetic instability is very common in *Streptomyces* species, and was one of the first reported properties (Beijerinck, 1913). Usually genetic instability has been detected as influencing easily scored phenotypes such as pigment production (Gregory and Huang, 1964), sporulation, auxotrophy (Redshaw et al., 1979) and antibiotic resistance (Freeman et al., 1977). In some cases genetic instability affects antibiotic production and can be a serious problem in industrial fermentations.

Genetic instability only affects particular genes in any one strain and usually involves irreversible deletion mutations (Cullum et al., 1986), although there is one well documented case of a reversible mutation (Danilenko et al., 1986). In many cases there is extreme DNA amplification of sequences adjacent to the deletions. The amplified DNA often accounts for over 10% of total DNA in the strains affected and the deletions may be very large (>500 kb).

Two types of amplification behaviour are seen (classification of Hütter et al., 1988):
(1) Type I : different DNA fragments are amplified in the different isolates. However, all (or most) of the amplified fragments lie in a particular chromosomal region.
(2) Type II: the same DNA fragment is amplified in all (or most) isolates.
Type I amplifications are commoner than Type II. It appears that Type II amplifications are associated with existing duplication of the amplifiable fragment in the wild type strain, whereas the

duplications for amplification must be created *de novo* in Type I strains.

The best investigated case of a Type I strain is the instability involving melanin production and streptomycin resistance in *S. glaucescens*. It was shown that both genes lie contiguous 800kb segment of DNA that is deleted in the mutants (Häusler et al., 1989; Birch et al., 1989). The amplified sequences seem to be more or less randomly chosen sequences from a 100kb region bordering the deletion. This strain was amenable to pulse field electrophoresis investigations.

The best investigated Type II system is of the chloramphenicol resistance and arginine auxotrophy in *S. lividans* 66. Both the first stage (to chloramphenicol sensitivity) and the second stage to arginine auxotrophy involve large deletions (Flett and Cullum, 1987; Altenbuchner and Cullum, 1985). The second stage also involves amplification of a 5.7kb sequence that is already present as a duplication in the wild type strain (Altenbuchner and Cullum, 1985). The strain is easy to transform – this allows integration of *in vitro* manipulated constructs into the unstable region (Altenbuchner & Cullum, 1987). However, *S. lividans* 66 is not amenable to pulse field electrophoresis, because of a DNA cutting activity (Zhou et al., 1988). A mutant lacking this activity can be used for pulse field electrophoresis, but there are certain problems when used for instability investigations (see later). In this paper we report a new Type I amplification system in *S. lividans* 66. A comparison of the Type I and Type II systems in the same strain will allow common properties to be deduced.

The important commercial antibiotic oxytetracycline is produced from *S. rimosus*. Work in Zagreb has established a genetic map using conjugation and protoplast fusion methods (Alacevic et al., 1973; Pigac and Alacevic, 1979: Pigac et al., 1982; Hranueli et al., 1983). The strain can be transformed efficiently and does not seem to show any restriction of *E. coli*-derived DNA (Pigac et al., 1986, 1988). Strains have been developed for high oxytetracycline production. When such strains are grown in complex liquid media, they grow as well dispersed mycelial fragments. About 2-3% of colonies derived from such mycelial fragments show a variant phenotype. However, about 80% of colonies derived from spores (on solid media) show such variants. The variants are altered in one or more phenotypes:

(1) Sporulation: the variants show little or no sporulation.
(2) Pigment formation: reduction or absence of dark brown pigment.
(3) Colony morphology: the parent colonies have a "pudding" (Kugel) morphology that is lost in the variants.
(4) Oxytetracycline production: variants show little or no production.
(5) Oxytetracycline resistance: sensitive variants arise.

These phenotypes are inheritable genetic changes. Different combinations of the phenotypes can occur and there is often a continuing "degeneration" of variants until they show all five phenotypes. These instability steps can occur in various orders and it is not clear whether the five phenotypic alterations correspond to independent mutation events. It is expected that analysis of the complex gentic instability system in *S. rimosus* using the methods developed for *S. lividans* 66 will show the relationship between strain degeneration in commercial strains and the model genetic instability systems.

RESULTS AND DISCUSSION

A Type I DNA amplification system in *S. lividans* 66

S. coelicolor A3(2) carries the transposable element IS466
(formerly the "agarase" element, Kendall and Cullum, 1986) close
to the agarase gene. *S. lividans* 66 strain TK64 was transformed
with the plasmid pMT664 (Birch, 1985) which consists of the 5kb
BamHI fragment carrying the agarase gene and IS466 (Kendall and
Cullum, 1986) cloned into the BglII site of the temperature-
sensitive replication plasmid pMT660 (Birch & Cullum, 1985).
Mutants resistant to the galactose analogue 2-deoxygalactose
were selected as in Kendall et al (1987) in the hope that some
would carry insertions of IS466 into the *gal* operon. Although
such events were not detected, it was observed that many of the
mutants (30-40%) contained high copy number DNA amplifications.
Most of the mutants showed a normal colony morphology without
any changes in pigmentation or sporulation.

DNA amplifications from 6 mutants are shown in Fig. 1 and
it can be seen that different mutants carry different
amplifications. None of the DNA amplifications showed homology
to the plasmid pMT664. Analysis of restriction patterns
suggested that some of the amplifications involved overlapping
DNA fragments and this was confirmed using Southern
hybridisation; the results of these experiments are summarised

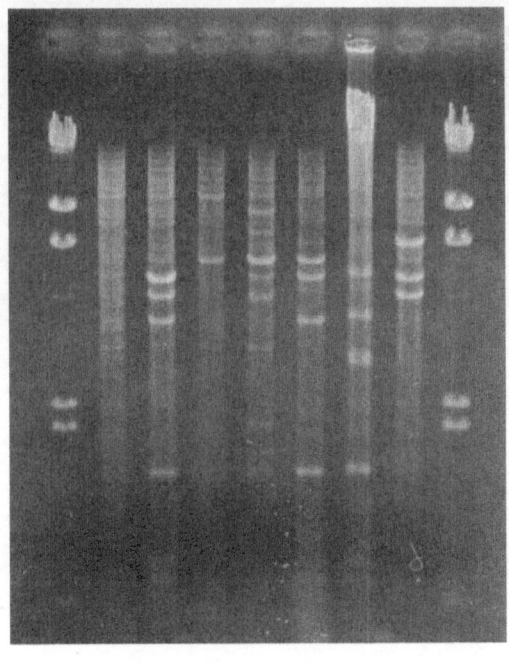

Fig. 1. Type I Amplifications
Tracks 1,9: DNA marker
Tracks 2-8: Total DNA
   restricted with BamHI
Track 2: Parent strain
Tracks 3-8: Mutants

Table 1.   DNA Amplifications in pMT664 Strains

Strain number	Length (kb)	Hybridisation with U7 U14 U17		
U7	21.9	+	–	+
U14	11.0	–	+	–
U17	13.6	+	–	+
U24	45.0	+	–	+

in Table 1. Further experiments may also localise non-overlapping amplifications to the same region and show whether flanking deletions are present as in other Type I systems.

The four mutants of Table 1 were propagated through three cycles of sporulation and the DNA was reexamined. None of them showed an alteration in amplification pattern, suggesting that the amplifications are quite stable.

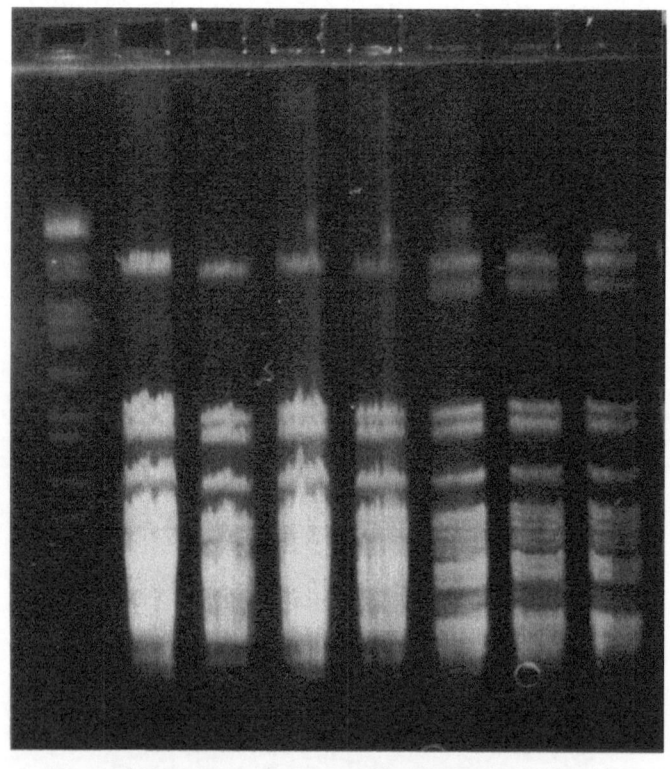

Fig. 2. Pulsed-field electrophoresis of
BfrI-digested *S. lividans* 66 DNA
Track 1: S. cerevisiae chromsomes
Tracks 2-5: ZX7 Cml^S mutant
Tracks 6-8: ZX7 parent strain

Cml and Arg Deletions in *S. lividans* 66 are linked

It was possible to isolate 16 cosmids from the Cml-deletion and 17 cosmids from the Arg-deletion (Flett et al, manuscript submitted). DNA from the mutant ZX7 (Zhou et al., 1988) was digested with several restriction enzymes and separated by pulsed-field-electrophoresis. This showed that the enzyme BfrI (recognition sequence CTTAAG) gave particularly large restriction fragments. Southern transfers from such gels were hybridised with digoxigenin-labelled DNA from representative cosmids from the Cml and Arg deletion regions. All cosmids used hybridised to the same large (945kb) BfrI fragment (data not shown) showing a physical linkage of the Arg and Cml deletions.

Spontaneous Cml^S mutants of ZX7 occur with similar frequencies to those in other *S. lividans* 66 strains. Fig. 2 shows BfrI digests of DNA from ZX7 and a Cml^S mutant. It can be seen that the second largest (945kb) BfrI band is missing in the mutant. No new band can be seen in the mutant. This might indicate that such a band is smaller than those well-resolved in Fig. 2, but this would be difficult to reconcile with the indications of the cosmid clones that the Cml-deletion is 250-300kb long (and that the Arg-deletion is of similar size). One possible explanation would be that the Cml^S strain is so unstable that a mixed population is always present so that single bands cannot be seen. Further hybridisation experiments are needed to resolve these questions.

The strain ZX7 is not ideal for studies on instability, because it is derived from a strain (JT46) deficient in intra-plasmid recombination (Chen et al., 1987). The results presented here, show that the Cml^S deletions still occur normally, but it would be desirable to construct a strain showing the Type II amplification, which could be used for pulsed-field electrophoresis.

Analysis of *S. rimosus*

The initial strategy to discover the molecular basis for genetic instability in *S. rimosus* was to screen variants for DNA amplifications and deletions. Amplifications in other *Streptomyces* species are easily detected by agarose gel electrophoresis of total DNA after digestion with various restriction enzymes. Preliminary experiments have not revealed any DNA amplifications in variants.

Large deletions should be revealed using pulsed-field electrophoresis after digestion with rarely cutting restriction enzymes. Two variants were chosen for initial analysis and to optimise the conditions for pulsed-field electrophoresis. They were MBV1 (Spo⁻ Pig⁻ Kug⁻ Otc⁻ Otc^R phenotype - see Introduction for definition of phenotypes) and MBV14 (Spo⁻ Pig$^{+/-}$ Kug$^+$ Otc$^{+/-}$ Otc^R), both of which were relatively stable strains. Mycelium from liquid cultures was immobilised in agarose blocks and used for DNA preparation. When undigested DNA was used for pulsed-field electrophoresis, a single band was seen that migrated together with yeast chromosomal markers of 370kb (Fig. 3(a)). When such a gel was subjected to normal gel electrophoresis (1 V/cm, 16 hr), there was no differential movement of the "370kb" band relative to the yeast chromosomal marker bands. This suggested that the band really does correspond to a giant linear plasmid rather than being a non-linear molecule (e.g. CCC-DNA). The 370kb band is present in both MBV1 and MBV14, showing that

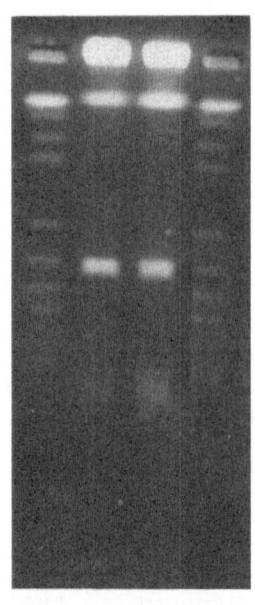 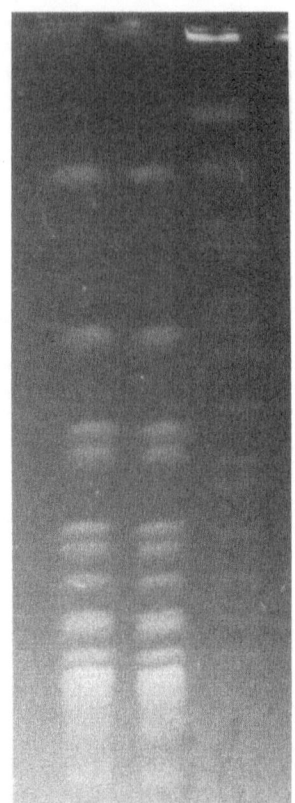

(a)   (b)

Fig. 3. Pulsed-field electrophoresis of DNA from
        S. rimosus mutants MBV1 and MBV14.
        (a) Undigested DNA.
            Tracks 2-3: MBV1, MBV14.
            Tracks 1,4: *S. cerevisiae* chromosomes.
        (b) DNA digested with XbaI.
            Tracks 1-2: MBV1, MBV14.
            Track 3: *S. cerevisiae* chromosomes.

no major change in size of the plasmid is involved in the
difference in phenotype (the gel system would resolve
differences of 10kb in size). Further analysis is needed to rule
out smaller changes in the plasmid between the variants.
    Several restriction enzymes were examined for their
suitability for analysing *S. rimosus*. A suitable enzyme must
give good digestion in agarose blocks and restrict the
chromosome into a relatively small number of large fragments.
XbaI and BfrI gave good results. Fig. 3(b) shows the results of
digesting DNA from the two strains with XbaI. In the size range
that is resolved by this gel (245-1100kb) no difference can be
seen in the restriction pattern. The well-resolved bands account
for about 5.3Mb of the genome. Further analysis using different

270

enzymes and electrophoresis conditions (e.g. altered pulse times) that resolve smaller sized fragments is needed to detect any differences. It is intended to establish these conditions and to continue analysis of instability in *S. rimosus*.

ACKNOWLEDGEMENTS

We would like to thank the EC for an International Scientific Cooperation grant (to JC and DH), the state of Rheinland-Pfalz for a Landesgraduiertenförderung Stipendium (to BG) and the Japanese government for a fellowship (to KM). We thank David Hopwood and Tobias Kieser for strains and Mr. S. J. Lucania of E. J. Squibb and Sons for the gift of thiostrepton.

REFERENCES

Alacevic, M., Strasek-Vesligaj, M., and Sermonti, G., 1973, The circular linkage map of *Streptomyces rimosus*, J. Gen. Microbiol., 77: 173-185.

Altenbuchner, J., and Cullum, J., 1985, Structure of an amplifiable DNA sequence in *Streptomyces lividans* 66, Mol. Gen. Genet., 201: 192-197.

Altenbuchner, J., and Cullum, J., 1987, Amplification of cloned genes in *Streptomyces*, Bio/Technology, 5: 1328-1329.

Beijerinck, M. W., 1913, Ueber Schröter und Cohn's Lakmus-micrococcus, Folia Microbiol., 2: 185-200.

Birch, A. W., and Cullum, J., 1985, Temperature-sensitive mutants of the *Streptomyces* plasmid pIJ702, J. Gen. Microbiol., 131: 1299-1303.

Birch, A., Häusler, A., Vögtli, M., Krek, W., and Hütter, R., 1989, Extremely large chromosomal deletions are intimately involved in genetic instability and genomic rearrangements in *Streptomyces glaucescens*, Mol. Gen. Genet., 217: 447-458.

Birch, A. W., 1985, Plasmid replication and recombination in *Streptomyces*, PhD Thesis, University of Manchester.

Chen, C. W., Tsai, J. F.-Y., and Chuang, S.-E., 1987, Intraplasmid recombination in *Streptomyces lividans* 66, Mol. Gen. Genet., 209: 154-158.

Cullum, J., Altenbuchner, J., Flett, F., and Piendl, W., 1986, DNA amplification and genetic instability in *Streptomyces*, Biotechnol. Gen. Eng. Rev., 4: 59-78.

Danilenko, V. N., Starodubtseva, L. I., and Navashin, S. M., 1986, Regulation of expression of kanamycin resistance and chloramphenicol resistance determinants in *Streptomyces lividans* 66, pp. 79-81, in "Biological, Biochemical and Biomedical Aspects of Actinomycetes", Szabo, G., Biro, S., and Goodfellow, M., eds., Akadémiai Kiadó, Budapest.

Flett, F., and Cullum, J., 1987, DNA deletions in spontaneous chloramphenicol-sensitive mutants of *Streptomyces coelicolor* A3(2) and *Streptomyces lividans* 66, Mol. Gen. Genet., 207: 499-502.

Freeman, R. F., Bibb, M. J., Hopwood, D. A., 1977, Chloramphenicol acetyl-transferase-independent chloramphenicol resistance in *Streptomyces coelicolor* A3(2), J. Gen. Microbiol., 98: 453-465.

Gregory, K. F., and Huang, J. C. C., 1964, Tyrosinase
    inheritance in *Streptomyces scabies*. I Genetic
    recombination, J. Bacteriol., 86: 1281-1286.
Häusler, A., Birch, A., Krek, W., Piret, J., and Hütter, R.,
    1989, Heterogeneous genomic amplification in *Streptomyces
    glaucescens*: Structure, location and junction sequence
    analysis, Mol. Gen. Genet., 217: 437-446.
Hranueli, D., Pigac, J., Smokvina, T., and Alacevic, M., 1983,
    Genetic interactions in *Streptomyces rimosus* mediated by
    conjugation and by protoplast fusion,
    J. Gen. Microbiol., 129: 1415-1422.
Hütter, R., Birch, A., Häusler, A., Vögtli, M., Madon, J., and
    Krek, W., 1988, Genome fluidity in Streptomycetes, in
    "Biology of Actinomycetes '88", Okami, Y., Beppu, T., and
    Ogawara, H., eds., Japan Scientific Societies Press, Tokyo.
Kendall, K., and Cullum, J., 1986, Identification of a DNA
    sequence associated with plasmid integration in
    Streptomyces coelicolor A3(2),
    Mol. Gen. Genet., 202: 240-245.
Kendall, K., Ali-Dunkrah, U., and Cullum, J., 1987, Cloning of
    the galactokinase gene (*galK*) from *Streptomyces coelicolor*
    A3(2), J. Gen. Microbiol., 133: 721-725.
Pigac, J., and Alacevic, M., 1979, Mapping of oxytetracycline
    genes on *Streptomyces rimosus* chromosome,
    Period biol, 81: 575-582.
Pigac, J., Hranueli, D., Smokvina, T., and Alacevic, M., 1982,
    Optimal cultural and physiological conditions for handling
    Streptomyces rimosus protoplasts,
    Appl. Environ. Microbiol., 44: 1178-1186.
Pigac, J., Vujaklija, D., and Gamulin, V., 1986, Structural
    segregation of a bifunctional vector pZG1 in *Streptomyces
    lividans* and *S. rimosus*, in "Biological, Biochemical and
    Biomedical Aspects of Actinomycetes", Szabo, G., Biro, S.,
    and Goodfellow, M., eds., Akadémiai Kiadó, Budapest.
Pigac, J., Vujaklija, D., Toman, Z., Gamulin, V., and Schrempf,
    H., 1988, Structural instability of a bifunctional plasmid
    pZG1 and single-stranded DNA formation in *Streptomyces*,
    Plasmid, 19: 222-230.
Redshaw, P. A., McCann, P..A., Pentella, M..A., and Pogell,
    B. M., 1979, Simultaneous loss of multiple differentiated
    features in aerial mycelium-negative isolates of
    Streptomyces, J. Bacteriol., 137: 891-899.
Zhou, X., Deng, Z., Firmin, J. L., Hopwood, D. A., and Kieser,
    T., 1988, Site-specific degradation of *Streptomyces
    lividans* DNA during electrophoresis in buffers contaminated
    with ferrous iron, Nucleic Acids Res., 16: 4341-4352.

ANALYSIS OF LARGE DELETIONS AND CHARACTERIZATION OF THE DELETION ENDPOINTS ASSOCIATED WITH AN AMPLIFIABLE DNA REGION IN STREPTOMYCES LIVIDANS

Wolfgang Piendl[1], Silvano Köchl[1], Fiona Flett[2] and John Cullum[3]

[1] Institut für Mikrobiologie (Med.Fak.), Univ. Innsbruck
[2] Department of Biochemistry and Applied Molecular Biology UMIST, Manchester
[3] LB Genetik der Universität Kaiserslautern

INTRODUCTION

The phenomenon, that Streptomyces species can lose spontaneously certain phenotypes at frequenzies between 10^{-3} and 10^{-1} has been recognized since at least 1913 (Beijerinck 1913). This genetic instability is very common in many Streptomyces species and can affect a variety of genes; however, only specific genes are affected in any one strain (reviewed by Cullum et al., 1986; Hütter and Eckhardt, 1988). Frequently antibiotic producing strains, including some of commercial importance, are subject to genetic instability: they lose the ability to produce antibiotics (e.g. tetracyclines) , i.e. they "degenerate". As plasmid-curing agents such as acriflavine and ethidium bromide or UV-irradiation increased the frequency of mutation drastically, several authors suggested that the loss of a plasmid caused the loss of antibiotic production and they concluded that genes (or regulatory genes) for antibiotic production are coded on plasmids.

Since DNA cloning methods have been developed for Streptomycetes, several unstable genes have been cloned and analysed. This has shown that the mutations usually result from deletion of the unstable gene and its flanking regions. There is no evidence for plasmid involvement in genetic instability and some unstable genes, e.g. the unstable tyrosinase gene (melC) and streptomycin resistance gene from S. glaucescens have been mapped to the chromosome (Crameri et al., 1983).

Sometimes deletions are accompanied by DNA amplifications where chromosomal sequences become amplified up to several hundred copies, which are arranged as tandem repeats. Thus in S. glaucescens 20% of melC/Str[S] strains showed intense DNA reiterations, which were heterogenous for stability, copy number, extent and DNA fragments involved. They are closely associated with polar deletions ranging from 270 to over 800kb (Birch et al., 1989; Häusler et al., 1989).

GENETIC INSTABILITY IN STREPTOMYCES LIVIDANS

Previous studies revealed a two-step process of instability in S. lividans. This strain spontaneously yields chloramphenicol sensitive (Cml^S) mutants at high frequency (0.1 - 1%) and the Cml^S mutants are very unstable and give rise to non sporulating (Amy^-) arginine auxotrophs (Arg^-) at frequencies of 10 - 25% of spores (Altenbuchner and Cullum, 1984). The Cml^S Arg^- double mutants have amplified the same 5.7 kb chromosomal element which is called AUD (amplifiable unit of DNA) to several hundred copies. In the parental strain the 5.7 kb AUD is already present as a tandem duplication and there are three copies of a 1 kb direct repeat flanking two copies of a 4.7 kb sequence (Altenbuchner and Cullum, 1985). For this amplifiable element the tandem structure seems to be a requirement for amplification, as Cml^S Arg^- mutants of S. lividans strain TK23, which has lost one of the two copies do not show a high copy number amplification (Dyson and Schrempf, 1987). The same is true for the closely related S. coelicolor A3(2) which contains only one copy of the 5.7 kb AUD (W. Piendl, unpublished; Flett et al., 1987).

J. Altenbuchner (1988) has discovered three more amplifiable elements in the wild type S. lividans 1326 which seem to be deleted in the plasmid-free derivative TK64 (Hopwood et al., 1983).

DELETIONS IN CML^S ARG^- STRAINS OF S. LIVIDANS

Ishihara et al. (1985) showed that in Cml^S Arg^- strains of S. lividans and S. coelicolor A3(2) the argG gene (the structural gene for argininosuccinate synthetase) was deleted. A plasmid carrying the cloned argG gene complemented the argG mutation and restored the Amy^+ phenotype. This finding was in accordance with the observation that in the amplified Cml^S Arg^- strains there was a deletion of neighbouring DNA sequences to one side of the amplified element. Sequences to the other side remain intact (Altenbuchner and Cullum, 1985). In fact chromosome walking experiments established that the argG gene is located 25 kb distant from the 5.7 kb AUD (Betzler at al., 1987).

Studies of Flett and Cullum (1987) demonstrated that chloramphenicol sensitivity in S. coelicolor A3(2) and S. lividans is associated with large chromosomal deletions.

To obtain further insight into the size and nature of these large deletions it would be attractive to use pulsed field gel electrophoresis. By careful choice of restriction enzyme it should be possible to investigate the arrangement of the deletions in the chromosome. Unfortunately S. lividans DNA contains a modification which makes it susceptible to double-strand cleavage during electrophoresis (Zhou et al., 1988) so the separation of large DNA fragments remains problematic.

Therefore, to determine the total length of the deletions associated with the Cml^S and Arg^- phenotype an alternative approach, a reverse-blotting procedure which relies on the identification of deletion clones by hybridisation of labelled total DNA from a deletion mutant with a wildtype cosmid bank, was chosen. Cosmid clones failing to give a hybridisation possess no homology with the deletion mutant DNA probe and should contain fragments which are part of the deletion (Birch et al., 1989).

A genomic cosmid library of S. lividans TK64 was screened by colony hybridisation using ^{32}P-labelled total DNA from the TK64 mutant Arg^- 1.1. It could be shown that a spontaneous Cml^S mutant of TK64 has deleted

$>$ 250 kb of chromosomal DNA. The $\underline{Cml}^S$ $\underline{Arg}^-$ mutant has a further deletion of $>$ 230 kb, including the $\underline{arg}G$ gene (F. Flett et al., manuscript in preparation). For the moment it is not clear wheather the $\underline{Arg}^-$ deletion is an extension of the Cml^S deletion or whether there are two or more independent deletions. The size of the deletion(s) is in the same range as that found in $\underline{S. \ glaucescens}$ (Birch et al., 1989). It represents the loss of approximately 10% of the total genome of $5-6 \times 10^3$ kb (Genthner et al., 1985).

Moreover we wanted to investigate if the deletions associated with the 5.7 kb AUD in different independent isolates of $\underline{Cml}^S$ $\underline{Arg}^-$ mutants of $\underline{S. \ lividans}$ were uniform, i.e. if they would share common deletion end-/start-points. We started to study the new junction bands formed in the deletion(s) by Southern hybridisation experiments using the 5.7 kb AUD element as a probe. Total DNA of 20 $\underline{Cml}^S$ $\underline{Arg}^-$ mutants with a high copy number amplification of the 5.7 kb AUD which were derived from 20 different Cml^S mutants of the parental strain TK64 (Altenbuchner and Cullum, 1984) was digested with four different restriction enzymes ($\underline{Bam}$ HI, $\underline{Bgl}$II, $\underline{Sac}$I, $\underline{Xho}$I) which cut the 5.7 kb AUD once (Fig.1). After gel

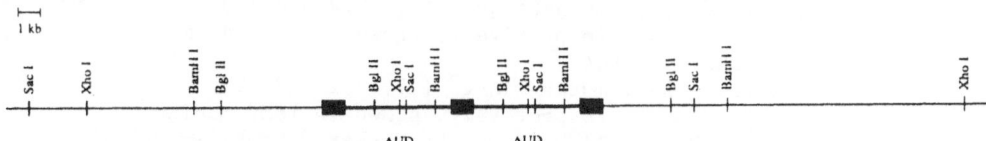

Fig. 1. Restriction map of the 5.7 kb AUD of $\underline{S. \ lividans}$ TK64 and its flanking regions. Only enzymes which cut the AUD once are mapped.

electrophoresis, the amplified 5.7 kb band was cut out of the gel (Fig.2), as the intense amplification (several hundred copies) precluded the identification of flanking fragments (one copy).

The alternative way, to look at strains with a very low level of amplification, where the flanking DNA would be detectable, was rejected, as during the deamplification secondary events might affect the original deletion.

For example, when total DNA from parental strain TK64 was cut with $\underline{Bam}$HI and probed with the 5.7 kb AUD, three bands were observed: (i) a 5.7 kb band representing the middle $\underline{Bam}$HI fragment of AUD, (ii) a 6.5 kb band representing the right hand flanking sequence and (iii) a 12.5 kb band, representing the left hand flanking sequence (Fig.1; Fig.3, track 1). In the amplified $\underline{Arg}^-$ variants we expected to find the 6.5 kb right hand fragment intact, as only deletions to the left of the amplification were found, the 5.7 kb amplified band and the left hand 12.5 kb band should be absent and replaced by the new deletion junction fragment.

In our first experiments all four restriction enzymes used ($\underline{Bam}$HI, $\underline{Bgl}$II, $\underline{Sac}$I, $\underline{Xho}$I) produced a new 11.4 kb fragment instead of the left hand flanking sequence in all 20 $\underline{Arg}^-$ strains tested. In similar experiments this 11.4 kb band was observed by another group and interpreted as "restriction site polymorphisms present in the amplified DNA" (Dyson et al., 1986). In our hands these 11.4 kb bands disappeared when we repeated the experiments with DNA we had purified once more. Thus, the bands were probably due to partial digestion.

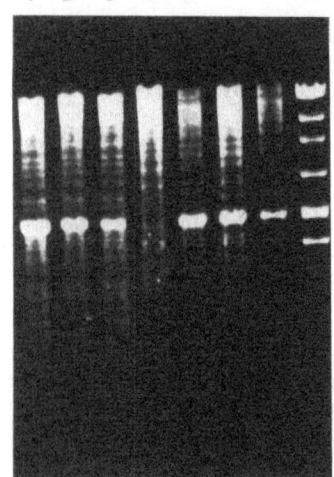
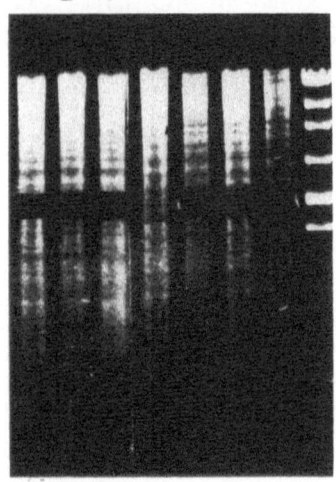

A

B

Fig. 2. A. Electrophoretic pattern of BglII digested DNA of
            Cml^S Arg$^-$ mutants of S. lividans TK64.
            Track 1-3: mutants Arg$^-$ 1.1, 3.2 and 7.3
                       respectively, grown in LB medium
            Track 4: TK64
            Track 5-7: mutants Arg$^-$ 1.1, 3.2 and 7.3
                       respectively, grown in YEME medium
            Track 8: molecular weight marker (EcoRI + BglII)
       B. Identical with 2.A., but the amplified band was cut
          out in all tracks.

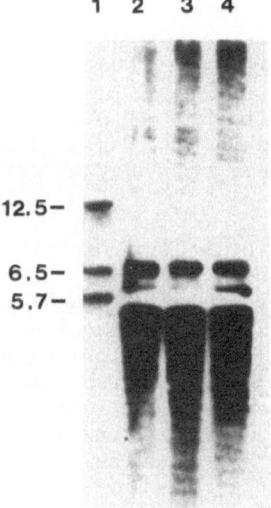

12.5—

6.5—
5.7—

Fig. 3. Hybridisation of the ^{32}P labelled 5.7 kb AUD fragment
        to Southern transfers of BamHI-digested total DNA from
        S. lividans strains.
        Track 1: TK64
        Track 2-4: Arg$^-$ 1.1, 3.2 and 7.3, respectively,
                   original DNA preparations.

Surprisingly we could not detect new junction bands of the left hand side using restriction enzymes BamHI, BglII, SacI or XhoI. The flanking sequences to the right were conserved as we had expected (Fig.3). Unfortunately we cannot detect fragments smaller than 6 kb in these experiments, as the amplified DNA undergoes partial double-strand cleavage during electrophoresis (Zhou et al., 1988) and the resulting smear would mask small bands. But it is very unlikely that all four bands arising from BamHI, BglII, SacI and XhoI digests carrying the new junction fragment are smaller than 6 kb. Therefore we favour another explanation why we were unable to detect new junction bands in 20 independent Arg⁻ isolates. We assume that the deletions are extremely instable and suggest a given population of chromosomes contains a very heterogenous set of junction fragments preventing identification of single bands.

In the course of these experiments we learned that the parental strain TK64, after protoplast regeneration and UV irridation, had not only lost the plasmids SLP2 and SLP3 (Hopwood et al., 1983), but had also undergone large chromosomal deletions. A 93 kb AUD element containing a mercury resistant determinant is completely deleted in TK 64 (Altenbuchner et al., 1988). To make sure that the results were not an artefact due to the choice of an unsuitable parent strain (TK64), we have looked for deletion junction bands in 20 independent (i.e. derived from 2o different Cml^s mutants) Arg⁻ isolates of the wildtype strain 1326. However the result was the same; in none of the original Arg⁻ mutants of the wildtype 1326 could new junction bands be detected.

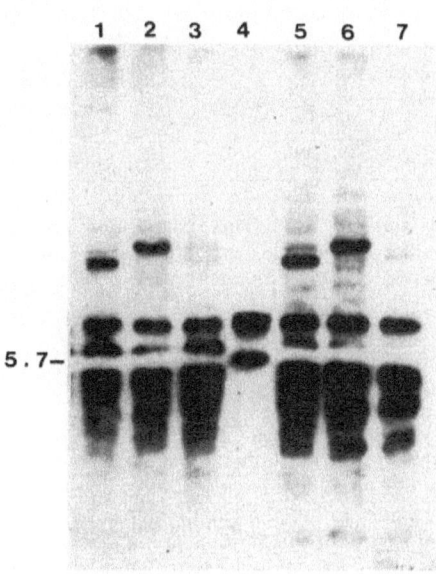

Fig. 4. Hybridisation of the ³²P labelled 5.7 kb AUD fragment to Southern transfers of BglII digestet total DNA from S. lividans strains.
Track 1-3: mutant Arg⁻ 1.1, 3.2 and 7.3 respectively, grown in LB medium (second DNA preparation)
Track 4: TK64
Track 5-7: mutant Arg⁻ 1.1, 3.2, and 7.3 respectively, grown in YEME medium (second DNA preparation)
The corresponding gel is shown in Fig.2.B.

Finally we succeeded in identifying deletion junction bands in two
Cml^s $\underline{Arg}^-$ mutants of parental strain TK64. After several rounds of subcul-
turing, mutants $\underline{Arg}^-$ 1.1 and 3.2 had been stored in LB/5%DMSO at -70°C
for several months. After this "treatment" we detected the expected stable
left hand junction bands in the new DNA preparations (Fig.4). We have
grown these and other $\underline{Arg}^-$ strains in different media, e.g. LB and YEME,
but the culture conditions chosen had no apparent influence on the deletions. In order to identify further deletion junctions in other $\underline{Cml}^s$
$\underline{Arg}^-$ mutants, they were subject to the same "treatment", but no more junction bands could be detected.

CLONING AND ANALYSIS OF THE DELETION JUNCTION AND THE CORRESPONDING
DELETION ENDPOINTS

To obtain information on the sequences involved in deletion formation
we decided to clone the two identified junction fragments from the $\underline{Arg}^-$
mutants 1.1 and 3.2 (Fig.4). The 11 kb $\underline{Bgl}II$ fragment of $\underline{Arg}^-$ 1.1 and
the 12 kb $\underline{Bgl}II$ fragment of $\underline{Arg}^-$ 3.2 were cloned into vector pBGS8. Comparison of the restriction maps of these two junction bands with the re-
striction map of the 5.7 kb AUD of the parental strain (Altenbuchner and
Cullum, 1985) revealed, that in both cases the deletion commenced in the
4.7 kb fragment of the amplified region (Fig.5). The two deletion endpoints
lie about 1 kb apart.

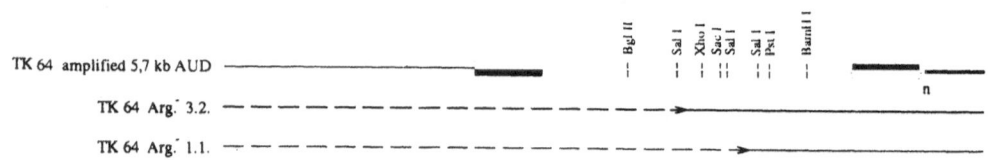

Fig. 5. Restriction map of the amplified 5.7 kb element and the
deletion end-points (→) of the $\underline{Arg}^-$ mutants 1.1 and 3.2.
------ deleted sequences.

Subclones that lacked any sequences homologous to the 5.7 kb element
were constructed from the junction fragment clones. These subclones were
used as hybridisation probes to isolate clones from a cosmid library of
TK64. The junction fragments from the $\underline{Arg}^-$ mutants 1.1 and 3.2 each hybridised to a different non-overlapping cosmid. Examination of the cosmids
shows that the two deletion end-points must be more than 10 kb apart on
the chromosome.

The deletion junctions in Arg^- 1.1 and Arg^- 3.2 and the corresponding
sequences in the parent strain (TK64) have been sequenced (Fig.6). The
ends of both deletions correspond to short direct repeats in the parental
strain sequence. The deletion had occured from a tetramer CCGG (Arg^- 1.1)
or a pentamer TCCGG (Arg^- 3.2) on the left side to an identical sequence
within the amplified region. In each case a single copy of the repeat
was left at the deletion junction. The sequence (T)CCGG occurs very fre-
quently in structural genes of $\underline{S.\ lividans}$ and other $\underline{Streptomyces}$ species,
thus it is not a specific "deletion-motif".

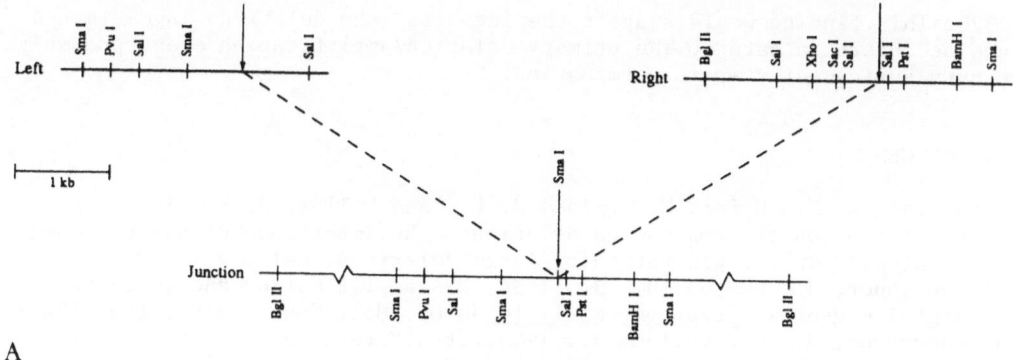

A

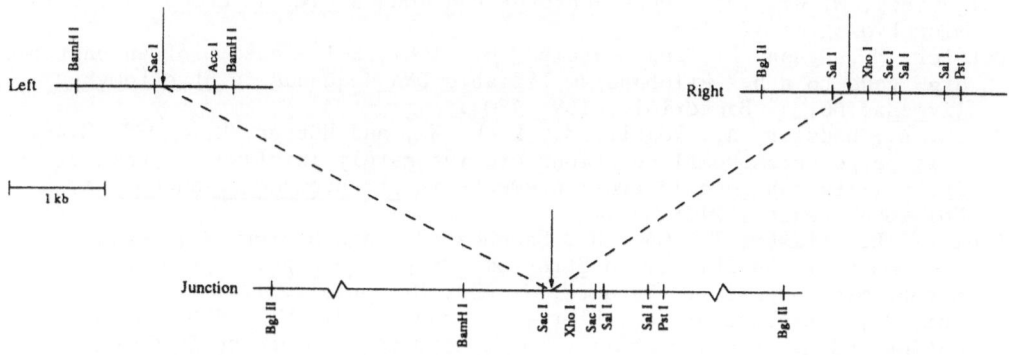

B

Fig. 6. Restriction maps of the chromosomal DNA fragments carrying
the right and left hand deletion end-point and the corresponding
junction fragment of the Arg⁻ mutant 1.1 (A) and 3.2 (B). Verti-
cal arrows denote the position of the deletion ends. Regions
surrounding the deletion end-point were sequenced.

Perfect and imperfect direct repeats can play a role in deletion
formation in several organisms. Such deletions are known to occur in E.
coli (Albertini et al., 1982) and Bacillus subtilis (Janniere and Ehrlich,
1987; Peters et al., 1988).

In Streptomyces, four small deletions, two occurring in plasmids
(Nakano et al., 1984; Kendall and Cohen, 1988) and two occurring in phage
ØC31 (Sinclair and Bibb, 1988) have been characterized at sequence level.
These deletions had occurred between short direct repeats (3 to 11 bp
long). No small direct repeats, but a quasi-palindromic sequence, which
precisely coincides with the deletion end-points, is involved in the formation of a large chromosomal deletion of S. glaucescens (Birch et al.,
1989).

Finally, we could show in hybridisation experiments that the two
deletions we had analysed in the Arg⁻ mutants 1.1 and 3.2 were larger
than the deletions in other Arg⁻ mutants derived from TK64 or wildtype

1326. This finding would support the idea that the deletions investigated
are secondary in nature. The primary deletion/amplification event probably
occurs during early spore germination.

REFERENCES

Albertini, A. M., Hofer, M., Calos, M. P., and Miller, J. H., 1982, On
    the formation of spontaneous deletions: the importance of short sequence
    homologies in the generation of large deletions, Cell, 29: 319.
Altenbuchner, J., and Cullum, J., 1984, DNA amplification and an unstable
    arginine gene in Streptomyces lividans 66, Mol. Gen. Genet., 195: 134.
Altenbuchner, J., and Cullum, J., 1985, Structure of an amplifiable DNA
    sequence in Streptomyces lividans 66, Mol. Gen. Genet., 201: 192.
Altenbuchner, J., Eichenseer, C., and Brüderlein, M., 1988, DNA amplifica-
    tion and deletion in Streptomyces lividans, in: Biology of Actinomycetes
    88, Y. Okami, T. Beppu and H. Ogawara, eds., Japan Scientific Societes
    Press, Tokyo.
Beijerinck, M. W., 1913, Über Schröter und Cohn's Lakmusmicrococcus, Folia
    Microbiologica, 2: 185.
Betzler, M., Dyson, P., and Schrempf, H., 1987, Relationship of an unstable
    argG gene to a 5.7 kilobase amplifiable DNA sequence in Streptomyces
    lividans 66, J. Bacteriol., 169: 4804.
Birch, A., Häusler, A., Vögtli, M., Krek, W., and Hütter, R., 1989, Extre-
    mely large chromosomal deletions are intimately involved in genetic
    instability and genomic rearrangements in Streptomyces glaucescens,
    Mol. Gen. Genet., 217: 447.
Crameri, R., Kieser, T., Ono, H., Sanchez, J., and Hütter, R., 1983,
    Chromosomal instability in Streptomyces glaucescens: mapping of
    streptomycin-sensitive mutants, J. Gen. Microbiol., 129: 519.
Cullum, J., Altenbuchner, J., Flett, F., and Piendl, W., 1986, DNA amplifi-
    cation and genetic instability in Streptomyces, Biotechnol. Genet.
    Eng. Rev. 4: 59.
Dyson, P., Betzler, M., Kumar, T., and Schrempf, H., 1986, Biochemical
    and genetic analysis of spontaneous genetic instability and DNA amplifi-
    cation of Streptomyces lividans, in: Fifth International Symposium
    on the Genetics of Industrial Microorganisms, M. Alacevic, D. Hranueli
    and Z. Toman, eds., Zagreb.
Dyson, P., and Schrempf, H., 1987, Genetic instability and DNA amplifica-
    tion in Streptomyces lividans, J. Bacteriol., 169: 4796.
Flett, F. and Cullum, J., 1987, DNA deletions in spontaneous chlorampheni-
    col-sensitive mutants of Streptomyces coelicolor A3(2) and Streptomyces
    lividans 66, Mol. Gen. Genet., 207: 499.
Flett, F., Platt, J., and Cullum, J., 1987, DNA rearrangements associated
    with instability of an arginine gene in Streptomyces coelicolor A3(2),
    J. Basic Microbiol., 27: 3.
Genthner, F. J., Hook, L. A., and Strohl, W. R., 1985, Determination of
    the molecular mass of bacterial genomic DNA and plasmid copy number
    by high pressure liquid chromatography, Appl. Environ. Microbiol.,
    50: 1007.
Häusler, A., Birch, A., Krek, W., Piret, J., and Hütter, R., 1989, Hetero-
    genous genomic amplification in Streptomyces glaucescens: structure,
    location and junction sequence analysis, Mol. Gen. Genet., 217: 437.
Hopwood, D. A., Kieser, T., Wright, H. M., and Bibb, M. J., 1983, Plasmids,
    recombination and chromosome mapping in Streptomyces lividans 66, J.
    Gen. Microbiol., 129: 2257.
Hütter, R. and Eckhardt, T., 1988, Genetic manipulation, in: Actinomycetes
    in biotechnology, M. Goodfellow, S. T. Williams and M. Mordarski, eds.,
    Academic Press, London.

Ishihara, H., Nakano, M. M., and Ogawara, H., 1985, Cloning of a gene from Streptomyces species complementing argG mutations, J. Antibiot., 38: 787.

Janniere, L., and Ehrlich, S. D., 1987, Recombination between short repeat sequences is more frequent in plasmids than in the chromosome of Bacillus subtilis, Mol. Gen. Genet., 210: 116.

Kendall, K. J., and Cohen, S. N., 1988, Complete nucleotide sequence of the Streptomyces lividans plasmid pIJ 101 and correlation of the sequence with genetic properties, J. Bacteriol., 170: 4634.

Nakano, M. M., Ogawara, H., and Sekiya, T., 1984, Recombination between short direct repeats in Streptomyces lavendulae plasmid DNA, J. Bacteriol., 157: 658.

Peters, B. P. H., de Boer, J. H., Bron, S., Venema, G., 1988, Structural plasmid instability in Bacillus subtilis: effect of direct and inverted repeats, Mol. Gen. Genet., 212: 450.

Sinclair, R. R., and Bibb, M. J., 1988, The repressor gene (c) of Streptomyces temperate phage ØC31: nucleotide sequence analysis and functional cloning, Mol. Gen. Genet., 213: 269.

Zhou, X., Deng, Z., Firmin, J. L., Hopwood, D. A., and Kieser, T., 1988, Site-specific degradation of Streptomyces lividans DNA during electrophoresis in buffers contaminated with ferrous ions, Nucleic Acids Res., 16: 4341.

AN INTRODUCTION

Jasenka Pigac

PLIVA Research Institute
41000 Zagreb
Yugoslavia

A prerequisite to the development of gene cloning and expression of homologous and heterologous genes in any organism is the identification of a suitable vector for the introduction and stable maintenance of cloned DNA in the host. During progressive application of rDNA technology in Streptomyces, the problems of efficient introduction of vectors by trans-formation of protoplasts were solved. However, the problems of stable maintenance of the cloned DNA in the host still persist. Streptomyces strains are often genetically unstable (Schrempf et al., 1987) and their mycelial growth habit hampers the study of plasmid instability. In general, two types of instability are observed and studied in gram-positive bacteria: one corresponding to the loss of the entire vector from the cell, the other to rearrangements, most often deletions of plasmid sequences. The first was named segregational, the second structural plasmid instability (Ehrlich et al., 1987). Recently attempts to study the problems of plasmid segregational instability in Streptomyces were reported. Because of their mycelial growth habit and irregular cell division, plasmids do not segregate into individual cells after replication. This fact makes investigation of plasmid inheritance difficult in Strepto-myces. Many vectors currently used in Streptomyces are derivatives of pIJ101 (Kieser et al., 1982). It is interesting to note that no evidence exists for a par locus on such plasmids which replicate via rolling circle, i.e. there are no membrane attachment sites that physically aid segregation of sufficient numbers of plasmids to be established in each daughter cell (Cesareni et al., 1987). In these plasmids, stability seems to be coupled with replication and not with a discrete par-like function (Gruss and Ehrlich, 1989).
Radnedge et al. (1989) described a method which allows rapid, multi-sample determination of plasmid copy number in order to evaluate stability in liquid culture. Using an optical scanning densitometer, undigested and digested total DNA samples, both segregational and structural stability of the vectors was measured. These experiments were scaled up for the use of 7 litre fermenters.

Another example of measuring segregational instability in Streptomyces using S. lividans TK21 containing representatives of high (pIJ702) and low (pIJ922) copy number plasmids has been described. There was a marked difference in plasmid segregation between spores and protoplasts, and between high and low copy number plasmids, in favour to spores and low copy number plasmid, respectively. Thus plasmid analysis of spores and copy number determinations could be used to investigate plasmid instability in serial subcultures using solid media and liquid batch and continuous culture (Simpson et al., 1990). In that respect a chemostat can be a suitable device to study the structural and functional stability of a Streptomyces plasmid. The chemostat provides an environment in which cell growth and division is continuous and the population size is held constant (Roth et al., this Symposium). The growth rate of the population, its density and other culture conditions can be independently manipulated. Among numerous possibilities the influence of additional DNA elements like plasmids, transposons and viruses on the fitness of their host cells as well as the segregational and structural stability of these elements in relation to the genetic background of the host and the culture conditions could be studied. Mathematical models of plasmid partitioning have also been developed and compared with experimental data (Müller et al., communicated at this Symposium).

For practical reasons, the effects on secondary metabolite production of a high copy pIJ702 and a low copy pIJ922 plasmids in S. thermoarchaensis transformants should be mentioned. The production of archaemycins but not oligomycin was depressed. The mechanism of this effect is still under investigation (Thomas et al., 1990).

The structural instability of recombinant plasmids in gram-positive bacteria is mainly the consequence of illegitimate recombination. The rearrangements of the DNA could be caused by the aberrant replication of a single-stranded DNA in the replication fork, called slipped mispairing (Ehrlich et al., 1987). Many vectors of gram-positive bacteria, including pIJ101 derivatives, accumulate ssDNA during their replication (for review see Gruss and Ehrlich, 1989). Their structural instability is more increased by inserts of foreign DNA as in the case of bifunctional constructs. To that group of plasmids belong pZG-bifunctional vectors (Pigac et al., 1988 and this Symposium).

Errors of DNA breakage and reunion (Franklin, 1967) might lead to deletions. In this model, enzymes that break and join DNA as part of their normal functions would do so at sequences that share little or no homology (Anderson, 1987). DNA topoisomerases modify the topological states of DNA by catalyzing specific types of coupled breakage/rejoining reactions (Wang, 1985).
Bacteria contain at least two DNA topoisomerase enzymes with complementary (i.e., opposing) activities (for reviews, see Gellert, 1981; Wang, 1985). Topoisomerase I and topoisomerase II (gyrase) can both reduce the super-helical density of ccc plasmid DNA, which is negatively supercoiled, by removal of negative supercoils, albeit by different mechanisms. On the other hand, in the presence of ATP, DNA gyrase can introduce negative supercoils into relaxed ccc DNA, a property not shared by topoisomerase I. Many antibiotics were discovered to inhibit the catalytic activities of gyrases in bacteria.

The classical gyrase inhibitors are synthetic quinolones. They interfere with the gyrase A subunit affecting its breaking and rejoining of the phosphodiester backbone of DNA (Gellert et al., 1977; Sugino et al., 1977) (for review, see Wolfson and Hooper, 1985). Novobiocin binds tightly to the DNA gyrase B moiety and coumarins inhibit energy transduction mediated by the B subunit, via the binding and hydrolysis of ATP (Mizuuchi et al., 1978; Sugino et al., 1978). Stoerl et al. (this Symposium) reported a peculiar new class of gyrase inhibitors, the antitumor-active non-intercalating DNA binding ligands. They proved to be equally effective inhibitors of the supercoiling reaction catalyzed by gyrases from Streptomyces and E. coli.
Moreover, as a group of dAdT-specific DNA binding ligands, they also showed a significant inhibitory effect on the relaxation reaction mediated by the topoisomerase I of S. noursei.

One of the goals of the cloning of desired genes is their successful expression in the host. Recently (Schauer et al., 1988) a new promoter probe vector for visualizing the expression of various streptomycete promoters lighting up during the differentiation of colonies of S. coelicolor was constructed. The system exploits the luxA and luxB operon coding for luciferase of the bacterium Vibrio harveyi cloned into a suitable Streptomyces vector.

The use of luciferase genes also enabled the analysis of developmental mutants in Streptomyces resulting from insertion of a derivative of the transposon TN4556 (isolated from S. fradiae) which simultaneously inactivates target genes and places the light-encoding luciferase genes under control of the exogenous transcription unit (Schauer et al., 1990).

Besides monitoring gene expression, Weiser et al. (communicated at this Symposium) constructed a vector for indirect measurement of the accuracy of translation in Streptomyces, based on the expression of the lux gene coding for luciferase isolated from the North American firefly Photinus pyralis.

Most of our efforts are usually directed to construct stable strains by controlling the stability of the genomic DNA, yet on the other hand we take advantage of infidelity of DNA replication which permits the selection of strains with novel traits. Though instabilities are often undesired, their study could be most rewarding, giving us powerful insights into the very nature of the stabilizing influences from which they have escaped.

REFERENCES

Anderson, P., 1987, Twenty years of illegitimate recombination. Genetics 115, 581:584.

Casareni, G., M. Anceschi, L. Castagnoli, F. Felici, M. Helmer Citterich, J. Hughes, D. Kirk, J. Murray, N. Rossi, M. Scarpa and L. Spinelli, 1987, E. coli plasmid stability and control of copy number. In: "Genetics of Industrial Microorganisms", Part A (M. Alačević, D. Hranueli and Z. Toman, eds.), p. 247:257. PLIVA, Zagreb.

Franklin, N. C., 1967, Extraordinary recombinational events in Escherichia coli. Their independence of rec$^+$ function. Genetics 55, 699:707.

Ehrlich, S. D., D. Brunier, L. Janniere, B. Michel, Ph. Noirot, M. A. Petit and H. te Riele, 1987, Structural instability of genes cloned in Bacillus subtilis. In: "Genetics of Industrial Microorganisms" Part B (M. Alačević, D. Hranueli and Z. Toman, eds.) pp. 93:96. PLIVA, Zagreb.

Gellert, M., 1981, DNA topoisomerases. Annu. Rev. Biochem. 54, 665:697.

Gellert, M., K. Mizuuchi, M. H. O'Dea, R. Itoh and J. Tomizawa, 1977, Nalidixic acid resistance: a second genetic character involved in DNA gyrase activity. Proc. Natl. Acad. Sci. USA 74, 4772:4776.

Gruss, A. and S. D. Ehrlich, 1989, The family of highly interrelated single-stranded deoxyribonucleic acid plasmids. Microbiol. Rev. 53, 231:241.

Kieser, T., D. A. Hopwood, H. M. Wright and C. J. Thompson, 1982. pIJ101, a multi-copy broad host-range Streptomyces plasmid: functional analysis and development of DNA cloning vectors. Mol. Gen. Genet. 185, 223:238.

Mizuuchi, K., M. H. O'Dea and M. Gellert, 1978, DNA gyrase: subunit structure and ATPase of the purified enzyme. Proc. Natl. Acad. Sci. USA 75, 5960:5963.

Pigac, J., D. Vujaklija, Z. Toman, V. Gamulin and H. Schrempf, 1988, Structural instability of a bifunctional plasmid pZG1 and single--stranded DNA formation in Streptomyces. Plasmid 19, 222:230.

Radnedge, L., R. Barallon, C. S. Davey, C. Jones, S. Zaman, J. M. Ward, C. Thomas and H. A. Richards, 1989, Plasmid vectors in Streptomyces. Abstracts of the 4th ASM Conference on the Genetics and Molecular Biology of Industrial Microorganisms, p. 29.

Schauer, A., H. Im, C. Sohaskey, A. Nelson, J. Daniel and G. Helt, 1990, Probing Streptomyces morphological development with TN4556 and luciferase. Abstracts of the UCLA Colloquium on the Molecular Biology of Streptomycetes, p. 121.

Schauer, A., M. Ranes, R. Santamaria, J. Guijarro, E. Lawlor, C. Mendez, K. Chater and R. Losick, 1988, Visualizing gene expression in time and space in the filamentous bacterium Streptomyces coelicolor. Science 240, 768:772.

Schrempf, H., P. Dyson, M. Betzler, T. Kumar and P. Groitl, 1987, Amplification and deletion of DNA-sequences in Streptomyces. In: "Genetics of Industrial Microorganisms" Part A (M. Alačević, D. Hranueli and Z. Toman, eds.), pp. 177:184. PLIVA, Zagreb.

Simpson, D-L., J. H. Cove, S. Baumberg, P. M. Rhodes and C. R. Bailey, 1990, Quantitative studies on plasmid stability in Streptomyces lividans. Abstracts of the UCLA Colloquium on the Molecular Biology of Streptomycetes, p. 96.

Sugino, A., N. P. Higgins, P. O. Brown, C. L. Peebles and N. R. Cozzarelli, 1977, Mechanisms of action of nalidixic acid: purification of Echerichia coli nalA gene product and its relationship to DNA gyrase and a novel nicking-closing enzyme. Proc. Natl. Acad. Sci. USA 74, 4767:4771.

Sugino, A., N. P. Higgins, P. O. Brown, C. L. Peebles and N. R. Cozzarelli, 1978, Energy coupling in DNA gyrase and the mechanism of action of novobiocin. Proc. Natl. Acad. Sci. USA 75, 4838:4842.

Thomas, D. I., J. H. Cove, S. Baumberg, C. A. Jones and B. A. M. Rudd, 1990, Plasmid effects on secondary metabolite production by Streptomyces thermoarchaensis. Abstracts of the UCLA Colloquium on the Molecular Biology of Streptomycetes, p. 121.

Wang, J. C., 1985, DNA topoisomerases. Annu. Rev. Biochem. 54, 665:697.

Wolfson, J. S. and D. C. Hooper, 1985, The fluoroquinolones: structure, mechanism of action and resistance, and spectra of activity in vitro. Antimicrob. Agents Chemother. 28, 581:586.

STRUCTURAL INSTABILITY OF BIFUNCTIONAL VECTORS IN STREPTOMYCES

Jasenka Pigac

PLIVA Research Institute

41000 Zagreb, Yugoslavia

SUMMARY

The molecular mechanisms responsible for plasmid structural istability in gram-positive bacteria and the data so far published on the same problem in Streptomyces are reviewed.

The usefulness of plasmid cloning vectors depends on their long term stability, which is sometimes affected by insertion of foreign DNA. Plasmid structural instability is frequently one of the problems hampering cloning experiments in gram-positive bacteria. Several studies of deletion end--points have indicated that plasmid rearrangements and deletions in gram--positive bacteria are often the consequence of illegitimate recombination (for review see Ehrlich et al., 1986; Anderson, 1987). Plasmid rearrangements (involving deletions) occur most frequently by recombination between short homologous sequences (3-30 bp long), but also between sequences of no homology (less than 3 bp). Two models have been proposed for illegitimate recombination between short homologous sequences. The first, called slipped mispairing, involves aberrant replication of single stranded DNA in the replication fork. The second, breakage and reunion model, implies the introduction of double stranded breaks and rejoining of the molecules between the direct repeats. Topoisomerases from prokaryotes and eukaryotes may be involved in this process. According to both models, deletions, duplications and translocations may occur (Ehrlich et al., 1986). Plasmid structural instability is often pronounced in bifunctional vectors, which consist of two replicons and qualitatively different DNAs. Most often they are used to shuttle DNA between gram-positive bacteria and Escherichia coli. If the replicon belonging to a gram-positive plasmid replicates by a rolling circle(RCR) mechanism, during which process single stranded molecules of plasmid DNA (ssDNA) accumulate, the structural instability is even more pronounced. The ssDNA plasmids represent an important family of replicons in gram-positive bacteria. Besides a plus origin and a replication protein (Rep), these plasmids possess a minus origin (M-O), which serves as an efficient initiation site, recognized by host factors, for the conversion of the circular plus-strand ssDNA to double-stranded DNA (dsDNA). The properties of a ssDNA plasmid in a foreign host may depend on whether the M-O is active in the particular host. In all hosts, a plasmid lacking an active M-O is still viable, but accumulates ssDNA. The minimal sequences of M-O in all described cases (Gruss and Ehrlich, 1989; Deng et al., 1988) are large (at least 130-220 bp) and contain imperfect palindromic structures (Radnedge et al., 1989). The plus origin, Rep protein activity and M-O recognition are involved in successful host adaptation (Gruss and Ehrlich, 1989). Insertion of certain DNA fragments, e.g. from E. coli, into these plasmids results in a shift in plasmid distribution from principally monomeric (for the original wild type) to principally multimeric (for hybrids) form. The formation of high molecular weight (HMW) multimers is a function of the replication of plasmids replicating via a ssDNA intermediate. It has not been observed for a plasmid which replicates as a dsDNA molecule (Gruss and Ehrlich, 1988). The HMW form could induce non-termination of plasmid replication, delayed completion of the second strand synthesis, and re-entry into the replication pool with the appearance of the deletion in the original construct (for the model see Gruss and Ehrlich, 1989).

The discovery of Streptomyces plasmid pIJ101, a multicopy plasmid with a broad host range eliciting lethal zygosis (Kieser et al., 1982), has increased the possibility of genetic engineering with respect to commercially important Streptomyces strains. This plasmid and its derivatives belong to the described group of plasmids of gram-positive bacteria, of which more than a dozen are already sequenced. They all replicate via a ssDNA intermediate. One of the most frequently used derivatives of pIJ101, either for cloning antibiotic biosynthesis genes or for construction of shuttle vectors, is pIJ702 (Katz et al., 1983). Plasmid pIJ702 is a broad host range multicopy plasmid, obtained by insertion of the mel or tyrosinase gene into pIJ350.

pIJ101 has an active second strand synthesis function (sti) (Deng et al., 1988), corresponding to M-O of Bacillus and Staphylococcus plasmids which replicate via a ssDNA intermediate. In many pIJ101 derivatives, like pIJ350 or pIJ702, the sti function is deleted,resulting in accumulation of ssDNA. Sequence analysis of the BclI-E fragment of pIJ101 (Deng et al., 1988; Radnedge et al., 1989) showed that the sti region stands out from the rest of the sequence because of its high degree of symmetry, with many sequences of 5-10 nucleotides occurring as direct or inverted repeats. The stability of this region of secondary structure and the presence of the GAGCGT sequence make sti act as the main initiation signal for the conversion of ss to ds forms of the plasmid (Radnedge et al., 1989). Replication of the leading strand (5'-3') proceeds anti-clockwise. The circular ss intermediates (Schrempf, personal communication) are converted to ds forms by a rate-limiting reaction, resulting in a copy number of c. 100.The addition of sti to the basic replicon provides an efficient signal for the initiation of the synthesis of the second (lagging) strand and leads to an increase in the number of ds copies of plasmids such as pIJ2743. The copy number of pIJ101 and its derivatives is influenced by sti and by an additional trans-acting function, cop, (Deng et al., 1988). It has been shown (Gruss and Ehrlich, 1989) that the plasmids with homologous plus origins also have corresponding homologies in their Rep proteins. A published sequence (Kendall and Cohen, 1988) of pIJ101 revealed a similar amino acid motif in the Rep protein, despite the high (72 %) G + C content of the Streptomyces plasmid (as opposed to 30 to 40 % G + C for Bacilli and Staphylococci). Thus far, homologies in the Rep proteins have been found to include only the ssDNA-accumulating plasmids (Gruss and Ehrlich, 1989). Classification of plasmids according to their mode of replication should prove useful in the construction of cloning vectors.

One of the first examples of intraplasmid recombination in Streptomyces has been reported by Nakano et al. (1984). The authors have observed a spontaneous tandem duplication of 900 bp in plasmid pSL1 in S. lavendulae. The spontaneous duplications in the S. lavendulae plasmid have been attributed to recombination between short (5 bp) direct repeats.

Some plasmid vectors become unstable in Streptomyces only after insertion of heterologous DNA. To study intraplasmid recombination in Streptomyces using convenient and well-defined starting material, the shuttle vector, pIJ132, containing a pair of direct repeats of the mel gene from pIJ702, was constructed. The construct was much more stable in E. coli than in S. lividans, in which homologous recombination between the repeats produced a single product, pIF138 (Chen et al., 1987).

Lee et al. (1986) used E. coli plasmid pUC12 alone or in conjunction with the hepatitis B viral surface antigen gene, and cloned them into the vector pIJ702. No S. lividans transformants stably maintaining the entire hybrid plasmids pWCL1 and pWTS2 were found. In each case deletions of the hybrid plasmids were observed. Radnedge et al. (1989) also constructed a shuttle vector based on E. coli pUC8 and Streptomyces plasmid pIJ702 to study the expression of the beta-lactamase gene in Streptomyces. Both orientations of pUC8 inserted at the BglII site of pIJ702 were unstable in S. lividans.

Chen et al. (1987) attributed particular instability to the BglII--PstI sequence in pIJ702, which obviously played an important role in the instability of the above-mentioned shuttle vectors in S. lividans. Consistent with this hypothesis was the observation of these authors that spontaneous deletions in the mel sequence usually had one end-point between the BglII and PstI sites (Chen et al., 1987).

As against the unstable pIJ702 shuttle vectors, several examples of constructs stable in Streptomyces were reported. Neesen and Volckaert (1989) described the construction of a new shuttle vector by cloning a small artificial E. coli replicon pGV462 into pIJ702 at SphI and SstI, which resulted in excision of 430 bp from the region preceding the tyrosinase (mel) gene of pIJ702. The shuttle vector pSKNO1 was stable in Streptomyces.

Radnedge et al. (1989) described the stable replication of plasmid pQR1 constructed from PstI digests of pBR325 and pIJ702. Jensen et al. (1989) reported the formation of the stable shuttle vector pSH obtained by ligating pIJ702 to pUC119 at the PstI site. The plasmid was used for successful cloning of the S. clavuligerus isopenicillin N synthase (IPNS) gene (Jensen et al., 1989).

Shuttle vectors pZG3.1 and pZG3.2 (communicated at this Symposium), as well as their ClaI-BclI deleted derivatives, were constructed by insertion of pBR328 into the PstI site of pIJ350, in both orientations. All plasmids extracted from transformant colonies less than 15 days old were completely stable in S. lividans and S. rimosus R6. However, deletions in all constructs were observed in some older colonies and upon subculture.

Other examples reported by us were the shuttle vectors pZG5 and pZG6. pIJ350 was linearized at its unique PstI site and ligated to PstI-cut Bluescribe M13⁻ or pUC18, respectively. Both vectors proved to be stable regardless of colony age or multiple subculture. Moreover, they were used for successful retransformation of three tRNA genes into S. rimosus.

However, when pBR322 was used for construction of shuttle vectors, they usually exhibited structural instability in Streptomyces (Schottel et al., 1981; Wohlleben et al., 1986). Upon transformation of S. lividans and S. rimosus with a simple shuttle vector pZG1, a high level of instability was detected (Pigac et al., 1988). The process of plasmid rearrangements and deletions occurred gradually, correlating with colony age. It was not unusual to find several plasmids with different deletions coexisting in the mycelium of a single transformant colony. Being aware that pIJ350 and pZG1 accumulate ssDNA and lack sti function, the shuttle vectors pZG4.1 and pZG4.2 were constructed (communicated at this Symposium). They were obtained by partial digestion of pIJ303 with PstI and ligation with PstI-linearized pBR322. pZG4.1 and pZG4.2 have pBR322 inserted at PstI(35) and PstI(13) of pIJ303, respectively. The presence of sti function and non-detectable amounts of ssDNA in the lysates of pIJ303 extracted from S. lividans (Schrempf and Pigac, 1986) could contribute to a reduction in the amount of ssDNA and consequently to deletions in the shuttle vectors. However, both vectors were rather unstable in Streptomyces.

On the contrary, other shuttle vectors based on pIJ101, but in combination with other E. coli replicons, proved to be stable in Streptomyces. Shareck et al. (1984) constructed pFSH102 by inserting E. coli plasmid pSAS1206, carrying a sulfonamide-resistance gene, into pIJ101 at PstI site. After initial deletions in both pSAS1206 and pIJ101 (c.4.6 kb), the remaining plasmid still possessing sti function replicated stably in Streptomyces, fully expressing sulfonamide resistance.

Stable replication in Streptomyces was also achieved with similar constructs based on pPFZ12, an in vivo deleted derivative of pIJ303 still carrying sti function, and pBR325 to create the bifunctional vector pPFZ54, which was used for construction of a streptomycete cosmid, pPFZ74 (Chambers and Hunter, 1984).

pIJ101 and its deleted derivatives are similar to the ssDNA family of plasmids in <u>Bacilli</u> and <u>Staphylococci</u> (Gruss and Ehrlich, 1989). Their mode of replication (RCR) via ssDNA, the high homology of the Rep proteins, and the <u>sti</u> function corresponding to M-O stimulate homologous and illegitimate recombination, and consequently rearrangements and deletions of the parental plasmids. The reported data also indicate that the <u>E. coli</u> part of the vector, as well as the site of the insertion of <u>E. coli</u> replicon into the streptomycete plasmid may affect the structural stability of the constructed shuttle vectors in <u>Streptomyces</u>.

An example of host influence on stabilization was observed when the pZG1 shuttle vector, being highly unstable in <u>S. lividans</u>, became almost completely stable in <u>S. rimosus</u> R6 on persistent subsequent retransformation, indicating that some minor changes in pZG1 did occur during propagation of the recombinant plasmid in <u>S. rimosus</u> R6 (Pigac et al., 1988).

Yet, some <u>Streptomyces</u> - <u>E. coli</u> shuttle vectors enabled the successful expression of homologous and heterologous genes in <u>Streptomyces</u>. In our earlier report (Pigac et al., 1988) we concluded that plasmid structural instability in <u>Streptomyces</u> is a sequential process comprising the change in plasmid structure and the replacement of the parental plasmid by the newly formed one. It is obviously correlated with transformant colony age, revealing the presence of the completely intact shuttle vectors in transformant colonies up to 10 days old, which could make them useful in cloning experiments.

At the moment the data can be discussed only in terms of experience, since there has been no systematic study of plasmid structural instability nor explanation of its mechanism in <u>Streptomyces</u>. Therefore we hope that this mini-review, revealing the state of knowledge on this subject, could be useful in further attempts to construct new shuttle vectors or to find new approaches to explain the mechanisms of this phenomenon in these commercially important bacteria.

ACKNOWLEDGMENT

The author wishes to thank Zora Toman for valuable suggestions and discussions. Part of the results presented at the Symposium (authors: J. Pigac, D. Vujaklija, Z. Toman, S. Durajlija, V. Gamulin) were included in this overview. The authors thank the Research Fund of SR Croatia, Federal Committee for Science, Technology and Informatics Programme P-22 and Federal Research Fund of SFR Yugoslavia GIBIT/SZNJ.

REFERENCES

Anderson, P., 1987, Twenty years of illegitimate recombination. Genetics 115, 581:584.

Chambers, A. E. and I. S. Hunter, 1984, Construction and use of a bifunctional streptomycete cosmid. Biochem. Soc. Trans. 12, 644:645.

Chen, C. W., J. F-Y. Tsai and S. Chuang, 1987, Intraplasmid recombination in Streptomyces lividans 66. Mol. Gen. Genet. 209, 154:158.

Deng, Z., T. Kieser and D. A. Hopwood, 1988, Co-integrate formation between Streptomyces plasmids stimulated by a sequence of the multi--copy plasmid. Mol. Gen. Genet. 214, 286:294.

Ehrlich, S. D., Ph. Noirot, M. A. Petit, L. Janniere, B. Michel and H. te Riele, 1986, Structural instability of Bacillus subtilis plasmids. In "Genetic Engineering" (J. K. Seatlow and A. Hollander, Eds.), Vol. 8, pp. 71:83, Plenum, New York.

Gruss, A. and S. D. Ehrlich, 1988, Insertion of foreign DNA into plasmids from gram-positive bacteria induces formation of high molecular -weight plasmid multimers. J. Bacteriol. 170, 1183:1190.

Gruss, A. and S. D. Ehrlich, 1989, The family of highly interrelated single-stranded deoxyribonucleic acid plasmids. Microbiol. Rev. 53, 231:241.

Jensen, S. E., B. K. Leskiw, J. L. Doran, A. K. Petrich and D. W. S. Westlake, 1989, Expression of the Streptomyces clavuligerus isopenicillin N synthase gene in Escherichia coli and Streptomyces lividans. Abstracts of the 4th ASM Conference on the Genetics and Molecular Biology of Industrial Microorganisms, p. 21.

Katz, E., C. J. Thompson and D. A. Hopwood, 1983, Cloning and expression of the tyrosinase gene from Streptomyces antibioticus in Streptomyces lividans. J. Gen. Microbiol. 129, 2703:2714.

Kendall, K. and S. Cohen, 1988, Complete nucleotide sequence of the Streptomyces lividans plasmid pIJ101 and correlation of the sequence with genetic properties. J. Bacteriol. 170, 4634:4651.

Kieser, T., D. A. Hopwood, H. M. Wright and C. J. Thompson, 1982, pIJ101, a multi-copy broad host-range Streptomyces plasmid: functional analysis and development of DNA cloning vectors. Mol. Gen. Genet. 185, 223:238.

Lee, Y-H. W., Z-Y. Tzecheng, S-C. Wang, W-L. Cheng and C- W. Chen, 1986, Structural stability of heterologous genes cloned in Streptomyces plasmid pIJ702. Biochem. Biophys. Res. Comm. <u>140</u>, 372:378.

Nakano, M. M., H. Ogawara and T. Sekya, 1984, Recombination between short direct repeats in Streptomyces lavendulae plasmid DNA. J. Bacteriol. <u>157</u>, 658:660.

Neesen, K. and G. Volckaert, 1989, Construction and shuttling of novel bifunctional vectors for Streptomyces spp. and Escherichia coli. J. bacteriol. <u>171</u>, 1569:1573.

Pigac, J., D. Vujaklija, Z. Toman, V. Gamulin and H. Schrempf, 1988, Structural instability of a bifunctional plasmid pZG1 and single-stranded DNA formation in Streptomyces. Plasmid <u>19</u>, 222:230.

Radnedge, L., R. Barallon, C. S. Davey, C. Jones, S. Zaman, J. M. Ward, C. Thomas and H. A. Richards, 1989, Plasmid vectors in Streptomyces. Abstracts of the 4th ASM Conference on the Genetics and Molecular Biology of Industrial Microorganisms, p. 29.

Schottel, J. L., M. J. Bibb and S. N. Cohen, 1981, Cloning and expression in Streptomyces lividans of antibiotic resistance genes derived from Escherichia coli. J. Bacteriol. <u>146</u>, 360:368.

Schrempf, H. and J. Pigac, 1986, Single stranded plasmid DNA in Streptomyces. Abstract of the Fifth International Symposium on Genetics of Industrial Microorganisms, p. 41.

Shareck, F., A. Sasarman and C. Vezina, 1984, Construction of a shuttle vector and expression in Streptomyces lividans of the sulfonamide-resistance gene direved from E. coli plasmid pSAS1206. Can. J. Microbiol. <u>30</u>, 515:518.

Wohlleben, W., G. Muth, E. Birr and A. Pühler, 1986, A vector system for cloning in Streptomyces and Escherichia coli. In "Biological, Biochemical Aspects of Actinomycetes" (G. Szabo, S. Biro and M. Goodfellow, Eds.), pp. 99:101. Academiai Kiado, Budapest.

POPULATION GENETIC PROCESSES IN CONTINUOUSLY CULTIVATED

STREPTOMYCES STRAINS REFLECT GENETIC INSTABILITIES

Dieter Noack and Rudolf Geuther

Centralinstitute of Microbiology and
Experimental Therapy, GDR Academy of Sciences.
Beutenbergstr. 11, 6900 Jena, GDR

INTRODUCTION

During the last few years the genetic instability of
Streptomycetes has been investigated very intensively. The
main progress in this field took place since molecular alterations and rearrangements within the chromosomal DNA of
these microorganisms have been detected. The study of related phenomena and processes are likely to be important in
understanding genome structure and are also important from
the commercial point of view as instability can cause strain
degeneration and the loss of proficiency to produce antibiotics as well as the loss of resistance against both the
strain's own antibiotics and other ones.

POPULATION GENETICS OF NOURSEOTHRICIN BIOSYNTHESIS AND
RESISTANCE

In order to study quantitatively the loss of antibiotic
biosynthesis and resistance we improved the chemostat method
of allowing the continuous cultivation of mycelially growing
Streptomycetes. This was achieved by a mechanical stirrer
which fractionates the mycelium into pieces of less then
100 μm of total length. By the aid of this method we investigated quantitatively the genetic alterations proceeding
in continuously cultivated Streptomyces populations in dependence on both the cultivation time and the number of generations respectively.
Fig. 1 demonstrates one of the characteristic experiments.

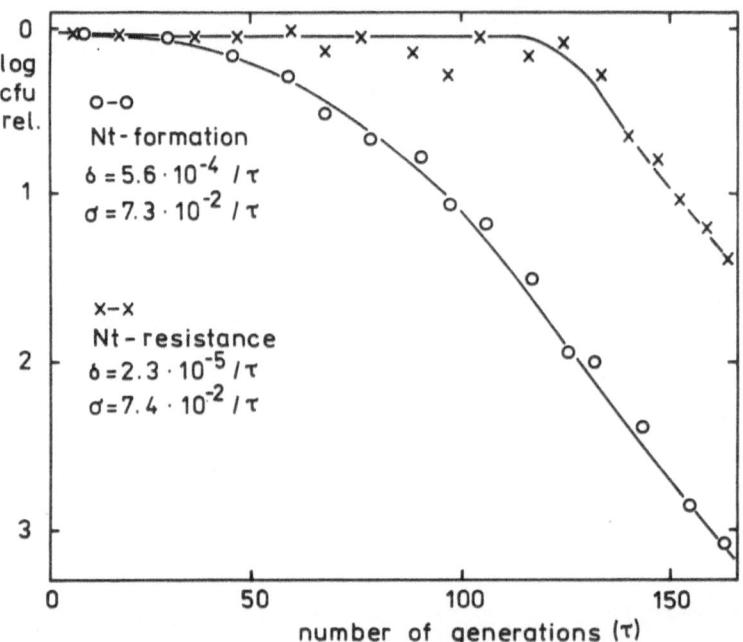

Fig. 1. Segregation kinetics of <u>Streptomyces noursei</u>. The nourseothricin producing strain was continuously cultivated in a chemostat with glucose as growth limiting substrate, with a dilution rate of D=0.12 h^{-1} and at 20°C. Samples were plated onto solid complex medium and the outgrown colonies were tested for their ability to form nourseothricin and for their resistance against nourseothricin. The logarithm of the relative colony forming units (cfu) were plotted over the number of generations proceeded in the chemostat. The genetic segregation rate δ and the selection pressure σ were calculated along the mathematical model of Noack et al. (1984).

A <u>Streptomyces noursei</u> strain, the producer of the strepto-thricin antibiotic nourseothricin which has been shown to be a useful ergothropicum, was cultivated in the chemostat equipped by the stirrer mentioned above. The selection pressure acting in the chemostat is determined by glucose as growth limiting substrate, by a dilution rate D=0.12 h^{-1} (equal growth rate) and by the cultivation temperature of 28°C. In distinct time intervals samples were harvested and tested for nourseothricin biosynthesis and nourseothricin resistance. The logarithm of relative colony forming units were plotted against the number N of generations estimated

by the aid of the equation $N = Dt/0.69$ with t the cultivation
time in terms of hours. Fig. 1 shows two curves one of them
designated Nt^+ for nourseothricin biosynthesis and the other
one designated Nt^r for nourseothricin resistance.
A mathematical model of the curves (Noack et al. 1984)
allows the calculation of two parameters. The first one is
the genetic segregation rate δ characterizing the frequency
with which the antibiotic productivity and the antibiotic
resistance respectively are lost during the time interval
of one generation and during the time interval of one doub-
ling of biomass respectively. As it can be seen the antibiotic productivity is lost ten times more frequent then the
antibiotic resistance. Therefor it is possible to select
clones which are resistent against nourseothricin and not
able to produce it.
The second parameter calculated from the experimental curves
is the selection pressure σ expressing the growth advantage
of one of both genotypes over the other one. The calculated
values of σ show that there is no significant growth advantage of one of the selected genotypes over the other one.
With other words both of them can grow with nearly the same
maximal specific growth rate under the cultivation condition
realized in the chemostat. It has been proved that all the
clones which have lost the antibiotic productivity and the
antibiotic resistance did not revert to the original geno-
type. This result shows that irreversible alteration of the
genetic information took place.

Population genetics of mycelium growth pattern

The selection pressure in the chemostat can be changed for
example by changing the growth limiting factor or by high
concentrations of phosphate or nitrogen a.s.o. In dependence
on the related selection pressure acting in the chemostat
quite different mutants of the originally inoculated strain
could be selected (Noack 1986; Roth et al. 1987). Recently
I tried to select mutants the mycelium of which fractionates
spontaneously giving rise to mycelium pieces comparable to
those resulting from the mechanical fractionation by the
stirrer mentioned above. This could be achieved by continuous cultivation of a Streptomyces noursei strain in a
stirrer-less chemostat operating with a dilution rate of

297

less then D=0.05 h^{-1} over at least 200 generation. After
this time the overflow of the culture suspension was used to
inoculate a second and a third stirrer-less chemostat vessel
and to cultivate each for further 50 generations under the
same conditions as before. The Streptomyces noursei population enriched thereafter contained identical clones which
exhibit an altered growth pattern of the mycelium.
This is characterized by
- very frequent fractionation of submerged mycelium,
- very low sedimentation of submerged mycelium,
- drastically inhibited pellet formation in liquid medium,
- drastically inhibited formation of aerial mycelium,
- very easy formation of protoplasts suitable for transformation with plasmid DNA and
- very easy release of cell contents such as DNA and proteins without the need of strong isolation procedures.
The last property was a prerequisite for isolating gyrase
and topoisomerase from Streptomyces mycelium as has been
demonstrated by Störl et al. (lecture presented at this
symposium). These proteins could never be isolated from
other usual Streptomyces mycelium.
By the same population genetic procedure mentioned above a
mutant derivative of Streptomyces lividans, named TC97,
could be isolated exhibiting much the same properties
described for Streptomyces noursei. This so-called "slowly
sedimenting" Streptomyces lividans derivative was investigated biochemically by Gräfe et al. (poster presented at
this symposium). They found that this derivative differs
from the parental strain by a drastical altered lipid composition of its cell wall. This result demonstrates that the
application of a distinct selection pressure during continuous cultivation provokes the enrichment of quite similar
derivatives originated from different Streptomyces species.
Revertants have never been observed so far.

Population genetics of amplification and deletion of plasmid DNA sequences

The last example presented here for the interrelation bet-

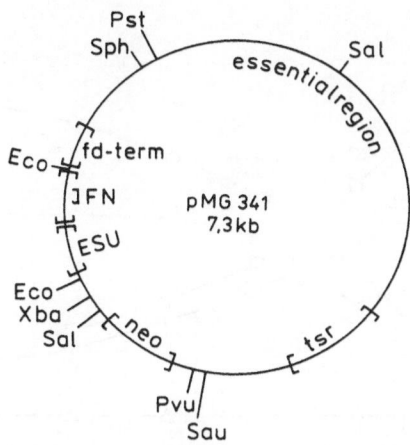

Fig. 2. Circular linkage map of a representative of the recombinant plasmid family pMG314. The determinant IFN for a
human interferon alpha 1, fused to an expression-secretion
signal ESU, was recombined with the vector plasmid pIJ487 at
the Eco site of the polylinker and transformed into Streptomyces lividans (Noack et al. 1988).

ween population genetic processes and genetic instability
concerns a recombinant plasmid family pMG341 and the interaction with its host, the slowly sedimenting derivative
Streptomyces lividans TC 97. Plasmid pMG341 is a recombinant between the promoter probe vector pIJ487 and an interferon determinant fused to the expression and secretion
signals derived from a Staphylococcus phage (Noack et al.
1988). The circular linkage map is shown in Fig. 2.
Immediately after cloning we observed a whole family of
plasmids differing one from another by the copy number of
an amplifiable sequence containing the neomycin resistance
gene of the vector. By restriction enzyme analysis it could
be determined that the amplified sequence amounts 1070 bp
including the coding region of the neo gene (Geuther et al.:
poster presented at this symposium).
Long time continuous cultivation of Streptomyces lividans
TC97 containing the plasmid family pMG341, using a liquid
medium with glucose in excess and ammonia as growth limiting
factor, revealed the enrichment of clones containing stably

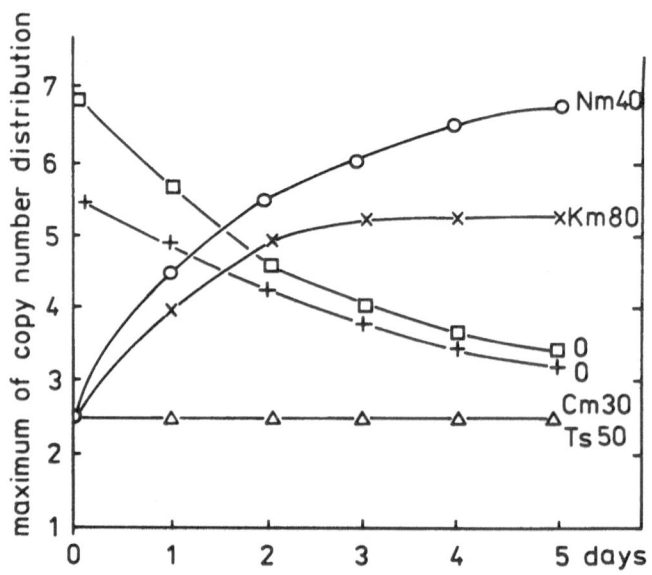

Fig. 3. Amplification and deamplification kinetics of the neo gene containing DNA sequence located onto plasmid pMG341. Streptomyces lividans TC97 bearing the plasmid family pMG341 was cultivated in liquid medium with the antibiotics neomycin (Nm), kanamycin (Km), chloramphenicol (Cm), thiostrepton (Ts) and without any antibiotics (O,O) respectively. The numbers 40, 80, 30, and 50 indicate their concentration in µg/ml. The plasmid content was followed by electrophoresis allowing the estimation of copy numbers of the amplified sequence. The maximum of the copy number distribution of the amplified DNA sequence was plotted over the cultivation time.

inherited plasmid family pMG341. Only after protoplasting, plasmid-free clones could be found with a frequency of less then 10^{-5} indicating a remarkable stabilization of inheritance functions.

However the tendency for amplification of the neo gene remained. Incubation with neomycin 40 µg/ml and 80 µg/ml respectively resulted after 5 days in the encrease of copy number. Because a whole family of plasmids is present within the biomass tested, we estimated this copy number which determines the maximum of the copy number distribution. This copy number is plotted against the cultivation time (Fig. 3) pointing to a dynamic process reaching a saturation level.

Transfer of the biomass thereafter into medium without antibiotic resulted in the restoration of the original copy
number distribution. Two copies of the amplifiable sequence
seem to represent a stable structure from which the amplification starts. Chloramphenicol and thiostrepton in concentrations of 30 µg/ml and 50 µg/ml respectively were
proved not to exert influence on the copy number distribution. This result demonstrates that the amplification process is only induced by an antibiotic the resistance to
which is encoded onto the amplifiable DNA sequence. The amplified structures are diluted out of the population after
it is transferred into a culture medium lacking the amplification inducing antibiotics.

However the alteration of the physical structure of the
plasmid is not restricted to the amplification of a DNA sequence containing the <u>neo</u> gene. A further stepwise enhancement of neomycin concentration up to 100 µg/ml within the
culture medium induces the subsequent deletion of a DNA sequence in the neighbourhood of the amplifiable sequence.
This deletion extends about 1600 bp and includes the determinants for IFN-ESU and the <u>fd</u>-terminator of the vector
plasmid. All the clones containing these deleted plasmids
exhibit a drastical encreased neomycin resistance by a factor of about 30 in comparison to those clones bearing only
"amplified" plasmids. This drastical enhanced neomycin resistance can be proposed to be the result of expression of
the neo gene by an expression unit located leftwards from
the deleted sequence, with other words leftwards of the deleted <u>fd</u>-terminator. The location of the amplified sequence
and the deleted one can be seen from the schematic representation in Fig. 4. The stem loop structure between both of
them results from the recombination of the IFN-ESU sequence
with the vector plasmid pIJ487 at the <u>Eco</u>-site of the polylinker. Its participation in the process of amplification
and deletion may be speculative until now.

In the frame of our population genetic studies we observed
that the "amplified" plasmid structure exhibits a selection
advantage over the "deleted" plasmid structure if a
Streptomyces population containing both plasmid types is
cultivated in a liquid medium lacking neomycin. After about
20 generations the "deleted" plasmid is diluted out of the

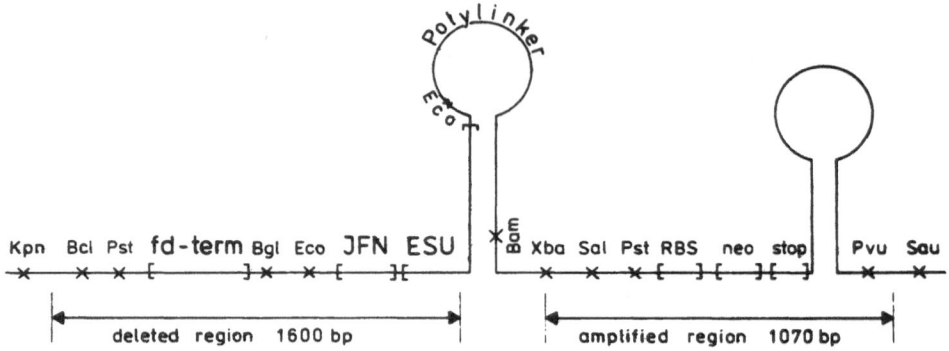

Fig. 4. Detailled restriction map of a partial sequence of
plasmid pMG341. The arrows indicate extent and location of
the amplifiable and deletable sequence. IFN, ESU, fd-term,
RBS, neo, and stop stand for determinants of human interferon alpha 1, expression-secretion-unit, fd-terminator,
ribosomal binding site, neomycin resistance and transcription stop respectively. The big stem-loop-structure is
possible due to the recombination of IFN-ESU with the vector
plasmid pIJ487 at the Eco-site.

population containing only the "amplified" plasmid type.
An interpretation for this observation can not be given until
now.

The whole set of molecular and population genetic processes
occuring within a Streptomyces lividans TC97 culture bearing
the recombinant plasmid family pMG341 is schematically
summarized with Fig. 5. Beyond a neomycin concentration of
about 50 µg/ml mainly amplifications and deamplifications
respectively take place. Above these concentration value a
deleted plasmid species appears and the amplified plasmid
species is diluted out. After transferring the Streptomyces
biomass into liquid medium lacking neomycin the processes
mentioned above proceed in opposite direction.

Concluding remarks

With the experimental results presented here I intended to
show that between population genetic processes and genetic
instabilities there exists not only a correlation but
mainly a strong interaction. The method of continuous
cultivation in the chemostat should be shown to be a useful
tool in studying population genetic processes and genetic
instabilities.

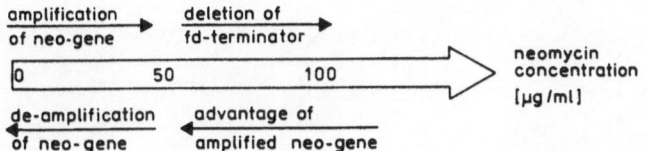

Fig. 5. Population genetic processes concerning the molecular structure of members of plasmid family pMG341 in
dependence on neomycin concentration present in the culture
medium. Enhancement of neomycin concentration from zero up
to 50 /ug/ml induces the amplification of the amplifiable
sequence containing the neo gene. Further enhancement of
neomycin concentration triggers the deletion of a DNA
sequence containing the IFN-ESU gene and the fd-terminator.
Finally the plasmid species containing amplifications are
diluted out of the population. Transfer of the biomass into
medium lacking neomycin induces processes with the opposite
direction as mentioned above.

References

Geuther, R., Noack, D. and Behnke, D.: An amplifiable unit
of DNA emerging from a recombinant Streptomyces plasmid.
Poster presented at this symposium.

Gräfe, U., Reinhardt, G. and Noack, D.: Lipid composition
of slowly sedimenting chemostat mutants of Streptomyces
lividans. Poster presented at this symposium.

Noack, D., Müller, G. and Roth, M.: Mathematical modelling
of genetic segregation kinetics obtained with chemostat
cultures of procaryotic microorganisms. Z.allg.Mikrobiol.
(1984) 24, 459-465.

Noack, D.: Directed selection of different mutants of
Streptomyces noursei using chemostat cultivation.
J.Basic Microbiol. (1986) 26, 231-239.

Noack, D., Geuther, R., Tonew, M., Breitling, R. and
Behnke, D.: Expression and secretion of interferon alpha 1
by Streptomyces lividans: use of staphylokinase signals and
amplification of a neo gene. Gene (1988)68, 53-62.

Roth, M.: Use of chemostat for selection of Streptomyces
chrysomallus mutants altered in the induction of D-glucose-
isomerase. Biotechnol. Letters (1987) 9, 855-860.

Störl, K., Störl, J. and Zimmer, Ch.: DNA topoisomerases
from Streptomyces and their inhibition by some antibiotic
and antitumor agents. Lecture presented at this symposium.

PARTITIONING OF PLASMIDS IN *STREPTOMYCES*:
SEGREGATION IN CONTINUOUS CULTURE OF A VECTOR
WITH TEMPERATURE-SENSITIVE REPLICATION

M. Roth, G. Müller, M. Neigenfind,
C. Hoffmeier and R. Geuther

Central Institute of Microbiology and Experimental
Therapy, Academy of Sciences of the GDR.
Jena, DDR-6900

INTRODUCTION

Streptomycetes are not only important producers of numerous antibiotics and enzymes, but they are also of increasing interest as alternative bacterial hosts for the production of eukaryotic and other heterologous gene products. Recombinant vector plasmids with sufficient structural and segregational stability under non-selective conditions are necessary for this purpose.

Stable plasmid inheritance is a result of complex processes: regulated replication of the plasmid DNA yielding a characteristic copy number and partitioning of the plasmid copies into daughter cells. These processes which have been extensively studied in unicellular bacteria are determined by the genotypes of both plasmid and bacterial host.

However, in chemostat studies it has been shown for different plasmid-host systems that, in addition, the stability of plasmid inheritance is strongly influenced by the growth conditions such as nutrient limitation and temperature (for review, see Dykhuizen and Hartl, 1983). In order to monitor plasmid stability repeated subculture is mostly used, but there are several disadvantages to this approach. The growing cells are exposed to continuously varying environmental conditions such as nutrient concentrations, pH, oxygen supply, etc., which may affect plasmid inheritance processes. Consequently repeated subculture experiments can yield only qualitative data on plasmid stability. Quantitative and reproducible results can only be obtained by investigations with continuous cultures.

Little is known about the mode of plasmid replication and, in particular, of partitioning in streptomycetes. Mycelia of streptomycetes consist of cells containing several nucleoids. The nucleoids and the plasmid copies do not segregate into individual daughter cells after replication. This situation complicates investigations of plasmid inheritance processes.

The process of plasmid partitioning is a deciding factor for stable plasmid maintenance. In order to examine plasmid partitioning in *Streptomyces* without interference by replication it would be favourable to use a plasmid with temperature-sensitive replication. Recently Muth et al. (1989) reported the construction of a family of cloning vectors with intermediate copy number based on the minimal replicon of the naturally temperature-sensitive plasmid pSG5 from *S. ghanaensis* (Muth et al., 1988). We are using in our experiments one of these vector plasmids, pGM4. It has a size of 4.75 kb and carries the aminoglycoside phosphotransferase gene from *S. fradiae* (*aphI*) and the thiostrepton resistance gene (*tsr*) from *S. azureus*.

In this paper we present first results of our attempts to investigate segregation kinetics of the vector pGM4 with chemostat cultures of *S. lividans* 66-PM7 (pGM4) under permissive and restrictive conditions.

The aims of our studies:

(a) Improvement of the experimental procedures to examine plasmid inheritance processes in streptomycetes,

(b) investigation of segregation kinetics of the plasmid pGM4, and

(c) mathematical modelling of the plasmid partitioning process in *Streptomyces* considering the mycelial growth and different possible modes of plasmid partition (Müller et al., 1990, this volume).

EXPERIMENTAL STRATEGY

Mycelial pellets of streptomycetes growing in continuous culture contain a large number of nucleoids and plasmid copies. Information about the distribution of plasmids within the mycelia can only be obtained if the mycelial pellets are divided into small pieces. We decided to fragment the mycelia by protoplast formation. By plating the protoplasts in parallel on antibiotic-containing and antibiotic-free regeneration media the fraction of plasmid-containing protoplasts can be estimated.

To study plasmid partitioning in *Streptomyces* we determine

(a) the kinetics of plasmid content of mycelia and protoplasts as a function of cultivation time in continuous culture under permissive and restrictive temperature (which stops plasmid replication),

(b) distribution parameters of mycelial pellets:
    - the hyphal length of the pellets,
    - the mean hyphal length of the pellets, and
    - the number of hyphal tips per pellet,

(c) the mean copy number of the plasmid in the population before the shift to restrictive temperature.

EXPERIMENTAL PROCEDURE

Strain

S. lividans 66-PM7 (pGM4) was used in the experiments.
It was constructed by transformation (standard protocol of
Hopwood et al., 1985) of pGM4 DNA (obtained from W. Wohlleben, Bielefeld, FRG.) into S. lividans 66-PM2, a plasmid-free
derivative of S. lividans 66-PM1 (pIJ2). Strain 66-PM2 was
selected from a chemostat at growth limitation by glucose
(Roth et al., 1985).

Continuous Culture

Segregation kinetics of the vector plasmid pGM4 were
examined according to the experimental procedure summarized
in Fig. 1. Continuous culture in chemostat was carried out as
described previously (Roth and Noack, 1982). Mineral salts
medium with growth-limiting concentration of ammonium chlor-
ide was used (Roth et al., 1985). It has been shown for
plasmids in E. coli (Noack. et al., 1981) as well as in
Streptomyces (Roth et al., 1985) that growth limitation by
glucose or phosphate led to faster loss of plasmids than that
by nitrogen. Therefore nitrogen limitation was chosen in our
experiments in order to avoid a strong influence of nutrient
limitation on plasmid replication and segregation before the
shift to restrictive temperature.
A low dilution rate of about 0.1 h^{-1} corresponding to a generation time of 6.9 h was selected to make sampling possible
over intervals of a few generations.
The preparation of stocks as inoculum for the experiments was
done as described previously (Roth et al., 1985) with the
alteration that neomycin (Nm) was added to the medium for
continuous culture at a concentration of only 2 μg ml^{-1}.
Experiments were started by inoculation of the batch culture
into the chemostat culture vessel filled with medium supplemented by Nm (2 μg ml^{-1}). Continuous flow of Nm-free medium
was initiated immediately. After 24 h of continuous culture
the first sample for platings was taken.

Estimation of the Fraction of Plasmid-containing Mycelia and Protoplasts

Platings of mycelia on YSA (Gräfe et al., 1981) and of
protoplasts on R2YE (Hopwood et al., 1985) were done in five
parallel sets per dilution on Nm-containing and Nm-free agar.
From each sample from continuous cultures 3 appropriate dilutions were plated. Very satisfactory results with respect to
the uniformity of colony numbers on parallel plates were obtained by inoculation of the protoplasts into R2 soft agar
and spreading on a basic layer of R2YE. Nm was added to both
agar layers of selective plates.

Measurement of Distribution Parameters of Mycelial Pellets

Mycelia withdrawn from chemostats were fixed by addition
of formaldehyde (3 %). After 15 minutes the mycelia were
washed 4 times with distilled water. A drop of the mycelia-
containing sample was spreaded on a grease-free slide and
dried at room temperature. Parameters of mycelial pellets
were determined by an automatic picture analyser (Quantimet

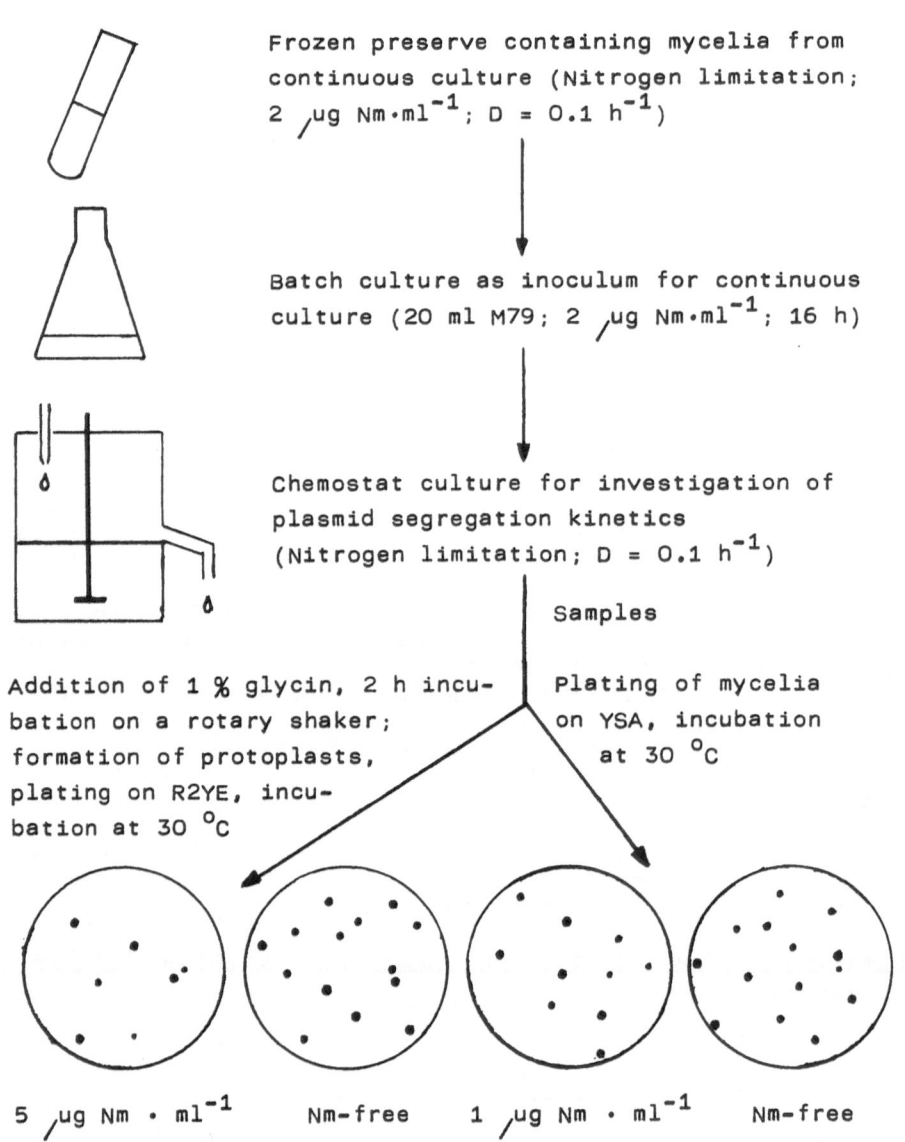

Frozen preserve containing mycelia from
continuous culture (Nitrogen limitation;
2 /ug Nm·ml^{-1}; D = 0.1 h^{-1})

Batch culture as inoculum for continuous
culture (20 ml M79; 2 /ug Nm·ml^{-1}; 16 h)

Chemostat culture for investigation of
plasmid segregation kinetics
(Nitrogen limitation; D = 0.1 h^{-1})

Samples

Addition of 1 % glycin, 2 h incubation on a rotary shaker;
formation of protoplasts,
plating on R2YE, incu-
bation at 30 $^{\circ}$C

Plating of mycelia
on YSA, incubation
at 30 $^{\circ}$C

5 /ug Nm · ml^{-1} Nm-free   1 /ug Nm · ml^{-1} Nm-free

Fraction of Nm resistant,   Fraction of Nm resistant,
plasmid-containing protoplasts   plasmid-containing mycelia

Fig. 1. Summary of the experimental procedure for the
examination of segregation kinetics in chemostat
cultures of the plasmid pGM4 from S. lividans
66-PM7 (pGM4)

308

720M, Image Analysing Computers Limited, Cambridge) which is coupled to a phase contrast microscope.

RESULTS AND DISCUSSION

In Fig. 2 the segregation kinetics of the vector plasmid pGM4 from *S. lividans* 66-PM7 (pGM4) during continuous culture under nitrogen limitation and at the restrictive temperature is shown. A rapid loss of pGM4 after shift to 37 °C was observed. Plasmid-free mycelial pellets appeared later than plasmid-free protoplasts as expected. Two platings of mycelia yielded lower fractions of Nm-resistant colonies compared with the course of the segregation curve. This accords with our experience that the results of platings of mycelia have

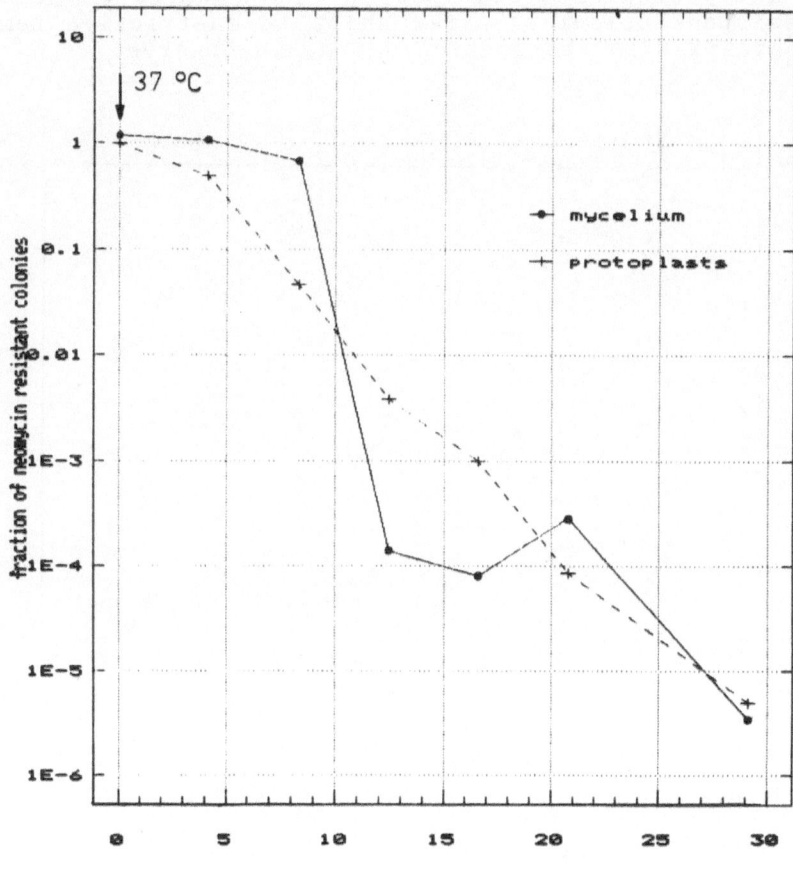

Fig. 2. Segregation kinetics of the vector plasmid pGM4 from an nitrogen-limited chemostat culture (dilution rate D was 0.12 h^{-1}) of *S. lividans* 66-PM7 at restrictive temperature. Samples of mycelia and protoplasts were plated on Nm-containing and Nm-free media in order to estimate the fraction of plasmid containing colonies in dependence on the cultivation time

in general higher errors than those of protoplasts. Only
slight increases in the Nm-concentration in selective plates
may lead to significant deviations of the results.

In Fig. 3 a similar experiment is presented. But in contrast to the first one after 28 generations the temperature
was changed to 30 °C. Unfortunately we obtained no result of
plating of protoplasts before temperature shift because of an
experimental error. Nevertheless, an increase of the fraction
of plasmid-containing protoplasts seems to occur. Similar
experiments will be repeated because they can provide information on spreading of plasmid copies within the mycelia.

In a third experiment (Fig. 4) the plasmid-containing
strain was cultivated 45 generations at 30 °C. The plasmid
was stably inherited during this time. Platings of both
mycelia and protoplasts resulted in fractions of about 100 %
of Nm-resistant colonies. After shift to restrictive temperature rapid plasmid segregation was again observed.

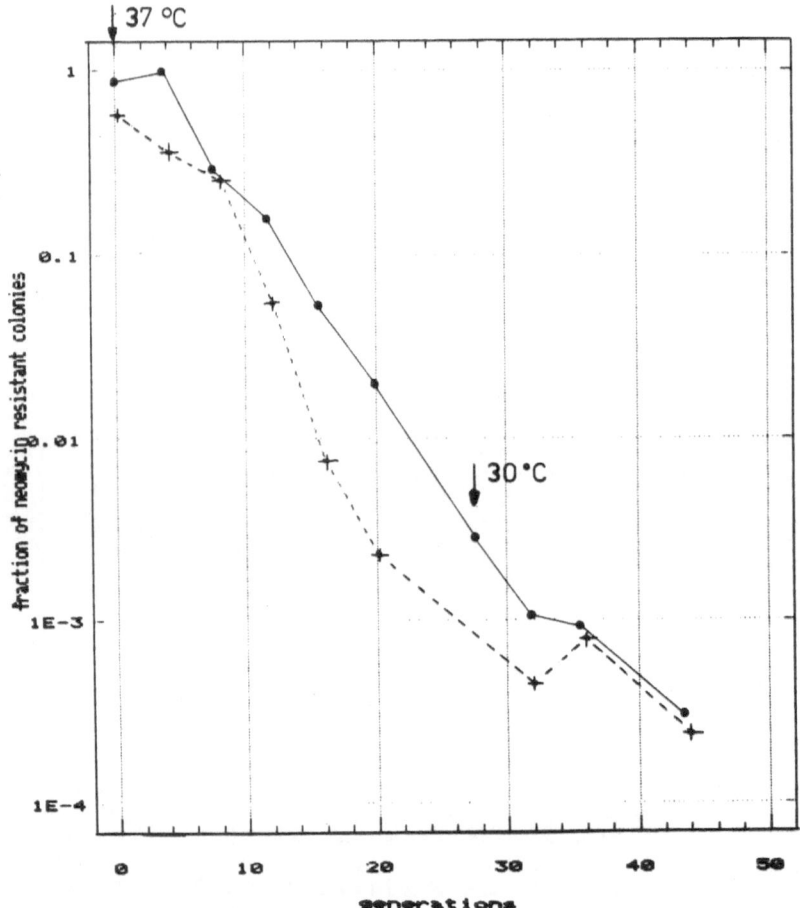

Fig. 3. Segregation kinetics as in Fig. 2 at
        D = 0.115 h^{-1}. After 28 generations
        at 37 °C the temperature was shifted to
        30 °C allowing replication of the
        plasmid pGM4. Symbols as in Fig. 2.

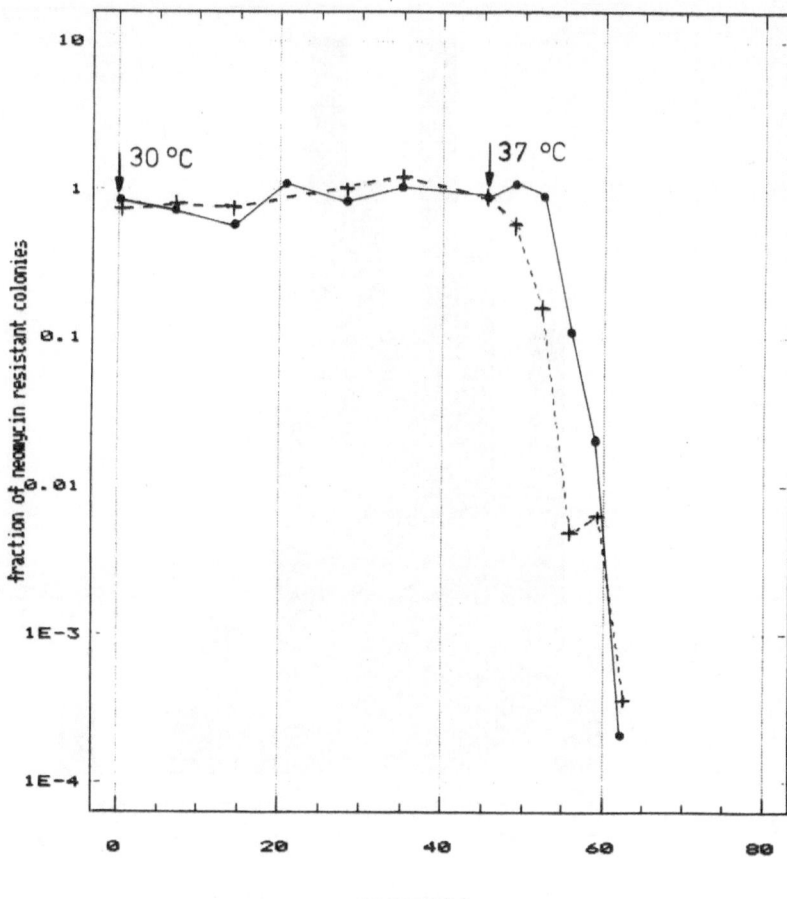

Fig. 4. Segregation kinetics as in Fig. 2 at
D = 0.1 h^{-1}. After 45 generations of
growth under permissive conditions the
culture was shifted to restrictive
temperature. Symbols as in Fig. 2.

In order to obtain an interpretation of these results
with respect to the mode of partition of the plasmid pGM4 in
S. lividans, complex mathematical models are under study (Müller et al., 1990). The models consider the mycelial growth
pattern of streptomycetes. Therefore data on distribution parameters of mycelial pellets are necessary for modelling. In
Fig. 5 an example of such data is given. The hyphal length of
the pellets withdrawn from chemostat culture seem to follow a
normal distribution.
Another important parameter for modelling is the plasmid copy
number. We are trying to estimate the mean copy number of
pGM4 in the population in the course of plasmid segregation
at restrictive temperature. But till now we have no results.

Concluding from the segregation kinetics of pGM4 under
restrictive conditions, it can be suggested that a partitioning of the plasmid takes place at increased temperature. If
the partition were impaired at restrictive temperature after

311

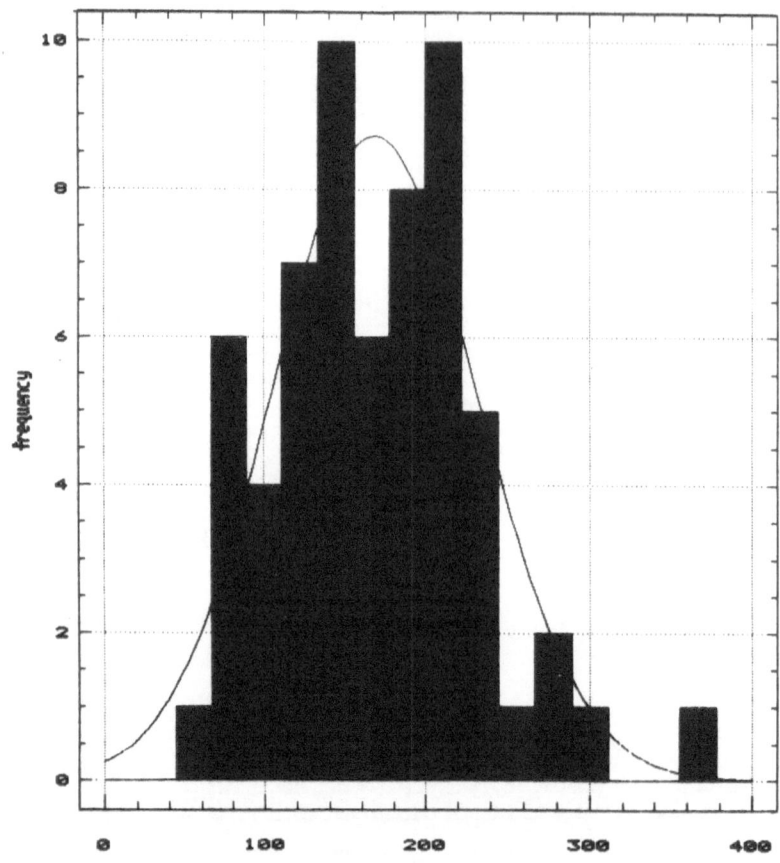

Fig. 5. Example of a frequency histogram of
the hyphal length of pellets withdrawn from a continuous culture of *S.
lividans* 66. The length of mycelia
(in um) prepared on slides was measured by the automatic picture analyser Quantimet 720M which is coupled
to a phase contrast microscope. The
curve represents the fitted distribution resulting from the experimental data

4 generations of growth only about 6 % of protoplasts should
carry plasmids. In contrast to this we observed in the 3
experiments 49, 36.4, and 50 % of Nm-resistant colonies regenerated from protoplasts at that time.

Additional experimental data and the results of analyses
of the mathematical models are necessary to characterize the
mode of partitioning of the vector plasmid pGM4 in *S. lividans*. However, the first results indicate that our experimental approach in combination with mathematical modelling should
be suitable for studying plasmid partitioning in *Streptomyces*.

Acknowledgements: The technical assistance of K. Perlet and E. Blanke is kindly acknowledged. For support of measurements with the automatic picture analyser we thank R. Blödner, C. Kühne and P. Mühlig. We are grateful to Dr. W. Wohlleben for providing the vector plasmid pGM4.

REFERENCES

Dykhuizen, D. E. and Hartl, D. L., 1983, Selection in chemostats, Microbiol. Rev., 47:150.

Gräfe, U., Roth, M., Christner, A. and Bormann, E.-J., 1981, Biochemical characteristics of non-streptomycin-producing mutants of *Streptomyces griseus*, I. Role of NAD(P)-glycohydrolase in cell differentiation, Z. Allg. Mikrobiol. 21:633.

Hopwood, D. A., Bibb, M. J., Chater, K. F., Kieser, T., Bruton, C. J., Kieser, H. M., Lydiate, D. J., Smith, C. P., Ward, J. M., Schrempf, H., 1985, Genetic manipulation of *Streptomyces*: a laboratory manual, John Innes Foundation, Norwich.

Müller, G., Stock, A., Löbus, J.-U. and Roth, M., 1990, Mathematical models of plasmid partitioning in unicellular bacteria and streptomycetes, in: "Genetics and Product Formation in *Streptomyces*", D. Noack, H. Krügel, S. Baumberg, eds., Plenum Publishing Corp., London.

Muth, G., Nußbaumer, B., Wohlleben, W. and Pühler A., 1989, A vector system with temperature-sensitive replication for gene disruption and mutational cloning in *Streptomyces*, Mol. Gen. Genet., 219:341.

Muth, G., Wohlleben, W., Pühler, A., 1988, The minimal replicon of the *Streptomyces ghanaensis* plasmid pSG5 identified by subcloning and Tn5 mutagenesis, Mol. Gen. Genet., 211:424.

Noack, D., Roth, M., Geuther, R., Müller, G., Undisz, K., Hoffmeier, C., Gaspar, S., 1981, Maintenance and genetic stability of vector plasmids pBR322 and pBR325 in *Escherichia coli* K12 strains grown in a chemostat, Mol. Gen. Genet., 184:121.

Roth, M. and Noack, D., 1982, Genetic stability of differentiated functions in *Streptomyces hygroscopicus* in relation to conditions of continuous culture, J. Gen. Microbiol., 128:107.

Roth, M., Noack, D. and Geuther, R., 1985, Maintenance of the recombinant plasmid pIJ2 in chemostat cultures of *Streptomyces lividans* 66 (pIJ2), J. Basic Microbiol., 25:265.

Wouters, J. T. M., Driehuis, F. L., Polaczek, P. J., Oppenraay, M.-L. H. A. van and Andel, J. G. van, 1980, Persistance of the pBR322 plasmid in *Escherichia coli* K12 grown in chemostat cultures, Antonie van Leeuwenhoek, 46:353.

MATHEMATICAL MODELS OF PLASMID PARTITIONING IN UNICELLULAR BACTERIA AND IN STREPTOMYCETES

G. Müller, A. Stock [*], J.-U. Löbus [*], and M. Roth

Central Institute of Microbiology and Experimental
Therapy, Academy of Sciences of GDR, Jena,
DDR-6900 and
*Friedrich Schiller University, Department of
Mathematics, Jena, DDR-6900

INTRODUCTION

Numerous experimental and theoretical treatments are known which describe plasmid inheritance processes in unicellular microorganisms (Novick and Hoppensteadt, 1978, Cullum and Broda, 1979, Müller et al., 1982, Seneta and Tavare, 1983, Müller, 1987). We have developed a Markov chain model which allows us to study extensively the segregation process from one generation to the next as a result of two processes, replication and partitioning of the plasmid copies. Different modes of plasmid replication and partitioning were modelled and relevant parameters were calculated. In comparison with these mathematical studies we have examined the process of plasmid partitioning in streptomycetes by using a complex mathematical model. The evolution of the plasmid copy number in the course of partitioning without replication was studied theoretically.
Mathematical model analysis incorporates
- the mycelial growth of the streptomycetes,
- the fragmentation of the mycelia as result of stirring,
- the distribution of the hyphal length and the total length and the number of hyphal tips of the mycelial pellets.
The results of the theoretical analysis have been compared with plasmid segregation kinetics from chemostat experiments at restrictive temperature with <u>Streptomyces lividans</u> 66 carring the vector plasmid pGM4 with temperature-sensitive replication (Roth et al., 1990).

PLASMID INHERITANCE IN UNICELLULAR BACTERIA

In unicellular bacteria plasmid inheritance processes have been studied extensively (Austin, 1988). Several quantitative aspects can be described by mathematical models. A Markov chain model allowed us to study the general properties of plasmid inheritance. This model described the whole process as a result of two processes:

(1) REPLICATION of plasmid copies
The replication matrix R contains the transition probabilities R_{ij} (index i is the number of plasmid copies in a single cell before and j after replication)

(2) PARTITIONING of plasmid copies as a result of cell division
The partition matrix Q contains the transition probabilities Q_{jk} (index j is the copy number in the mother cell after replication and k is the copy number in the daughter cell after cell division)

Therefore the process of plasmid inheritance from one generation to the next will be described by the matrix product

$$\pi = R * Q .$$

Starting from an initial distribution of plasmid copies in a population (distribution vector p(0)) the evolution of the copy number distribution in each generation can be calculated iteratively

$$p(n) = p(0) * \pi^n .$$

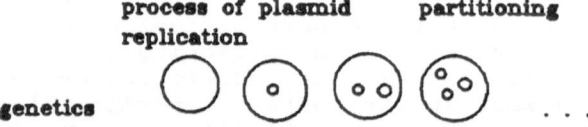

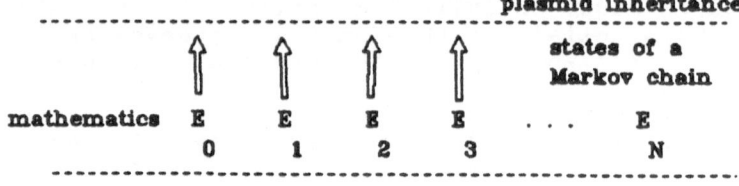

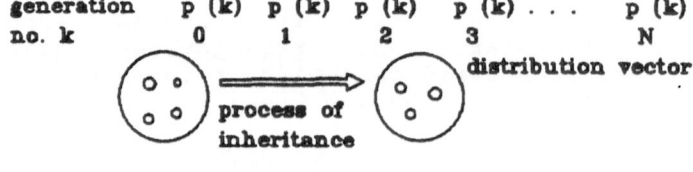

Fig. 1. Plasmid inheritance in unicellular micro-organisms (Scheme of the model)

The modelling framework employed here is suitable for studying the following aspects of the segregation process:
- Is there any asymptotic distribution of the copy number in the population ?
- The influence of initial distribution (p(0)) and of model parameters on the asymptotic distribution
- How long does a population need to reach steady state(s) ?
- Kinetics of segregation rate
- How many generations are necessary to reach a constant segregation rate ?
- Kinetics of the mean copy number of a population
- Influence of the initial conditions and parameters on the mean copy number
- Which mode of replication and partition are probable ?

MODEL OF PLASMID PARTITIONING IN STREPTOMYCETES

Streptomycetes grow in a complex branching pattern. The principal element of the mycelium is the hyphae, which is divided by cross walls into sections (Kretschmer, 1989, Reichl et al., 1990). The mycelium pellet grows by elongation at the hyphal tips and by branching. In opposite to unicellular bacteria little is known about

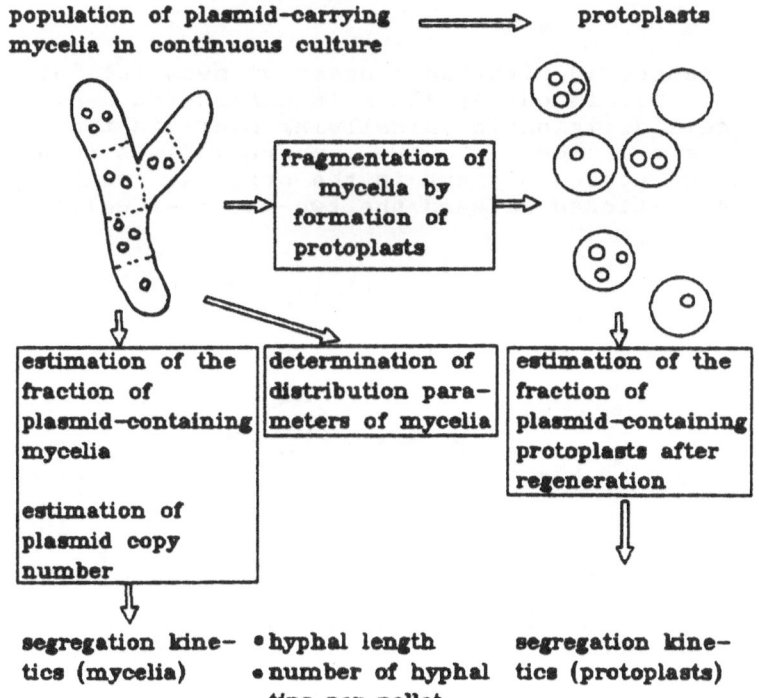

Fig. 2. Scheme of the experimental strategy to study plasmid inheritance processes in streptomycetes

the plasmid inheritance processes in streptomycetes (Roth et al., 1985) . Because of the mycelial growth of streptomycetes without regular cell division, plasmids do not segregate into individual cells after replication.

In our experimental studies on plasmid partitioning we used the vector plasmid pGM4 in S. lividans 66-PM7. Furthermore we enlarged in the theoretical studies the Markov chain model described here. The temperature-sensitive replication of the plasmid pGM4 allows the investigation of the partitioning process without interference by replication (Fig.2).

MODEL ASSUMPTIONS

(1)   The growth and segregation process starts from one mycelial pellet. This original pellet contains an initial number of plasmid copies.
       Plasmid replication is stopped by shift to restrictive temperature.

(2)   In the course of mycelial growth the hyphae of Streptomyces mycelia are septated by cross walls and branching occurs. This compartment between two cross walls is regarded as a single cell in the model.

(3)   The whole length of the mycelium pellet reaches a critical value and by the influence of stirring in the chemostat it divides into two parts and the process (2) starts again (Fig.3).

(4)   Simultaneously with the process of mycelial growth the growth within one mycelium is considered. Similar to the cell division in unicellular bacteria the cellular units are divided into two compartments in each case. Plasmid copies present in the original cellular unit are partitioned between the two daughter cell units.

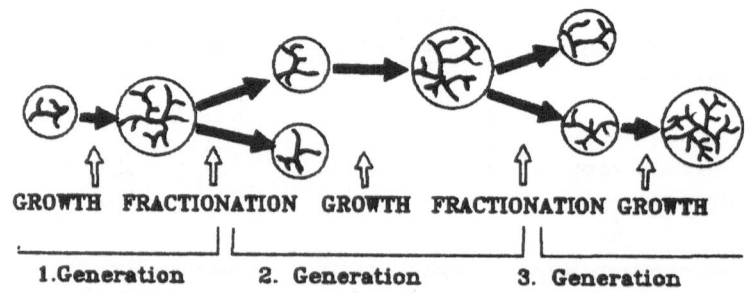

Fig. 3. Schematic pattern and terminology used for Streptomyces growth

PARAMETERS AND MATHEMATICAL QUANTITIES USED

(a) Distribution parameters of the mycelial pellets (Fig. 5 in Roth et al., 1990)
(b) Plasmid copy number (N)
(c) Probability of branching in the mycelia (p_1)
(d) Mean value and variance of the critical number of cell units in one mycelium generation (l_A).
(e) Parameter of the binomial partitioning of plasmid copies in the course of compartment formation

We used a homogeneous Markov chain model for the description of copy number along a path of mycelial growth and fragmentation (Fig. 3). This is similar to the model for unicellular bacteria. The plasmid copy number can be regarded as a state of the Markov chain and one unit of time in the model corresponds to one generation of growth of the mycelia. In order to compute the transition probability $p_{i,j}$ reaching state j (the number of plasmid copies is j) conditioned to starting at state i the parameters introduced in (a)-(e) were needed. A paper dealing with the details of the mathematical model is in preparation.

REFERENCES

Austin, S. J., 1988, Plasmid partition, Plasmid, 20:1.
Cullum, J., Broda, P., 1979, Rate of segregation due to plasmid incompatibility, Genet. Res. Camb., 33:61.
Kretschmer, S., 1989, Septation behaviour of the apical cell in *Streptomyces granaticolor* mycelia, J. Basic Microbiol., 29:587.
Müller, G., 1987, Modelluntersuchungen zur Segregationsproblematik plasmidbesitzender Mikroorganismen, Dissertation, Jena.
Müller, G., Noack, D., Schorcht, R., Gaspar, S., Herenyi, L., 1982, Mathematical modelling of segregation processes in microbial populations containing a single plasmid species, Acta Physica Acad. Sci. Hungaricae, 53:253.
Noack, D., Müller, G., Roth, M., 1984, Mathematical modelling of genetic segregation kinetics obtained with chemostat cultures of procaryotic microorganisms, J. Basic Microbiol., 24:459.
Noack, D., Roth, M., Geuther, R., Müller, G., Undisz, K., Hoffmeier, C., Gaspar, S., 1981, Maintenance and genetic stability of vector plasmids pBR322 and pBR325 in *Escherichia coli* K12 strains grown in a chemostat. Mol. Gen. Genet., 184:121.
Novick, R. P., Hoppensteadt, F., 1978, On plasmid incompatibility, Plasmid, 1:421.
Reichl, U., Yang, H., Gilles, E.-D., Wolf, H., 1990, An improved method for measuring the interseptal spacing in hyphae of *Streptomyces tendae* by fluorescence microscopy coupled with image processing, FEMS Microbiol. Letters, 67:207.
Roth, M., Müller, G., Neigenfind, M., Hoffmeier, C., Geuther, R., 1990, Partitioning of plasmids in Streptomyces: Segregation in continuous culture of a vector

with temperature-sensitive replication, in:
"Genetics and Product Formation in Streptomyces",
D. Noack, H. Krügel, S.Baumberg, eds., Plenum Publishing Coop., London.

Roth, M., Noack, D., Geuther, R., 1985, Maintenance of the recombinant plasmid pIJ2 in chemostat cultures of <u>Streptomyces lividans</u> 66 (pIJ2), J. Basic Microbiol., 25:265.

Seneta, E., Tavare, S., 1983, Some stochastic models for plasmid copy number, Theor. Pop. Biol., 23:241.

DNA TOPOISOMERASES FROM STREPTOMYCES AND THEIR INHIBITION BY SOME ANTIBIOTIC AND ANTITUMORACTIVE AGENTS

K. Störl, J. Störl and Ch. Zimmer

Central Institute of Microbiology and Experimental
Therapy, Academy of Sciences of the GDR
Jena, GDR

INTRODUCTION

It has been recognized that DNA supercoiling is an essential factor in the maintainance of the chromosomal state and cellular growth of bacteria.[1-3] Supercoiling plays an important role in the interaction with a large number of proteins and hence influences the cellular processes such as replication, recombination, transcription and repair. In prokaryotes, DNA supercoiling is controlled by the action of two enzymes, DNA gyrase and DNA topoisomerase I. Gyrase, a type II topoisomerase produces negative DNA supercoils and can remove negative as well as positive supercoils.[1-3] Topoisomerase I relaxes negatively supercoiled DNA and counteracts the supercoiling activity of gyrase.[2,3] Studies on DNA topoisomerases, their enzymatic properties and inhibitory effects of various agents have been reported for different microorganisms.[1-4] Topoisomerases from E.coli are the most extensively investigated enzymes.[1,3,5]

In our studies on topoisomerases we have focused on Streptomycetes because of a number of different striking features which are characteristic for the behaviour and the complex cycle of the morphological development of these organisms. Streptomycetes are grampositive bacteria and show a mycelial growth by forming hyphae. They undergo a differentiation process, sporulate and may produce antibiotics in a secondary metabolism. As Eubacteria they contain nucleoids composed of condensed chromosomal DNA with supercoiled domains.[6] The genome is about three times that of E.coli containing extremely GC-rich DNA (72 mole-% G+C). Since the condensed chromosomal structure in nucleoids of Streptomycetes is associated with proteins[7,8] and DNA supercoiling is an important element in the control of gene activity we have isolated DNA gyrase and topoisomerase I from the antibiotic nourseothricin producing strain Streptomyces noursei and investigated their in vitro behaviour as well as their sensitivity against antibiotics and antitumor agents.

MAIN PROPERTIES OF TOPOISOMERASES FROM STREPTOMYCES NOURSEI

DNA topoisomerase I and DNA gyrase were purified simultaneously from the same crude extract using heparin agarose chromatography.[9] They showed the characteristic relaxing and supercoiling activities, respectively, which are typical for the reaction of these two types of enzymes (for reviews cf.[1-3]). The relaxing activity of topoisomerase I depends on Mg^{2+} with a pH range from pH 7.6 to pH 8 and a relatively broad temperature range. The enzyme has a molecular mass of about 120 kDa, its reaction is inhibited by single-stranded DNA and spermidine concentrations $>0.5mM$; positively supercoiled DNA is not accepted as a substrate.[9]

The gyrase of S.noursei requires ATP for its supercoiling activity and the reaction is stimulated by potassium ions and spermidine. Like the E.coli enzyme it converts positive DNA supercoils into negative ones in the presence of ATP. DNA gyrase from Streptomyces noursei was found to consist, in pairs, of two different subunits with an apparent molecular mass of 103 kDa and 85 kDa, respectively.[9]

SENSITIVITY OF TOPOISOMERASES FROM STREPTOMYCES NOURSEI AGAINST BIOLOGICALLY ACTIVE AGENTS

The best known specific inhibitors which block DNA gyrases are novobiocin and related compounds, called coumarins and the class of quinolones.[2,3,10] The inhibitory mechanism is generally believed to occur by drug binding to the enzyme (for details see reviews[2,3,10]), but the quinolones most probably bind stronger to DNA in the complex formed with the topoisomerase.[11,12] The effectiveness of gyrase inhibitors may be very different for enzymes from various organisms suggesting a different sensitivity of topoisomerases which is an unsolved issue at present. In table 1 the effect of inhibitors on DNA gyrase from S.noursei is compared with that from E.coli and M. luteus. Data are presented by the lowest drug concentration required for a partial inhibition of the supercoiling activity of gyrase (IC_i). It is immediately evident that the Streptomyces enzyme exhibits a higher resistance against the gyrase inhibitors than the E.coli enzyme. Very similar the gyrase from M. luteus shows a relatively low sensitivity to 4-quinolones (table 1) as reported by Zweerink and Edison.[13] The origin of this enhanced resistance of Streptomyces gyrase against these inhibitors is unclear at present. One speculative explanation could be some differential effects in the molecular interactions between DNA and the enzymes from different origin. The local binding behaviour of gyrases could be different which may induce changes in DNA secondary structure leading to a significant modulation of the affinity for the drug. From our results it appears that the molecular mechanism of the antigyrase active quinolones is very complex in nature. We also found that topoisomerase I from S. noursei is significantly inhibited by some quinolones, e.g. by ciprofloxacin and temafloxacin (not shown). These results agree with the findings of Tabary et al.[14] reported for the E.coli enzyme. From these data one is tempted to suggest that not only the enzyme in the DNA-topoisomerase reaction but also DNA possess a

Table 1. Inhibition of DNA supercoiling activity
of different gyrases by some antimicrobial agents

agent	IC_1^a (μg/ml)		
	Str.noursei	E.coli	M.luteus[b]
Novobiocin	5	0.5	
Coumermycin	20	0.5	
Ciprofloxacin	50	1	8
Temafloxacin	100	10	
Oxolinic acid	200	2.5	625
Pefloxacin	200	5.0	312
Ofloxacin	200	10.0	156
Nalidixic acid	300	100	625

[a]The lowest concentration of drug that resulted
in partial inhibition of supercoiling activity
of gyrase was estimated by agarose gel electrophoresis.
[b]values published by Zweerink and Edison.[13]

target function for quinolones. Shen et al.[12] demonstrated
that the quinolone antibiotics show a higher affinity to specific sites on DNA created by gyrase rather than to sites on
the enzyme itself.

In this context it was of special interest to examine a
group of DNA binding drugs, that binds tightly to the minor
groove by a nonintercalating mechanism. We investigated a
variety of minor groove binders which show more or less a specificity for an abundance of dA·dT base pairs and are known
as antibiotics, antiviral or antitumor agents (for review see
ref.[15]). Most of this class of DNA binding drugs affect other
DNA processing enzymes, e.g. the reaction of RNA and DNA po-
lymerases[15,16] and some eukaryotic topoisomerases.[17] Details
of the structures of the nonintercalating drugs are given in
reference.[15]

The results of table 2 demonstrate that most of the DNA
minor groove binding ligands effectively inhibit the supercoiling activity of S.noursei gyrase as measured by the IC_i
and IC_c values. The inhibitory effects are similar for E.coli
gyrase (not shown). The effectiveness of the antigyrase activity of minor groove binders decreases in the order Dst-3,
Nt > Im$_3$,Im$_2$ > SN-6999 > Hoechst 33258 > bis-Nt > SN 18071.
Interestingly, we also observed an inhibitory effect of the
groove binding drugs on the relaxation reaction catalyzed by
Streptomyces topoisomerase I with some differences in their
decreasing order of the inhibitory ability from Dst-3 > bis-
Nt > SN-18071 > SN-6999 > Im$_3$ > Nt (data not shown). These
data clearly suggest that DNA minor groove binders can be regarded as a new class of gyrase inhibitors and other topoiso-

Table 2. Inhibition of DNA supercoiling activity of gyrase from <u>Streptomyces noursei</u> by nonintercalating DNA binding ligands

	μg/ml	
agent	IC_1 [a]	IC_c [b]
Distamycin-3	0.5	5
Netropsin	1	5
Lex: Im_3	2	10
Lex: Im_2	2	50
SN 6999	5	50
Hoechst 33258	10	50
Bis-netropsin-5	15	50
SN 18071	20	>300

[a] IC_1 is defined as in table 1.

[b] IC_c means that concentration of drug that causes a full inhibition of the enzymatic reaction

merases such as topoisomerase I in agreement with results reported for eukaryotic topoisomerase II.[17] Since most of the nonintercalating DNA binding drugs show no significant binding tendency to enzyme proteins[15] the blocking of the topoisomerase reaction is related to the DNA-bound drug. The mechanism and differential inhibitory effects of various groove binders is more complex and cannot be explained unequivocally at present. Although all DNA ligands used in this study show more or less an affinity to dA·dT base pairs this preference is not a priori a determinant molecular factor of the strong inhibitory effect on the topoisomerase reaction, e.g. the lowering of the dA·dT base pair affinity associated with the structural modification from Nt to lexitropsin Im_3 does not decrease the inhibition to a great extent. There is also no relation to the binding strength and site size of the minor groove binders (for corresponding data see ref.[15]). From the variation of the potential inhibitory effects of the minor groove binders (table 2) together with their DNA binding properties[15] we may tentatively assume that the nature of the ligand and geometrical factors in the minor groove interaction play a role for the ability of inhibition of topoisomerases. The effects may be different for blocking of the gyrase and topoisomerase I. In case of the anti-gyrase action of the nonintercalators some possible molecular factors seem to be important for their effectiveness: a moderate length of the ligand and its adaptability in the minor groove including the capability to form hydrogen bonds as well as van der Waals contacts; the inducibility of topological changes in DNA upon minor groove binding could be another determinant responsible for their potential to block the gyrase reaction. These considerations are, however, speculative at present and require further studies on this subject.

ACKNOWLEDGEMENT

We thank Dr. Linus Shen (Abbott Laboratories) for support and providing a sample of temafloxacin.

REFERENCES

1. J. C. Wang, DNA topoisomerases, Annu. Rev. Biochem. 54: 665 (1985).
2. K. Drlica, Biology of bacterial DNA topoisomerases, Microbiol. Rev. 48:273 (1984)
3. Ch. Zimmer, K. Störl, and J. Störl, Microbial DNA topoisomerases and their inhibition by antibiotics, J.Basic Microbiol. in press (1990).
4. K. Drlica and S. Coughlin, Inhibitors of DNA gyrase, Pharmac. Ther. 44:107 (1989).
5. M. Gellert, DNA topoisomerases, Annu. Rev. Biochem. 50:879 (1981).
6. O. G. Stonington and D. E. Pettijohn, The folded genome of Escherischia coli isolated in a protein-DNA-RNA complex, Proc. Natl. Acad. Sci. USA 68:6 (1971).
7. E. Sarfert, Ch. Zimmer, J. Gumpert, and H. Lang, Folded chromosome structure and DNA-binding protein of Streptomyces hygroscopicus, Biochim. Biophys. Acta 740:118 (1983).
8. E. Sarfert, V. Sedova, H. Triebel, H. Bär, and Ch. Zimmer, DNA binding protein from Streptomyces hygroscopicus: detection of binding by gel retardation, sedimentation and effects on the transcriptional activity in vitro, Biomed. Biochim. Acta 9:633 (1989).
9. K. Störl, J. Störl, Ch. Zimmer, M. Roth, and D. Noack, Isolation and properties of DNA topoisomeraseI and DNA gyrase from Streptomyces noursei, J. Basic Microbiol. in press (1990).
10. H.-P. Vosberg, DNA topoisomerases; enzymes that control DNA conformation, Curr. Top. Microbiol. Immunol. 114: 19 (1985).
11. L. L. Shen and A. G. Pernet, Mechanism of inhibition of DNA gyrase by analogues of nalidixic acid: The target of the drugs is DNA, Proc. Natl. Acad. Sci. USA 82: 307 (1985).
12. L. L. Shen, L. A. Mitscher, P. N. Sharma, T. J. O'Donnell, D. W. T. Chu, C. S. Cooper, T. Rosen, and A. G Pernet, Mechanism of inhibition of DNA gyrase by quinolone antibacterials: A cooperative drug- DNA binding model, Biochemistry 28:3886 (1989).
13. M. M. Zweerink, and A. Edison, Inhibition of Micrococcus luteus DNA gyrase by norfloxacin and 10 other quinolone carboxylic acids, Antimicrob. Agents Chemother. 29:598 (1986).
14. X. Tabary, N. Moreau, C. Dureuil, and F. Le Goffic, Effect of DNA gyrase inhibitors pefloxacin, five other quinolones, novobiocin and chlorobiocin on E. coli topoisomeraseI, Antimicrob. Agents Chemother. 31:1925 (1987).

15. Ch. Zimmer and U. Wähnert, Nonintercalating DNA - binding ligands: Specificity of the interaction and their use as tools in biophysical, biochemical and biological investigations of the genetic material, <u>Prog. Biophys. Molec. Biol.</u> 47:31 (1986).

16. D. C. Straney and D. M. Crothers, Effect of drug-DNA interactions upon transcription initiation at the lac promoter, <u>Biochemistry</u> 26:1987 (1987).

17. J. M. Woynarowski, R. D. Sigmund, and T. A. Beerman, DNA minor groove binding agents interfere with topoisomeraseII mediated lesions induced by epipodophyllotoxin derivative VM-26 and acridine derivative m-AMSA in nuclei from L1210 cells, <u>Biochemistry</u> 28:3850 (1989).

Luciferase, 57, 59-61
lux gene, 57-61

Maltose metabolism, 216
Markov chain model, 315, 318, 319

NAD glycohydrolase, 15
nat1, 185-190
Nitrogen assimilation (Ntr), 246, 247
Nonsense mutants, 65

otc cluster, 105
Oxytetracycline, 105, 137

pat, 174
Phosphinothricyl-alanyl-alanine, 171
Plasmid family, 309, 310
  pGM4, 296-302, 315, 318
  partitioning, 295, 296, 301, 302, 315-318
  segregation, 297, 298, 301, 315
  copy number, 295, 296, 301, 315-319
  with ts replication, 296, 315, 318
Polyamine formation, 37
Polyketides, 105, 117
Polyketide synthase, 117
Population genetics, 305, 307, 308, 245
ppGpp, 29
Primary metabolism, 35
Promoter region, 41, 199, 239
Protease, 230
Protein sequence matrix alignment, 188, 179
Pullulanase, 215
Pulsed field gel electrophoresis, 248

Reading frame, 53, 61
Regulation of differentiation, 53, 59, 60
Repeated sequences, 265, 266, 257, 278
Resistance
  to aminoglycosides, 163
  to β-lactams, 162, 195
  to chloramphenicol, 246, 269, 253, 274
  to macrolides, 163
  to neomycin, 297-300, 309-312
  to nourseothricin, 185-190, 306
  to oxytetracycline, 246, 248, 265, 266, 268, 269
  to polyketides, 162
  to streptomycin, 54-56, 60, 61

to streptothricin, 164, 185
to tetracyclines, 162
to thiostrepton, 58, 59, 77-79, 296
Reverse blotting, 274
Rhizobiaceae, 174
rRNA genes, 70

SAT, 185
SCP2, 105
Secretion vector, 235
Segregation rate, 307
Selection
  pressure, 306-308
  advantage, 307
Serine hydroxyamate, 30-32
Sigma factor, 5
Signal sequence, 238
Slowly sedimenting mycelium, 308
Southern hybridization analysis, 77, 79
Sporulation, 3, 14-16, 19, 22, 24, 29
STAT, 185
Strain degeneration, 266
Streptomyces
  glaucescens, 166
  griseus, 165, 166
  hygroscopicus, 174
  lavendulae, 185
  noursei, 185
  viridochromogenes 161, 171
Streptomycin
  formation, 37
  phosphotransferases, 162
  production genes, 165
Stringent response, 29
Supressor mutants, 65

Tandem promoter, 229
Tetracycline, 105, 117, 129, 137
Tetracycline dehydrogenase, 137
TGA termination codon, 54, 57-61
Topoisomerase I, 321-324
Translation, 53, 57, 60, 61
Translational accuracy, 53, 54, 59, 60
Transposable element IS466, 267
tRNA genes, 47, 65, 68, 70, M21/1
Tylosin, 145

Undecylprodigiosin, 31
Upstream sequence, 198

whi genes, 4

Xylan hydrolysis, 212
Xylanase enzymes, 207, 210